D0753605

Cluster Analysis for Researchers

Cluster Analysis for Researchers

H. Charles Romesburg

Lifetime Learning Publications
Belmont, California
a division of Wadsworth, Inc.
London, Singapore, Sydney, Tokyo, Toronto, Mexico City

To Lafayette College, University of Pittsburgh, and Utah State University

Designer: Rick Chafian
Copyeditor: Kirk Sargent
Illustrator: John Foster
Composition: Science Typographers, Inc.

Printed in the United States of America

1 2 3 4 5 6 7 8 9 10—88 87 86 85 84

Library of Congress Cataloging in Publication Data

Romesburg, H. Charles
Cluster analysis for researchers.

Bibliography: p.
Includes index.
1. Cluster analysis. I. Title.
QA278.R66 1984 519.5′3 83-24835
ISBN 0-534-03248-6

Contents

Preface

PURPOSE OF THIS BOOK

This book explains and illustrates the most frequently used methods of *hierarchical cluster analysis* so that they can be understood and practiced by researchers with limited backgrounds in mathematics and statistics.

Widely applicable in research, these methods are used to determine clusters of similar objects. For example, ecologists use cluster analysis to determine which plots (i.e. objects) in a forest are similar with respect to the vegetation growing on them; medical researchers use cluster analysis to determine which diseases have similar patterns of incidence; and market researchers use cluster analysis to determine which brands of products the public perceives in a similar way. In all fields of research, there exists this basic and recurring need to determine clusters of objects.

This book will ground you in the basic methods of cluster analysis and guide you in all phases of their use. You will learn how to recognize when you have a research problem that requires cluster analysis, how to decide upon the most appropriate kind of data to collect, how to choose the best method of cluster analysis for your problem, how to obtain a computer program to perform the necessary calculations, and how to interpret the results.

INTENDED AUDIENCES

This is primarily a self-study text for researchers and graduate students in all fields of the physical sciences, the social sciences, the life sciences, as well as those in planning, management, and engineering. In addition, this is a reference book for applied statisticians and mathematicians, especially those who want to learn how researchers are using cluster analysis to solve a variety of research problems. Finally, this book will serve as a text, or a supplemental text, in applied statistics and mathematics programs for students enrolled in courses that aim to help them improve their abilities in using mathematical methods in *real* applications.

SPECIAL FEATURES

As well as examining the mathematics of cluster analysis, this book analyzes the *qualitative* aspects of cluster analysis, which help researchers to make the best choices of both data and methods for use in their research problems.

Nearly one hundred applications of cluster analysis are described in ways that nonspecialists will understand. Drawn from diverse research, these applications illustrate qualitative aspects of choosing and using methods in which researchers must be proficient if they are to make valid applications. By studying this wealth of examples, your ability to creatively use cluster analysis and other mathematical methods will improve dramatically.

ORGANIZATION OF MATERIAL

The material is divided into four parts and two appendices as follows:

Part I presents the basic calculations of cluster analysis and illustrates a variety of its uses. You will see how it can be used to a) develop interesting scientific questions; b) create research hypotheses; c) test research hypotheses; d) make general-purpose and specific-purpose classifications; and e) facilitate planning, management, and engineering.

Part II presents alternative methods for tailoring cluster analysis to specific applications. These include methods for standardizing the data, for estimating similarities among the objects to be clustered, and for clustering the objects. Also included is a chapter that shows how to report the results found using cluster analysis.

Part III shows how to use cluster analysis to make classifications. While Part I introduced this function of cluster analysis, Part III proceeds to supply considerable detail.

Part IV discusses and illustrates how to make the subjective decisions required to frame and validate applications. It shows how researchers are guided by norms within their research communities, both when they choose methods and when they validate applications.

Appendix 1 describes books and articles about cluster analysis and multivariate methods so that you can continue to learn beyond the basic methods presented in this book.

Appendix 2 describes computer programs for cluster analysis, including those in the widely available SAS and BMDP statistical packages, and including the CLUSTAR and CLUSTID programs available from Lifetime Learning Publications as a companion to this book.

There are exercises at the end of key chapters with answers given at the back of the book. For self-study, you should work the exercises and also independently rework the examples in the text.

ACKNOWLEDGMENTS

Development of the examples of applications was partly supported by a National Science Foundation grant (NSF/SER-8160606) for "Development of a Course in the Principles of Mathematical Applications, with Emphasis on Cluster Analysis." I thank the reviewers of earlier versions of the manuscript for their comments, and I also thank Lana Barr for typing it.

H. Charles Romesburg

Cluster Analysis for Researchers

A Road Map to This Book

OBJECTIVES

This book is an introduction to cluster analysis for researchers. It

- Presents the most often used methods of hierarchical cluster analysis so that they can be understood and applied by researchers with limited backgrounds in mathematics and statistics.
- Shows a variety of uses of cluster analysis that span the biological sciences, the social sciences, and the natural sciences, as well as planning and management.
- Explains how to use cluster analysis validly in research.

This chapter discusses

- What cluster analysis is about.
- What level of understanding you can expect to gain from this book.
- How you should use this book to meet your needs.

1.1 WHAT CLUSTER ANALYSIS IS ABOUT

Cluster analysis is a generic name for a variety of mathematical methods, numbering in the hundreds, that can be used to find out which objects in a set are similar. For example, if we gathered a set of pebbles from a stream shore, noted their attributes of size, shape, and color, and sorted similar pebbles into the same piles, we would be physically performing a cluster analysis. Each pile of similar pebbles would be a cluster.

Mathematical methods of cluster analysis accomplish this mathematically. Instead of sorting real objects, these methods sort objects described as data. Objects with similar descriptions are mathematically gathered into the same cluster. In fact, if we cluster-analyzed a set of pebbles physically and then again by using a mathematical method of cluster analysis, we should obtain essentially the same set of clusters.

For a variety of research goals, researchers need to find out which objects in a set are similar and dissimilar. The best known of these research goals is the making of classifications. One reason that cluster analysis is so useful is that researchers in all fields need to make and revise classifications continually. For example, hydrologists need to make classifications of streams and watersheds. Psychiatric researchers need to make classifications of mental patients based on psychiatric test scores in order to improve their understanding of mental illness and plan for its treatment. Business researchers need to make classifications of companies in the course of doing market research. An endless list of other examples could be drawn up because classifications are essential building blocks in all fields of research.

In addition to making classifications, there are several other research goals for which researchers need to find out which objects are similar. The discussions in this book explain how cluster analysis can be used to help create interesting scientific questions that spur further research, to help create research hypotheses, and to test research hypotheses. You can also learn in this book how to use cluster analysis as an aid when making decisions in planning and management. Studying real applications that have a variety of purposes and contexts, such as those described in this book, can enable you to prepare yourself for using cluster analysis in your own research.

1.2 THE METHODS OF CLUSTER ANALYSIS IN THIS BOOK

The methods in this book were selected on the basis of a literature search of applications of cluster analysis. The search utilized the DIALOG, BRS, and SDC computerized data bases of abstracts of journal articles covering virtually all fields of application. Using the key phrase CLUSTER ANALYSIS, the search retrieved several thousand abstracts of articles published in the late 1970's and early 1980's. Most of these applications used certain methods of hierarchical cluster analysis. These methods have therefore been made the core of this book; lesser-used methods are covered in less depth or omitted entirely.

Methods of hierarchical cluster analysis follow a prescribed set of steps, the main ones being: (1) collect a data matrix whose columns stand for the objects to be cluster-analyzed and whose rows are the attributes that describe the objects; (2) optionally standardize the data matrix; (3) using the data matrix or the standardized data matrix, compute the values of a resemblance coefficient to measure the similarities among all pairs of objects; (4) use a clustering method to process the values of the resemblance coefficient, which results in a diagram called a **tree**, or **dendrogram**, that shows the hierarchy of similarities among all pairs of objects. From the tree the clusters can be read off.

These steps, which are dealt with in the next chapter, are constant. But within these steps you have the freedom to choose among alternative ways of standardizing the data matrix, of choosing a resemblance coefficient, and of choosing a clustering method. The upshot is that you have wide latitude in tailoring a cluster analysis to your specific application.

Based on their frequency of use by researchers, this book stresses certain methods of standardizing the data matrix, certain resemblance coefficients, and certain clustering methods. For example, it stresses the **UPGMA** clustering method because in hierarchical cluster analysis it is used the most often. It discusses several lesser-used clustering methods in enough detail to do them but does not give them equal treatment. Examples of these are the **single linkage**, the **complete linkage**, and **Ward's** clustering methods. Finally, it mentions in passing other clustering methods that are even less frequently used than these—for example, the **centroid** clustering method, which comprised less than two percent of the application articles found in the literature search.

The literature search also showed that applications of hierarchical cluster analysis outnumber applications of nonhierarchical cluster analysis by more than ten to one. Therefore, this book mentions only briefly some of the nonhierarchical methods, but other literature is cited in which you can learn more about them. Because the basic concepts that apply to finding clusters are the same, much of what you learn about hierarchical cluster analysis can also help you to understand nonhierarchical cluster analysis. (Henceforth in this book, it should be assumed that "hierarchical cluster analysis" is what is meant by "cluster analysis.")

1.3 SPECIAL FEATURES OF THIS BOOK

The mathematics of cluster analysis is well established and is treated in several books—two well-known ones are *Numerical Taxonomy* (Sneath and

Sokal, 1973) and *Cluster Analysis for Applications* (Anderberg, 1973). The first of these focuses on the philosophy of using numerical methods to make biological classifications, and the description of cluster analysis it gives is intermixed with a discussion of taxonomic philosophy. The second of these books presents the mathematics of cluster analysis but does not illustrate their calculations, nor does it contain applications that illustrate how cluster analysis is used in research. Consequently, there is a need for a book that explains cluster analysis as it is used in virtually all fields, not just numerical taxonomy, and that gives illustrations of calculations and applications. This book is aimed at satisfying this need.

A second difference between this book and others is that it presents cluster analysis within the framework of a research philosophy. Not only does it present the mathematics of cluster analysis, but it also shows how to frame and validate applications. By "frame" we mean how to choose correctly the objects and attributes for the data matrix, the method of standardization, the resemblance coefficient, the clustering method—in short, how to make the subjective judgments that transcend the objective mathematics of cluster analysis. And by "validate" we mean how to decide whether the cluster analysis has in fact achieved the intended research goal.

In particular, this book emphasizes that there are six different research goals you can address by using cluster analysis, and that you must identify and clearly state your research goal before you use cluster analysis. It is also emphasized that you must state the criteria that you will later use to frame and validate your cluster analysis. How these criteria are guided by norms within your field, how the norms differ among fields of science, and particularly how their applications in science differ from their applications in planning and management—all of these are important topics in this book.

Whether you are a graduate student or an experienced researcher, this material can help you improve the quality of your research in general, by heightening your awareness of the relation between the ends of research and the means of research. And it can also help you understand how researchers use unwritten norms to make countless subjective research decisions, whether your research uses cluster analysis, regression analysis, or computer simulation modeling.

Another special feature of this book is the inclusion of case studies, called "Research Applications," selected from the application literature of cluster analysis. They encompass a variety of research goals and contexts and give an overall picture of how cluster analysis is used—a kind of second sense to guide you as you use cluster analysis in your research. Some of these Research Applications serve as model examples, and describe the application of cluster analysis in detail from start to finish. Others are shorter and leave some details implicit while highlighting other important features, such as the nature of the objects and attributes that were used.

Finally, some end with a brief listing of other applications of cluster analysis that are similar to, or complement, or augment, the main application. By obtaining the original articles and studying them, you can further your understanding.

1.4 HOW TO USE THIS BOOK

This book is broken into the following four parts, plus appendices:

Part I. Shows the basic steps and calculations of cluster analysis; it also gives the six kinds of research goals that you can use cluster analysis to address and illustrates these with applications. Read Part I to gain an overview of cluster analysis: what it is and what it can be used for. At the conclusion of Part I you should be able to understand, at a general level, most applications articles.

Part II. Gives the details necessary to use cluster analysis in your research, alternative methods for standardizing the data, alternative resemblance coefficients, and alternative clustering methods. It also describes what information you should include in your own application reports or articles so that other researchers can understand how you have used cluster analysis.

Part III. Because cluster analysis is important for making classifications, this special use is given detailed discussion here.

Part IV. Gives the principles for framing and validating applications. This part teaches you how to select the correct methods of cluster analysis for your research application, and how to validate your choices.

Appendix 1. Gives a reading list of books and articles on cluster analysis to help you learn beyond this book. These materials contain more detail about hierarchical cluster analysis, and can also extend your understanding to nonhierarchical cluster analysis and other methods of multivariate analysis.

Appendix 2. Describes computer programs for doing cluster analysis. Among those discussed are the CLUSTAR and CLUSTID computer programs, which do all the methods presented in this book, and the cluster analysis routines in the popular statistical packages such as BMDP and SAS.

Exercises are given at the end of certain chapters (answers are given in the back of the book). Working these should reinforce your understanding.

PART I

Overview of Cluster Analysis

When I say "similarity exists," it is this fact about the world, not a fact about language, that I mean to assert. The word "yellow" is necessary because there are yellow things; the word "similar" is necessary because there are pairs of similar things. And the similarity of two things is as truly a nonlinguistic fact as the yellowness of one thing.

BERTRAND RUSSELL
An Inquiry into Meaning and Truth

Much research depends on the estimation of the similarities between pairs of things. Cluster analysis is the most basic method for estimating similarities. Therein lies its use and appeal.

Applications of cluster analysis are found in virtually all professions. Some aid science, and others aid planning and management. An archaeologist uses cluster analysis to piece together evidence supporting a hypothesis about how a prehistoric Indian culture lived, for example; a historian uses it to show that sculptors of ancient Greece were influenced by Egyptian sculptors' style; and an industrial engineer uses it to find the best layout for a factory's machines.

Cluster analysis is easily communicated. Its aims are directly and simply stated. Its language is ordinary arithmetic. And it is easily learned. It is one quantitative method anyone can do, and do well.

Part I begins with an explanation of the basic calculations of cluster analysis that are required to estimate similarities between pairs of things. Then, a variety of applications in science, and in planning and management are presented; all required that researchers estimate similarities.

Basics—The Six Steps of Cluster Analysis

OBJECTIVES

When stripped of detail, the skeleton of any subject is the part that cannot be removed without destroying the subject itself. Once the skeleton is seen, details can be added and understood in relation to each other. So it is with cluster analysis. Its skeleton comprises six steps.

- Obtain the data matrix.
- Standardize the data matrix.
- Compute the resemblance matrix.
- Execute the clustering method.
- Rearrange the data and resemblance matrices.
- Compute the cophenetic correlation coefficient.

All else is supplementary detail.

2.1 STEP 1—OBTAIN THE DATA MATRIX

Cluster analysis begins with a **data matrix**. The data matrix we will first cluster-analyze is the following.

Data matrix

		Object				
		1	2	3	4	5
Attribute	1	10	20	30	30	5
	2	5	20	10	15	10

The columns of the data matrix stand for objects, the things whose similarities to each other we want to estimate. Its rows stand for attributes, the properties of the objects.

For example, the objects could be people; the attributes, a set of their responses to a psychiatric test. Or the objects could be plots of land in a forest; the attributes, counts of the tree species growing on the plots. Or the objects could be fossils; the attributes, their dimensions. That the objects and attributes can be almost anything gives cluster analysis endless variety.

The cluster analyst's objective is to find out which objects are similar and dissimilar to each other. Which people are similar, which plots of land are similar, which fossils are similar—and in general, which objects are similar? By finding out which ones are similar, the analyst automatically finds out which are dissimilar.

With a small data matrix like the one just shown, there is ordinarily no need for cluster analysis. We can simply look at the data matrix and find the similar and dissimilar objects. Two objects that have about the same values, attribute for attribute, are more similar than two objects that do not. But for a large data matrix—one with, say, a hundred objects and a hundred attributes—visual inspection fails us and cluster analysis becomes useful.

For the sake of interest, our data matrix needs a context. Imagine that the five objects are plots of farm land. Two attributes describe their important properties: (1) water-holding capacity (cm^3 of H_2O/cm^3 of soil, expressed as a percent); and (2) the percent by weight of soil organic matter. Our objective is to find the two most similar plots, because we wish to treat one with fertilizer while leaving the other, a control plot, untreated. It is therefore crucial that this experiment use two similar plots; that way, any differential response of plant growth can be attributed to the treatment and not to initial differences among the plots themselves.

2.2 STEP 2—STANDARDIZE THE DATA MATRIX

This step is optional. Standardizing the data matrix converts the original attributes to new unitless attributes. As Chapter 7 shows, standardizing requires only simple arithmetic. Without going into detail now other than to say that we have used equation 7.1, one of the alternative standardizing functions in Chapter 7, we convert the original data matrix to the following **standardized data matrix**:

Standardized data matrix

		Object				
		1	2	3	4	5
Attribute	1	−0.79	0.90	0.96	0.96	−1.22
	2	−1.23	1.40	−0.35	0.53	−0.35

Like the original data matrix, it has five objects and two attributes. But its values have changed and some are even negative.

The main calculations of the cluster analysis can be done on either the data matrix or on the standardized data matrix. Standardizing recasts the units of measurement of attributes as dimensionless units. It does not matter whether the attributes of the data matrix are measured in a mixture of units (in inches, meters, pounds, grams, or whatever). Standardizing strips the identity from each attribute, changes its numerical value, and recasts it in dimensionless form.

To discuss the reasons for wanting to standardize the data matrix, and to show how it is done, would at this point only confuse you. Therefore, in your mind think of Step 2 as optional. By the time you get to Chapter 7 you will know how to make use of standardization. For now, we will cluster-analyze the original data matrix rather than the standardized data matrix.

2.3 STEP 3—COMPUTE THE RESEMBLANCE MATRIX

A **resemblance coefficient** measures the overall resemblance—the degree of similarity—between each pair of objects. A resemblance coefficient is a mathematical formula. Its value is computed by entering, for a given pair of objects, the values from their columns in the data matrix (or standardized data matrix if it is being used). The formula gives one value that represents how similar the pair of objects are. (In Part II it is shown that there are many different resemblance coefficients to choose from.)

The data matrix given in Section 2.1 contains five objects. This means there are ten values of the resemblance coefficient to compute, one value for each pair of objects. We store the ten values in a matrix called the **resemblance matrix.**

The diagram below shows the computed resemblance matrix for our example:

Resemblance matrix

First object in pair

Second object in pair	1	2	3	4	5
1	—	—	—	—	—
2	18.0	—	—	—	—
3	20.6	14.1	—	—	—
4	22.4	11.2	5.00	—	—
5	7.07	18.0	25.0	25.5	—

It is square. Each column identifies the first object in a pair; each row identifies the second object. The cell formed by the intersection of a column and row contains the value of the resemblance coefficient for the given pair of objects. Moreover, the concept of resemblance is symmetric: the resemblance between objects 2 and 1 is the same as between objects 1 and 2. For this reason, only the lower-left half of the matrix contains values.

A **resemblance coefficient** is always one of two types: (1) a **dissimilarity coefficient** or (2) a **similarity coefficient**. Their difference is a matter of "direction." The *smaller* the value of a *dis*similarity coefficient, the *more similar* the two objects are. The larger its value, the more dissimilar they are. Conversely, the *larger* the value of a *similarity* coefficient, the *more similar* the two objects are. And the smaller its value, the more dissimilar they are.

For example, two objects whose dissimilarity coefficient has a value of 5.0 are more similar than two whose value is 6.0. And two whose similarity coefficient is 12.0 are more similar than two whose value is 8.0. Thus the difference between a similarity coefficient and a dissimilarity coefficient is merely a difference in which way the scale runs.

In this example, the resemblance matrix contains the ten values of a dissimilarity coefficient called the **Euclidean distance coefficient**. As its name implies, it measures the literal distance between two objects when they are viewed as points in the two-dimensional space formed by their attributes.

Let us take the original data matrix, treat the two attributes as coordinates that fix the positions of the objects, and plot the objects in this space.

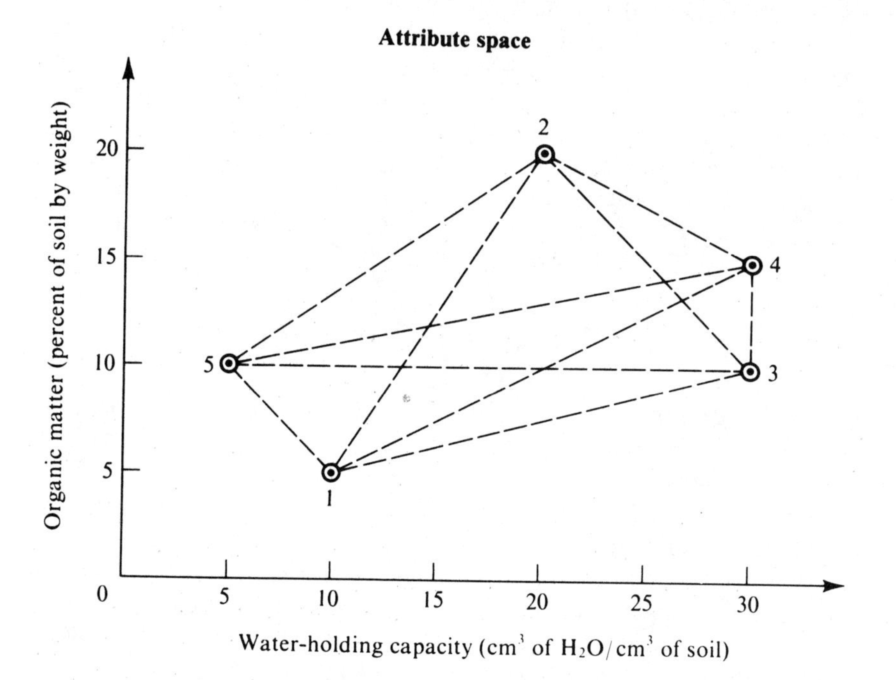

Then we can use dashed lines, connecting the ten pairs of objects, to show the distances between pairs. These distances will next be computed by using the Pythagorean theorem. From plane geometry recall that the Pythagorean theorem says, "the hypotenuse of a right triangle is equal to the square root of the sum of the squares of the other two sides." Pictorially, it looks like this:

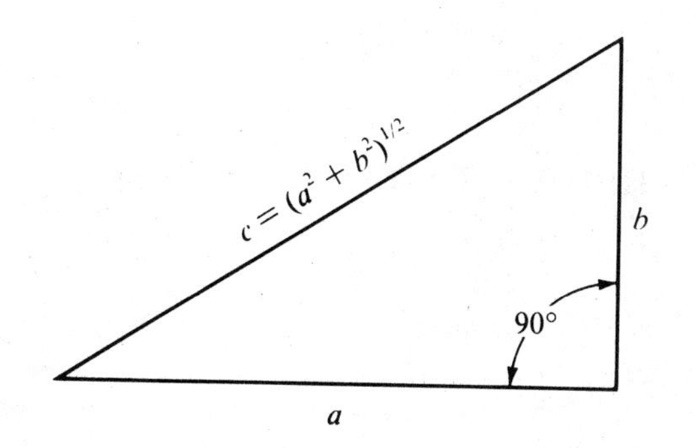

Note that the exponent "1/2" means to take the square root of the quantity inside the parentheses it is attached to. In the remainder of this book the exponent "1/2" is used to mean the square root. Thus the length of the hypotenuse of the above triangle is $c = \sqrt{(a^2 + b^2)} = (a^2 + b^2)^{1/2}$.

Take a pencil now and lightly make a dot on the attribute space plot shown above at the point (5, 5)—that is, where the values of both attributes are 5. With a ruler, draw a vertical line upward from the horizontal baseline to object 5, and a horizontal line rightward from the vertical line to object 1. These lines, with the already-drawn dashed line connecting objects 1 and 5, define a right triangle. Letting the symbol e_{15} stand for the Euclidean distance coefficient for objects 1 and 5, the Pythagorean theorem gives the following length for the dashed line from 1 to 5:

$$e_{15} = \left[(10 - 5)^2 + (5 - 10)^2\right]^{1/2} = 7.07.$$

This value is written in the resemblance matrix in the cell formed by the intersection of column 1 and row 5. By convention the smaller subscript is written first—that is, as e_{15} and not e_{51}. In general, e_{jk} is the Euclidean distance coefficient for objects j and k, where j is the smaller of the two object identification numbers.

In this way we construct ten imaginary right triangles, one for each pair of objects. We then compute the ten values of the Euclidean distance coefficient between the pairs and enter them into the resemblance matrix. (To test your understanding, compute one of the other e_{jk} values.)

The dissimilarity coefficient e_{jk} can be no smaller than 0.0, its value when objects j and k are maximally similar—that is, when they are identical. Identical objects occupy the same point in the attribute space: one plots on top of the other, and their distance apart is zero. But even when objects j and k plot very near to each other and e_{jk} has a very small value, say $e_{jk} = 0.01$, because the objects do not occupy the same point they are no longer identical. They are nearly identical, however, and we would call them "highly similar." The larger e_{jk} is, the less similar the objects j and k are, and the more we must search for apt adjectives that describe their similarity. As e_{jk} increases, the similarity of any two objects j and k decreases through a progression of adjectives like: highly similar → mildly similar → mildly dissimilar → highly dissimilar.

2.4 STEP 4—EXECUTE THE CLUSTERING METHOD

When we have found the values of the Euclidean distance coefficient for all pairs of objects, we will be able to produce a map of sorts, called a **tree**, that shows at a glance the degrees of similarity between all pairs of objects.

A resemblance matrix is turned into a tree by using a **clustering method**. A clustering method is a series of steps that incrementally "eats away" at the resemblance matrix, reducing it in size. At the same time, it develops the tree from the e_{jk} values that are eaten away. At the last step, the resemblance matrix disappears completely while the final touches are applied to complete the tree.

A **cluster** is a set of one or more objects that we are willing to call similar to each other. A cluster can be as few as one object, if we are willing to call no other objects similar to that object. Or it can be as many as all of the objects in the data matrix, if we are willing to call all of them similar to each other. It may seem strange to use the word "willing," but that is exactly the right word. To call two or more objects similar, we must be willing to neglect some of the detail that makes them nonidentical. We must be tolerant of some of their differences.

At the start of the steps of the clustering method, each object is regarded as being in a separate cluster; so, in our example with five objects, there will be five clusters at the start. Each clustering step will merge the two most similar clusters that exist at the start of the step, reducing the number of clusters by one.

The clustering method that we will use is called the **unweighted pair-group method using arithmetic averages**, which convention abbreviates as UPGMA. Of the clustering methods that are discussed in Chapter 9, it is the one most widely used. Using the resemblance matrix we have computed, the following steps show how the UPGMA clustering method is computed.

Step 1. Merge clusters 3 and 4, giving 1, 2, (34), and 5 at the value of $e_{34} = 5.0$.

Before doing this, observe that there are five clusters, each containing one object and having the identification number of the object it contains—1, 2, 3, 4, and 5. No two clusters are as similar as clusters 3 and 4 because no other value of e_{jk} is smaller than 5.0. Step 1 merges the two most similar clusters—clusters 3 and 4.

It is possible that two or more pairs of clusters tie for being the most similar pair at a clustering step. Although this seldom happens in practice, if it does occur we should select one of the tied pairs at random and merge the two clusters that it contains.

As a result of step 1, clusters 3 and 4 have disappeared. In their places a new cluster, written as *(34)*, has appeared. The clusters 1, 2, and 5 remain. They were unaffected by the clustering step. Bear in mind that the merging occurred at the value of $e_{34} = 5.0$.

Think of what this means. At the start of step 1 we were not willing to call any pair of clusters (objects) similar. What is the smallest amount by which we would have to relax our tolerance—increase our blindness to differences between objects—in order to be willing to call two separate clusters the same: The answer is 5.0, and the two clusters are *3* and *4*. They became the new cluster *(34)*.

Step 1 leaves the attribute space looking this way:

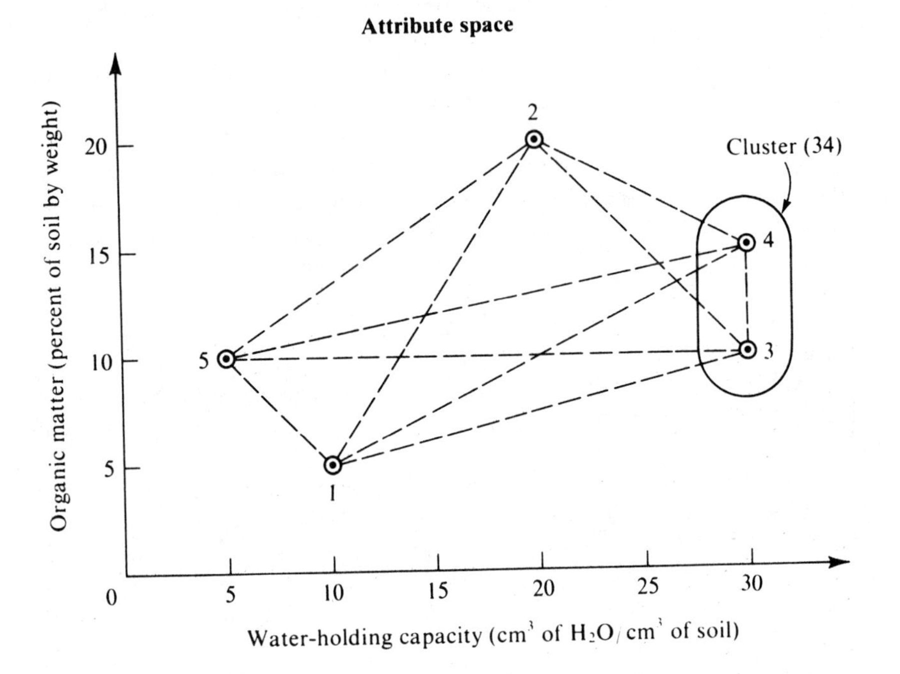

The new cluster (34) is circled; clusters 1, 2, and 5 remain as before.

Before going to step 2, we must revise the resemblance matrix. It is now smaller, "eaten away" by one row and one column. It shows the values of e_{jk} between all pairs of the four clusters that result.

	1	2	5	(34)
1	—	—	—	—
2	18.0	—	—	—
5	7.07	18.0	—	—
(34)	21.5	12.7	25.3	—

Here is where its values come from. Because step 1 did not affect clusters 1, 2, and 5, the values of the Euclidean distance coefficient between them remain unchanged and are transcribed from the initial resemblance matrix: we transcribe $e_{12} = 18.0$, $e_{15} = 7.07$, and $e_{25} = 18.0$. However, the values between clusters 1 and (34), between clusters 2 and (34), and between clusters 5 and (34) must be recalculated. To do this, we follow the dictates of the UPGMA clustering method and compute the averages of the values between the individual objects in one cluster and those in the other cluster, namely:

$$e_{1(34)} = \tfrac{1}{2}(e_{13} + e_{14}) = \tfrac{1}{2}(20.6 + 22.4) = 21.5;$$

$$e_{2(34)} = \tfrac{1}{2}(e_{23} + e_{24}) = \tfrac{1}{2}(14.1 + 11.2) = 12.7;$$

$$e_{5(34)} = \tfrac{1}{2}(e_{35} + e_{45}) = \tfrac{1}{2}(25.0 + 25.5) = 25.3.$$

The values of all e_{jk} used in the averaging come from the initial resemblance matrix.

Step 2. Merge clusters 1 and 5, giving 2, (34), and (15) at the value of $e_{15} = 7.07$.

Here is how step 2 is decided. We scan the revised resemblance matrix, looking for the smallest value. It is $e_{15} = 7.07$. Thus, we merge clusters 1 and 5 at this value. This leaves three clusters: 2, (34), and (15).

With step 2 completed, the attribute space looks like this.

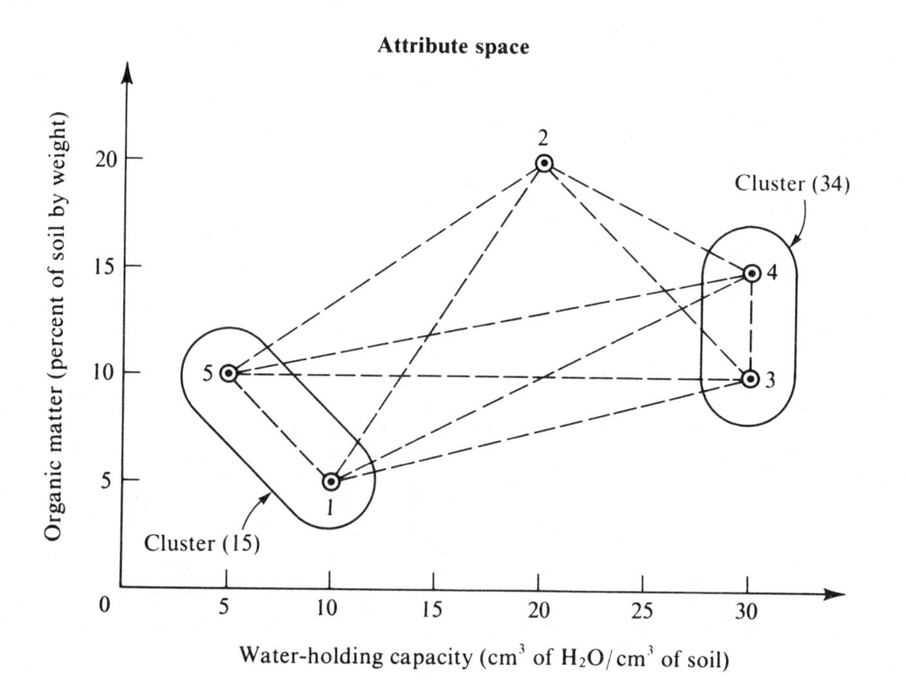

Object 2 is in a cluster by itself; objects 3 and 4 are in cluster (34); objects 1 and 5 are in cluster (15).

Before going on to the next clustering step, a new revised resemblance matrix must be computed:

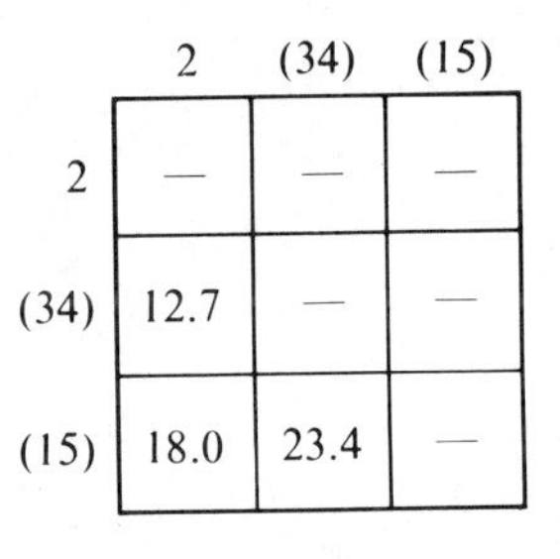

	2	(34)	(15)
2	—	—	—
(34)	12.7	—	—
(15)	18.0	23.4	—

Its values are computed as follows. Step 2 left the Euclidean distance between clusters 2 and (34) unchanged; thus, $e_{2(34)} = 12.7$ is transcribed from the previous revised resemblance matrix. The two remaining values, $e_{2(15)}$ and $e_{(15)(34)}$, are calculated by averaging the Euclidean distances between the individual objects that are in different clusters, namely:

$$e_{2(15)} = \tfrac{1}{2}(e_{12} + e_{25}) = \tfrac{1}{2}(18.0 + 18.0) = 18.0;$$

$$\begin{aligned} e_{(15)(34)} &= \tfrac{1}{4}(e_{13} + e_{14} + e_{35} + e_{45}) \\ &= \tfrac{1}{4}(20.6 + 22.4 + 25.0 + 25.5) \\ &= 23.4. \end{aligned}$$

The attribute space plot makes it clear that four values of e_{jk} must be averaged to get $e_{(15)(34)}$: there are four pairwise distances between the objects in clusters (15) and (34), shown as four dashed lines.

Step 3. Merge clusters 2 and (34), giving (15) and (234) at the value of $e_{2(34)} = 12.7$.

Step 3 is decided by scanning the newly revised resemblance matrix and looking for the smallest value. It is 12.7, indicating that clusters 2 and (34) are the most similar and therefore to be merged. Again, we revise the attribute space:

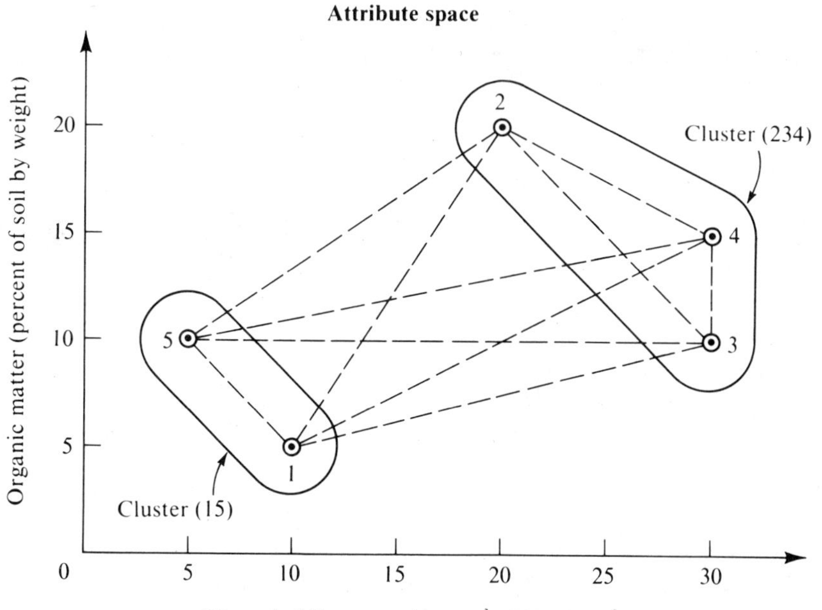

Before approaching the final step, we must revise the resemblance matrix for the last time. Step 3 left two clusters, (15) and (234). There is only one value of Euclidean distance remaining, the one between the two clusters, which is $e_{(15)(234)}$. We obtain it as before, by averaging the Euclidean distances between the objects in the one cluster and those in the other. The attribute space plot shows that six distances must be averaged, namely:

$$e_{(15)(234)} = \tfrac{1}{6}(e_{12} + e_{13} + e_{14} + e_{25} + e_{35} + e_{45})$$
$$= \tfrac{1}{6}(18.0 + 20.6 + 22.4 + 18.0 + 25.0 + 25.5) = 21.6.$$

The revised resemblance matrix now becomes:

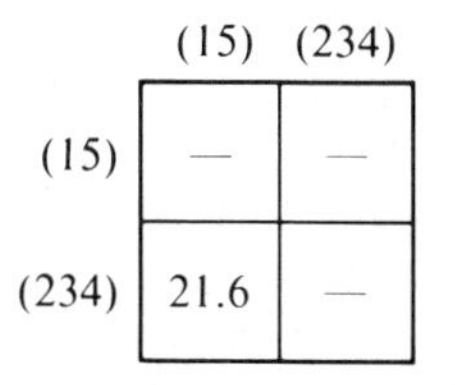

	(15)	(234)
(15)	—	—
(234)	21.6	—

Step 4. Merge clusters (15) and (234), giving (12345) at the value of $e_{(15)(234)} = 21.6$.

This last step is automatic since the smallest entry in the newly revised resemblance matrix is the only entry. The attribute space plot now shows one cluster, (12345), containing all five objects:

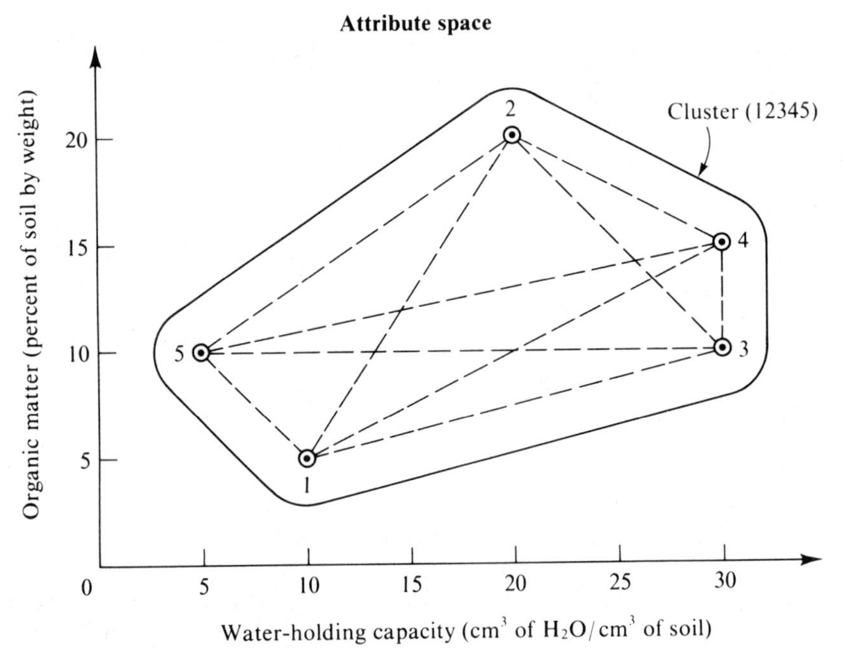

We can depict what the four clustering steps have accomplished in another way. With each clustering step, a tree has been growing. It is now fully grown and appears on page 21.

The tree's left-hand vertical scale gives the value of the Euclidean distance coefficient, e_{jk}. The five objects are the plots of farm land, and they number the tree's branch tips. The lines in the tree running upwards from the branch tips are joined with horizontal lines at the values of e_{jk} given by the clustering steps. For example, the tree shows that at step 1 clusters 3 and 4 were merged at the value of $e_{34} = 5.0$; that is, the branches containing objects 3 and 4 join at a value of 5.0.

Both the initial resemblance matrix and the tree show the similarity between any pair of objects. However, when we look at the resemblance matrix we can only pick out the similarity between individual pairs. We

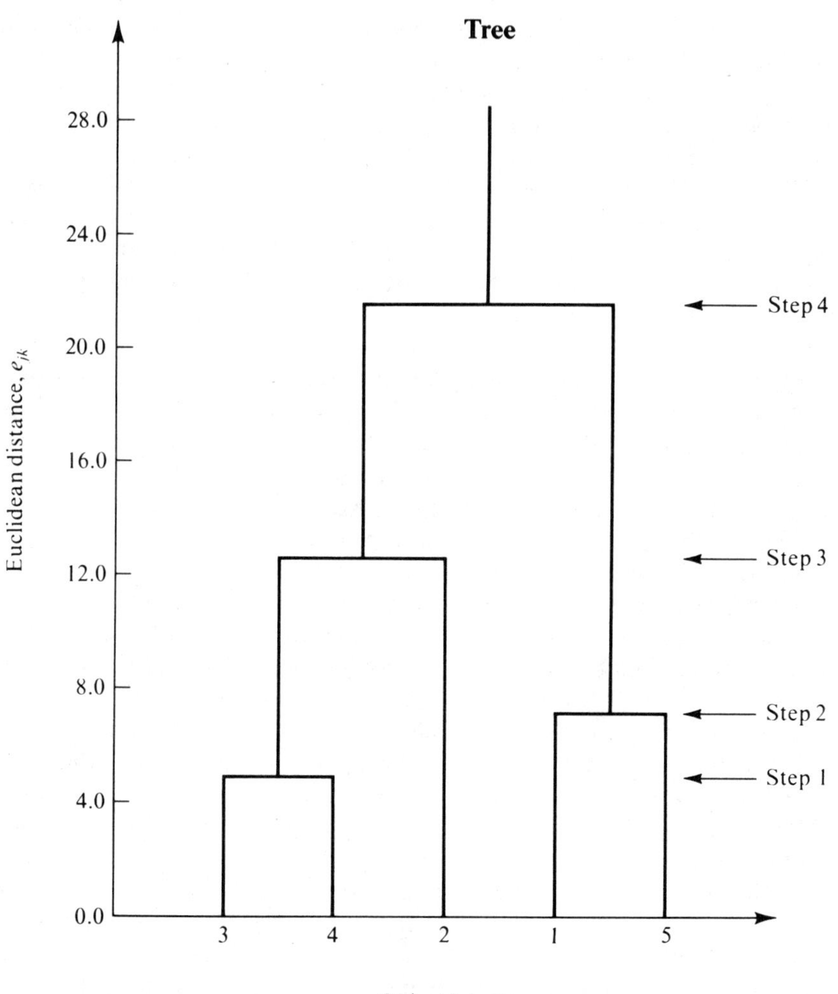

cannot easily grasp the relation of similarity among the many pairs. The value of the tree is that it lets us grasp this relation. To illustrate this, consider the tree. The value of the similarity of objects j and k is the value of e_{jk} at which their branches join. For example, the similarity of objects 2 and 3 is $e_{23} = 12.7$, because the branches connecting objects 2 and 3 join at this level. Likewise, the similarity of objects 3 and 5 is $e_{35} = 21.6$.

The similarity between any two objects j and k, as measured by their e_{jk} value in the resemblance matrix, may not agree exactly with their e_{jk} value obtained from the tree. This is because the clustering method must distort, to some degree, the similarity structure of the resemblance matrix in order to produce the tree. The sixth step of cluster analysis, explained in Section 2.6, shows how to assess the amount of distortion, to decide if it is tolerable.

How would the tree be used in the context of our original problem of finding two similar soil plots, one to be treated with fertilizer and the other to be left untreated as a control? The tree shows that plots 3 and 4 are the most similar. Should the researcher decide they are similar enough for his purpose, then one could be given the fertilizer treatment and the other could be left as the untreated control. Inasmuch as we found that plots 3 and 4 were the most similar as soon as step 1, why is the whole tree needed to answer the question? One reason is that it lets us see the similarity between plots 3 and 4 in relation to the similarities between all pairs of plots. Similarity is relative. Restricted to plots 3 and 4, the resemblance coefficient $e_{34} = 5.0$ does not say much, because there is no way of telling whether the value 5.0 stands for a high or low degree of similarity. The tree tells us how similar the most similar pair of objects is compared to the similarities of other pairs of objects. A second reason for generating the tree is that it lets us change our minds. We may decide we want three similar soil plots; two for treatment, one for control. In that case the tree suggests plots 2, 3, and 4.

The map given by the tree is a hierarchical ordering of similarities that begins at the tree's bottom where each object is separate, in its own cluster, and similar only to itself. As we move to higher levels of e_{jk} we become more tolerant and allow differences between objects to exist while still proclaiming them to be similar. As we reach the tree's top we are tolerant of all differences in detail, and all objects are put into one cluster and we look on them as similar to each other.

That is to say, the concept of similarity exists on a variable scale of resolution. Each of us, according to our needs and circumstances, sets the right level of resolution. We are always caught between the opposite extremes of wanting complete detail about nature and yet wanting economy of detail too. Complete detail comes at the expense of cramming our brains with so many facts we cannot think. Little or no detail gives our brains too little to think about.

We can view evolution as establishing the point of tradeoff between too much and too little detail. Imagine two prehistoric hunters, one at each pole of these extremes. As a tiger charges, the first hunter stores every detail about it and its environment. By the time he processes this detail and arrives at an optimal decision, it is too late. Meanwhile, the second hunter cannot decide whether he is being charged by a tiger or a large yellow butterfly. He is still thinking about it, searching for some detail that can make a difference in what he will do, when.... The hunters that survived and passed their genes to us were a third type cast between these extremes. They carved sensory experience into classes that had enough detail to call tigers and butterflies dissimilar animals, while still leaving mental room for thinking and deciding how to deal with these classes.

The extremes of full detail and no detail have decided disadvantages. The optimal compromise lies somewhere between at a point that changes from one research problem to another. Part of our adventure into cluster analysis will be to understand how these points of optimal compromise can be determined.

With step 4 we reached our main objective, the tree. Two more steps, dealt with in the next two sections, take us to important secondary objectives.

2.5 STEP 5—REARRANGE THE DATA AND RESEMBLANCE MATRICES

This step, which requires no computation, rearranges the values in the data matrix and in the resemblance matrix, so as to make the similarities between objects stand out as given in the tree. To illustrate, let us recopy the data matrix and the resemblance matrix for our example:

Data matrix

Attribute \ Object	1	2	3	4	5
1	10	20	30	30	5
2	5	20	10	15	10

Resemblance matrix

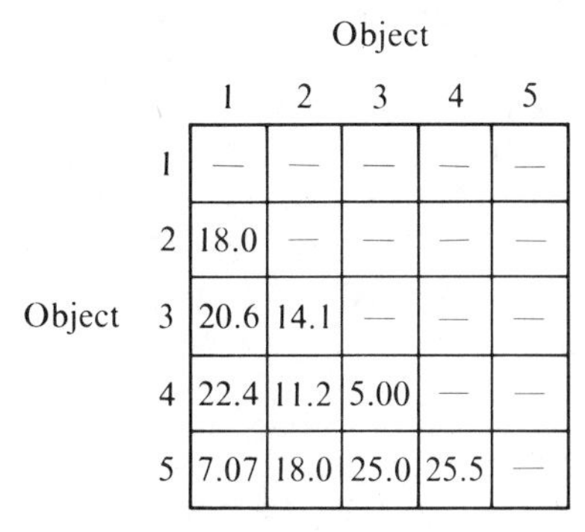

Object	1	2	3	4	5
1	—	—	—	—	—
2	18.0	—	—	—	—
3	20.6	14.1	—	—	—
4	22.4	11.2	5.00	—	—
5	7.07	18.0	25.0	25.5	—

Recall that we arbitrarily assigned identification numbers to objects and therefore had no way to know whether or not the most similar objects were adjacent to each other in these matrices. However, the most similar objects tend to have adjacent positions across the tree's bottom: there the order is 3, 4, 2, 1, 5. Rearranging the data and resemblance matrices is a matter of writing the numbers according to their order in the tree:

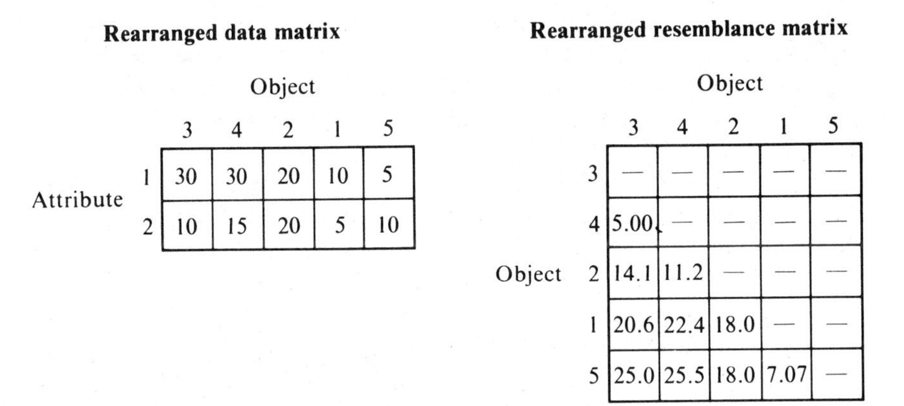

Rearranged data matrix

Attribute \ Object	3	4	2	1	5
1	30	30	20	10	5
2	10	15	20	5	10

Rearranged resemblance matrix

Object	3	4	2	1	5
3	—	—	—	—	—
4	5.00	—	—	—	—
2	14.1	11.2	—	—	—
1	20.6	22.4	18.0	—	—
5	25.0	25.5	18.0	7.07	—

In the **rearranged data matrix**, similar objects 3 and 4, and similar objects 1 and 5 are side by side and can be easily compared and contrasted. In the **rearranged resemblance matrix**, the most similar pairs of objects tend to lie near the diagonal; the similarity between pairs tends to decrease away from the diagonal.

This organization helps to highlight the interobject similarities in the data themselves. While the tree suggests similarities, we should go back to the original data matrix and the original resemblance matrix, and rearrange them to confirm the nature of the similarities.

On small matrices like the ones in our example the power of rearranging to help reveal the similarities between objects is hardly noticeable. But in examining large matrices—those with hundreds of objects and attributes—the unaided mind can become overloaded and the benefits of rearranging are then striking and appreciated.

2.6 STEP 6—COMPUTE THE COPHENETIC CORRELATION COEFFICIENT

A map is not exactly like the terrain it represents. In the same way, a tree is not exactly like the data matrix it represents. We need to know how well the tree represents the data matrix. The **cophenetic correlation coefficient,** $r_{X,Y}$, gives a partial answer: it measures how well the tree and the resemblance matrix "say the same thing." While it would be best if we could compare the tree (the output of a cluster analysis) with the data matrix (the input to a cluster analysis), this comparison is not feasible. So we do the next best thing and compare the tree and the resemblance matrix.

To compute $r_{X,Y}$ we begin by converting the tree to its equivalent, the **cophenetic matrix**. The comparison is then made between the resemblance matrix (recopied on the left, below) and the cophenetic matrix (shown on the right).

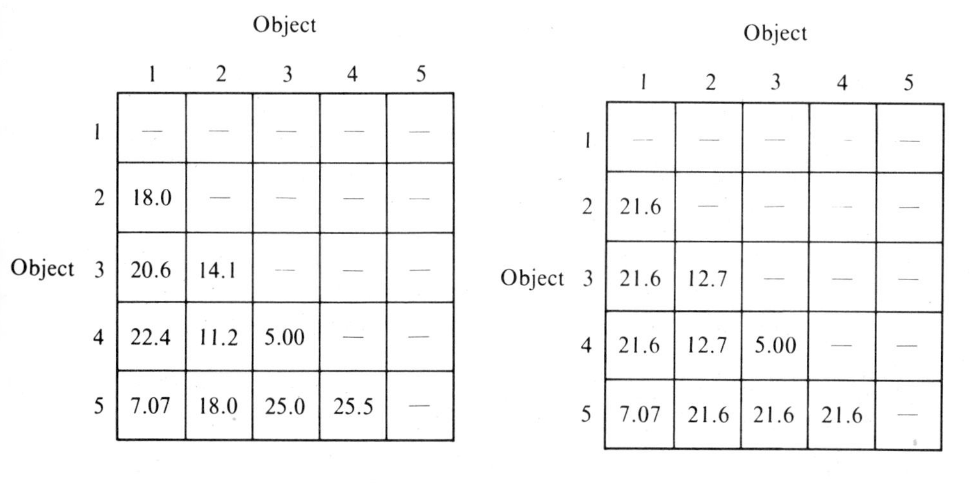

Resemblance matrix

Object	1	2	3	4	5
1	—	—	—	—	—
2	18.0	—	—	—	—
3	20.6	14.1	—	—	—
4	22.4	11.2	5.00	—	—
5	7.07	18.0	25.0	25.5	—

Cophenetic matrix

Object	1	2	3	4	5
1	—	—	—	—	—
2	21.6	—	—	—	—
3	21.6	12.7	—	—	—
4	21.6	12.7	5.00	—	—
5	7.07	21.6	21.6	21.6	—

The values that appear in the cophenetic matrix come from the tree. To illustrate, the value $e_{23} = 12.7$ is the largest value of e_{jk} found in tracing a path from object 2 upwards through the tree (shown in Section 2.4) and downwards again to object 3. This is the level at which objects 2 and 3 were merged into the same cluster at step 3. All remaining values in the cophenetic matrix are obtained in the same way—that is, by tracing the path connecting each pair of objects to its highest point in the tree.

The tree and its cophenetic matrix are two different forms of the same thing. If we have a tree we can always compute its cophenetic matrix. And should we lose the tree, the cophenetic matrix contains all the information needed to reconstruct the tree. Since we can go from one to the other without loss of information, the two are equivalent representations of the similarity structure between pairs of objects.

To affect the comparison, we unstring the resemblance matrix and the cophenetic matrix into two lists, one for each matrix, where each row corresponds to the same (j, k)-cell. The X list and the Y list contain respectively the values in the resemblance and cophenetic matrices. Each list is ten items long. The following are the two unstrung lists for the example:

Cell	X	Y
(2, 1)	18.0	21.6
(3, 1)	20.6	21.6
(4, 1)	22.4	21.6
(5, 1)	7.07	7.07
(3, 2)	14.1	12.7
(4, 2)	11.2	12.7
(5, 2)	18.0	21.6
(4, 3)	5.00	5.00
(5, 3)	25.0	21.6
(5, 4)	25.5	21.6

Next we compute the Pearson product-moment correlation coefficient, $r_{X,Y}$, between the lists X and Y. For this particular use we rename it the **cophenetic correlation coefficient**. Its formula, which is given in most elementary statistics texts and is programmed into many scientific pocket calculators, is

$$r_{X,Y} = \frac{\Sigma xy - (1/n)(\Sigma x)(\Sigma y)}{\left\{\left[\Sigma x^2 - (1/n)(\Sigma x)^2\right]\left[\Sigma y^2 - (1/n)(\Sigma y)^2\right]\right\}^{1/2}}. \tag{2.1}$$

The component sums (indicated by Σ) are of the n terms in the two lists X and Y. To calculate $r_{X,Y}$ for our example, we therefore first calculate

$$\Sigma xy = (18.0)(21.6) + (20.6)(21.6) + \cdots + (25.5)(21.6) = 3{,}193.5;$$

$$\Sigma x = 18.0 + 20.6 + \cdots + 25.5 = 166.9;$$

$$\Sigma y = 21.6 + 21.6 + \cdots + 21.6 = 167.1;$$

$$\Sigma x^2 = (18.0)^2 + (20.6)^2 + \cdots + (25.5)^2 = 3{,}248.6;$$

$$\Sigma y^2 = (21.6)^2 + (21.6)^2 + \cdots + (21.6)^2 = 3{,}196.9.$$

Putting these sums into equation 2.1 gives

$$r_{X,Y} = \frac{3{,}193.5 - (1/10)(166.9)(167.1)}{\left\{\left[3{,}248.6 - (1/10)(166.9)^2\right]\left[3{,}196.9 - (1/10)(167.1)^2\right]\right\}^{1/2}} = 0.93.$$

This is a little less than perfect concordance ($r_{X,Y} = 1.0$); but it is far above the point of no concordance ($r_{X,Y} = 0.0$); and it is out of the range of a negative concordance ($-1.0 \le r_{X,Y} < 0.0$). Our value indicates that the tree represents well the interobject similarity structure inherent in the resemblance matrix.

In essence, the clustering method is interposed between its input—the resemblance matrix—and its output—the tree. The coefficient $r_{X,Y}$ is an index that tells us how much the clustering method distorts the information in its input in order to produce its output. We cannot state exact guidelines that define how much distortion is tolerable. However, most fields would accept that when $r_{X,Y}$ is large (thus approaching concordance), say 0.8 or more, the distortion is not great.

Other indicies for assessing the distortion are discussed in Section 14.3, although in practice the cophenetic correlation coefficient is the index most researchers prefer to use.

SUMMARY

Cluster analysis is a method for displaying the similarities and dissimilarities between pairs of objects in a set. It consists of six steps and uses only simple arithmetic.

For the first step, the researcher assembles a set of objects, defines and measures their attributes, and enters the data into a data matrix. At the second step, which is optional, the researcher standardizes the data and enters it into a standardized data matrix. The third step requires choosing a resemblance coefficient and computing its value for each pair of objects, using either the data matrix or the standardized data matrix. The resemblance coefficient is a measure of how similar each pair of objects is.

With the fourth step, the researcher chooses a clustering method and uses it to transform the resemblance matrix into a tree. The tree makes it easy to see the similarities and dissimilarities between all pairs of objects.

The next two steps are auxiliary. The fifth step rearranges the objects in the original and the standardized data matrices, and in the resemblance matrix, so that similar objects tend to be near each other. This helps the researcher to find similarities and dissimilarities visually among the objects of the input data. The sixth and final step computes an index called the cophenetic correlation coefficient. It measures the agreement between the resemblance matrix (the input to the clustering method) and the tree (the output of the clustering method). That is, it measures how much of the similarity structure in the resemblance matrix is retained in the tree.

EXERCISES

(Answers to exercises are given in the back of the book.)

2.1. Using the data matrix given below, do the following and satisfy yourself that you know how to do a cluster analysis:
(a) Make an attribute space plot of the five objects.
(b) Using the Euclidean distance coefficient, e_{jk}, and the UPGMA clustering method, perform a cluster analysis. Draw the tree.
(c) Compare your tree to your attribute space plot.
Data Matrix:

	Object				
Attribute	1	2	3	4	5
1	2	9	4	10	3
2	2	5	3	2	10

2.2. Suppose you are doing a cluster analysis of six objects by using (1) the Euclidean distance coefficient, e_{jk}, for the resemblance coefficient, and (2) the UPGMA clustering method. You calculate the initial resemblance matrix, and it is:

	Cluster					
Cluster	1	2	3	4	5	6
1	—	—	—	—	—	—
2	6.1	—	—	—	—	—
3	3.2	7.3	—	—	—	—
4	7.3	10.0	4.1	—	—	—
5	7.1	2.2	8.9	12.0	—	—
6	7.8	4.0	7.3	8.2	6.1	—

When you perform the cluster analysis (you aren't asked to do this as part of the exercise), the fourth clustering step is

Step 4. Merge 4 and (13), giving (256) and (134) at the value of $e_{4(13)} = 5.7$.
Calculate the last resemblance matrix value, $e_{(256)(134)}$.

General Features of Cluster Analysis

OBJECTIVES

For the expedience of showing how cluster analysis is done, we have so far omitted some details or sketched them only lightly. In this chapter we fill them in, first, by discussing in general terms

- How cluster analysis is used in numerical taxonomy to make classifications.
- The different scales of measurement that can be used to measure attributes.
- How the scales of measurement constrain the choice of resemblance coefficients that can be used.

Then, we see

- How to cluster-analyze attributes as well as objects.
- How computers are used to do cluster analysis.

3.1 RELATION OF CLUSTER ANALYSIS TO NUMERICAL TAXONOMY

Numerical taxonomy is a subject that uses numerical methods to make classifications of objects. Every profession has its objects of interest for which classifications must be made. For example, biology is concerned with classifying forms of life, medicine is concerned with classifying diseases of patients, and geology is concerned with classifying rocks and ores.

In addition to cluster analysis, numerical taxonomists use other multivariate methods such as factor analysis, multidimensional scaling, principal component analysis, and principal coordinate analysis. But cluster analysis is the method most often used. Unlike other multivariate methods that are based on mathematical deduction and expressed through matrix algebra, cluster analysis is an algorithmic series of steps that uses matrices for tidiness but does not use them algebraically for calculations.

Because of the mathematical simplicity of cluster analysis, many assume that it may be inferior to the other, more complex multivariate methods. However, no scientific study has ever shown that "mathematical simplicity" equates to "inferiority" and that the more complex a method is the better it must be. In truth, each multivariate method, whether simple or complex, has potential for revealing insight lacking in another multivariate method. Since there is no guarantee that what works best for one research problem will work best for another, Shepard (1980) and Aspey and Blankenship (1978) recommend that a data matrix be analyzed by using several multivariate methods to maximize the insight gained.

This advice applies not only to the use of cluster analysis in numerical taxonomy, but also to its use for the other research goals described in this book. Thus, if you are unfamiliar with multivariate methods, a good strategy is to learn cluster analysis first and let it be a springboard to learning other methods.

One cautionary note: when reading the literature of cluster analysis as applied in numerical taxonomy, you will find that a variety of terms are used to describe the same thing. For example, numerical taxonomists often refer to objects as "OTU's," which stands for "operational taxonomic units." Some researchers call objects "individuals," or "cases," or "entities," or "subjects," or "data units." Similarly, attributes are called "variables," or "features," or "descriptors," or "characters," or "characteristics," or "properties." Trees are called "dendrograms" or "phenograms."

For numerical taxonomy, as well as for other research goals, cluster analysis is used as a *descriptive* method for gauging the similarities of objects in a sample. Usually the sample is chosen nonrandomly for its interest. Any conclusions the researcher ascribes to a larger population from which the sample was obtained must be based on analogy, not on inferential statistics.

It follows that in cluster analysis we shall not be testing statistical hypotheses about the tree: whether or not the tree we got is typical of all possible trees we could have gotten does not concern us. The nonrandom sample of objects prevents this. Thus, we must rely on informed judgment to assess the risk of extrapolating similarity relations found in the tree to a more general domain. Common sense is important here. If a sample of objects, although nonrandom, seems typical of the larger population of objects, and if the sample is large, then this subjectively reduces the risk of incorrect extrapolation.

Even though this may seem an unscientific way of proceeding, informed subjective judgments do play a crucial role in all of research. When used with nonrandom samples, the utility of cluster analysis and other multivariate methods is not seriously lessened because these methods do not provide a quantitative estimate of the risks of extrapolating.

3.2 USING CLUSTER ANALYSIS TO MAKE CLASSIFICATIONS

Now let us sketch how numerical taxonomists use cluster analysis to make classifications. To begin, a **classification** is a set of mutually exclusive "pigeonholes" that partition an attribute space. Each pigeonhole is a class, and the average properties of its attributes function as the address of the class. Objects whose attributes match the address of a class fall within the pigeonhole and belong to the class.

Let us use the data matrix presented in Chapter 2 to make a classification of plots of farm land. While this example has too few objects and attributes for a real application, nonetheless it can illustrate the principles.

To begin, let us regard the five plots as a representative sample from a larger population of plots for which we want a classification. Figure 3.1 reproduces the tree we obtained in Section 2.4, with the addition that we have cut its branches with a horizontal line drawn at $e_{jk} = 16.0$. This severs the tree into two branches: one is cluster (234) and the other is cluster (15). Assume that we have decided that these clusters are archetypes of two classes of our classification.

The number of classes we want the classification to have determines where we should cut the tree. For example, to get three classes we should cut the tree at, say, $e_{jk} = 10.0$. Deciding where to cut the tree resolves the tradeoff between the desire for detail (many classes) and the desire for generality and simplicity (few classes). The decision is subjective.

Next we let the average attribute measurements of the objects in each cluster be the defining properties of each class. When we average the values

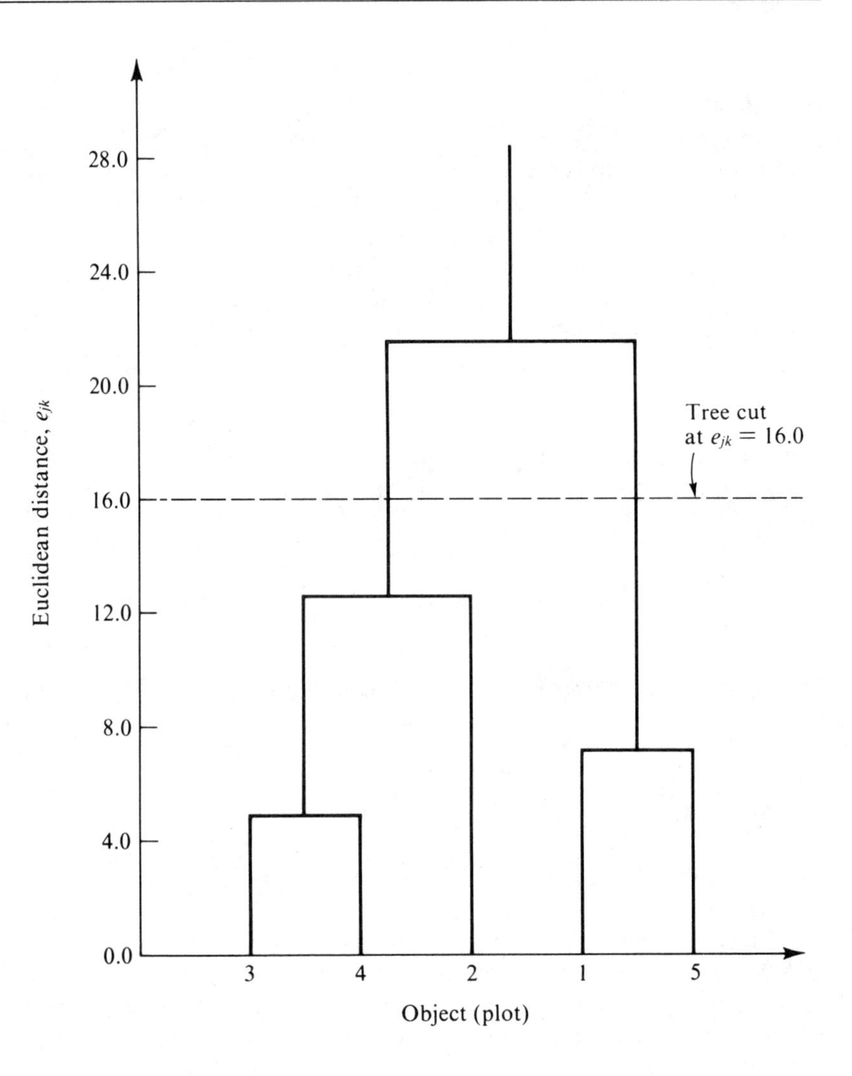

Figure 3.1. An example in which the tree is cut to form a classification.

in the data matrix given in Section 2.1, we get:

Attribute	Cluster	
	(234)	(15)
1	(20 + 30 + 30)/3 = 26.7	(10 + 5)/2 = 7.5
2	(20 + 10 + 15)/3 = 15.0	(5 + 10)/2 = 7.5

Classifications are made to be used, and the act of using them is called **identification.** To illustrate, suppose we have a new plot of land, from the same "population" but not one of the five plots used to make the classification, and suppose we want to identify which of the two classes it belongs to. Labeled *U* for "unknown," its attribute measurements are as follows:

Attribute	*U*
1	20
2	17

Because *U* more nearly agrees with the average measurements of cluster (234) than it does with those of cluster (15), it is identified as a type "(234)." To see this another way, you can plot *U* in the attribute space shown on page 19 and confirm that *U* plots nearer to cluster (234) than to cluster (15).

Obviously a classification must exist before an object can be identified. For example, a single strange object that has fallen from outer space cannot be meaningfully identified because there is no classification more refined than the one that contains the crude class of "strange fallen object." But should enough of these strange objects fall and be collected, they can be compared and sorted into piles of similar objects. The average properties of the piles will define the classes of a classification. Then, when a new object falls it can be identified as a member of one of the classes. Because it will belong as a member of a familiar class, it no longer will be strange.

This, in essence, is how numerical taxonomists use cluster analysis to make classifications, and identify objects into classifications. In Part III classification and identification are covered in more detail.

3.3 SCALES OF MEASUREMENT FOR ATTRIBUTES

Attributes can be measured on several scales—for example, on a continuous quantitative scale (such as temperature) or on a discrete qualitative scale (such as sex). Both scales are used in practice.

A scale of measurement is a rule for assigning numbers to represent different states of sensory data. There is no single all-purpose scale of measurement because there is no single amount of detail about nature that will satisfy the needs of all researchers. Typically researchers use four degrees of detail, each a different scale of measurement. In order of increasing detail, these are: (1) nominal scale; (2) ordinal scale; (3) interval scale; and (4) ratio scale.

A nominal scale measures unordered classes. For example, the binary classes of sex (0 = male, 1 = female) and the multistate classes of color (0 = red, 1 = blue, 2 = yellow, and so on) are measured on nominal scales.

An ordinal scale measures ordered classes—for example, the first three finishers in a horse race (1 = win, 2 = place, 3 = show). Sociological and market research studies often use ordinal scales in questionnaires that ask subjects to rank their preferences for products. Whereas the numbers used to measure a nominal scale merely identify the classes, in the way that Social Security numbers identify people, the numbers used with ordinal scales are the natural integers "1, 2, 3, . . . ," and each is one whole unit greater than its neighbor to the left in the sequence.

Interval and ratio scales are much alike. They are continuous, and the difference between any two scale values stands for a difference in the magnitude of sensory data. However, on an interval scale the value of zero is arbitrarily defined, whereas on a ratio scale the value of zero is inherently fixed by reality. For example, temperature in degrees centigrade is measured on an interval scale. It is continuously ordered in magnitude, and the zero point of 0°C is set by definition with arbitrary reference to the freezing point of water. On the other hand, weight in kilograms is measured on a ratio scale. Although it too is continuously ordered in magnitude, its zero point of 0 kg is a bottom limit that is not arbitrary (the absence of any weight).

More detailed discussions of these scales can be found in Stevens (1946), Siegel (1956), Ware and Benson (1975), Agnew and Pyke (1978), and Lawson (1980).

Usually a researcher has a choice of scales for measuring an attribute. Consider, for example, the measurement of the amount of ponderosa pine on a forested plot. On the coarsest scale—the nominal scale—the researcher would record the presence or absence of pine trees, coded with the numbers, 1 = present and 0 = absent. On the finest scale—the ratio scale—the number of pine trees, coded 0, 1, 2, 3, . . . , would be recorded.

Generally it takes less sophisticated instruments, and consequently costs less, to measure on a coarser scale. Thus, deciding the best scale involves a tradeoff of benefits and costs. The finer the scale, the more detail with which we can measure the states of sensory data and the closer we can get to a true picture of nature. But detailed measurements cost more to obtain. In the end the best tradeoff comes down to answering the question, "Is the more detailed picture worth the added cost?" Although better research goes hand in hand with increased detail, a law of diminishing returns often applies. Thus, nominal scales are cost effective for some research. And for much research, nominal scales are the only possible choice.

The scales of measurement used for the attributes constrain the choice of a resemblance coefficient. One class of resemblance coefficients can be used when attributes are measured on ordinal, interval, or ratio scales. We call such attributes **quantitative attributes**. A data matrix with quantitative attributes can contain a mixture of the three scales. For example, some attributes may be measured on ratio scales, some on interval scales, some on ordinal scales. Chapter 8 discusses resemblance coefficients for quantitative attributes.

Another class of resemblance coefficients is used when the attributes are measured on nominal scales. We call these **qualitative attributes**. These resemblance coefficients are discussed in Chapter 10.

Then there are special resemblance coefficients for use when all of the attributes are measured on ordinal scales. These are discussed in Chapter 11.

Finally, there is the case where some of the attributes of the data matrix are quantitative and some are qualitative. Data matrices with mixtures of the two kinds of attributes can be handled by using the special resemblance coefficients presented in Chapter 12.

3.4 CLUSTER-ANALYZING ATTRIBUTES

We can cluster-analyze attributes by transposing the data matrix and making the objects the rows and the attributes the columns. We can then compute a value of the resemblance coefficient for each pair of attributes and store these values in a resemblance matrix. When cluster-analyzed, the tree will have one branch for each attribute and will show the similarities among attributes.

By convention in the literature of multivariate analysis, a technique for investigating the association among objects is called a Q-technique, and a technique for investigating the association among attributes is called an R-technique (Sneath and Sokal, 1973, pp. 115–116). Our terminology will closely follow this. We will call a cluster analysis of objects a **Q-analysis**, and a cluster analysis of attributes an **R-analysis** (for a mnemonic, "R" stands for the rows of attributes).

For some applications, both a Q-analysis and an R-analysis of the same data matrix can be informative. For example, suppose a group of people answer a questionnaire. In the data matrix, the people are the objects, the questions are the attributes, and the responses (for example, 1 = yes, 0 = no) are the data. A Q-analysis will show which people gave similar and dissimilar responses, and will define classes of people having similar attitudes. An R-analysis will show which questions received similar responses, and may be useful for locating redundant questions that tap the same underlying attitudes. Note that for a data matrix with t objects and n attributes, the following statistics hold:

	Type of analysis	
	Q-analysis	R-analysis
Number of values in the initial resemblance matrix	$t(t-1)/2$	$n(n-1)/2$
Number of clustering steps	$t-1$	$n-1$

This means that for a data matrix having $t = 20$ objects and $n = 200$ attributes, a Q-analysis will be a relatively short set of calculations compared to an R-analysis.

Before closing, let us note a possible point of confusion about how the data in a matrix are displayed. This book follows the convention of Sneath and Sokal (1973) and makes objects the columns and attributes the rows. Another convention, found for example in some of the cluster analysis routines of statistical packages such as SAS and BMDP, reverses this by making objects the rows and attributes the columns. Either way, a Q-analysis always means objects are clustered, and an R-analysis always means attributes are clustered.

3.5 COMPUTERS AND CLUSTER ANALYSIS

Computers make cluster analysis routinely possible. One early cluster analysis of 97 specimens of bees described by 122 attributes was performed by technicians who used desk (mechanical) calculators, and it required 160 man-hours to obtain a tree (Sokal and Michener, 1958). Today this problem would be considered average in size, and a researcher using a computer could obtain its solution in a few minutes.

For a Q-analysis of objects, the number of calculations required grows exponentially with the number of objects t. There will be $t(t - 1)/2$ values of the resemblance coefficient to compute and store in the initial resemblance matrix, and $t - 1$ clustering steps will be needed to obtain the tree. Because attributes are dealt with only at the time the initial resemblance matrix is computed, the total time spent on calculations is influenced to a lesser extent by the number of attributes. Doubling the number of attributes will double the time required to compute the values in the initial resemblance matrix, but the actual work of obtaining those values is but a small portion of all the calculations. So the total time required to obtain the tree will be less than double. Naturally this logic reverses on all points for an R-analysis of attributes.

Computer programs for cluster analysis are widely available. Appendix 2 describes some of these programs, several of which are part of commonly available statistical packages—for example, BMDP and SAS. Many universities, government agencies, and corporations make use of these statistical packages. At several places in this book, outputs from the CLUSTAR cluster analysis program are shown, and the preface describes how you can obtain it and its manual.

SUMMARY

Cluster analysis is one of the multivariate methods used in numerical taxonomy, a field that uses numerical methods to make classifications. When numerical taxonomists use cluster analysis, they cut the tree into clusters, which become the archetypical classes of a classification. After doing this, and after the average values of the attributes of the classes are calculated, objects of unknown identity can be identified to see which class they belong to.

Attributes can be measured on different scales. Qualitative attributes are measured on nominal scales. Quantitative attributes are measured on ordinal, interval, and ratio scales. Whether a data matrix contains qualitative or quantitative attributes, or a mixture of the two, determines the type of resemblance coefficient that can be used.

A cluster analysis of objects is called a Q-analysis; a cluster analysis of attributes is called an R-analysis. In applications, research goals arise for which each is appropriate.

Computer programs for cluster analysis ease the burden of extensive numerical calculations. Although the methods of cluster analysis presented in this book are mathematically simple, the computations are lengthy. A computer program is required to help ensure that the results will be free of error and cost effective.

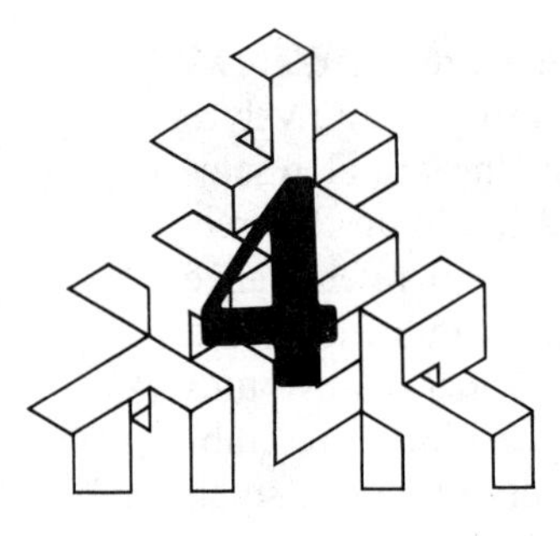

Applications of Cluster Analysis in Retroductive and Hypothetico-Deductive Science

OBJECTIVES

Cluster analysis can be used to create knowledge about natural and historical processes. In particular, this chapter describes how you can use cluster analysis

- To create a scientific question.
- To create a research hypothesis that answers a scientific question.
- To test a research hypothesis to decide whether it should be confirmed or disproved.

We will call each of the uses listed on page 38 a "Research Goal." Briefly described, they are as follows:

Research Goal 1: Create a Question. Here we collect a data matrix, cluster-analyze it to obtain a tree, examine the tree (and rearranged data and resemblance matrices), and hope that a good question comes to mind. In other words, we explore the data through cluster analysis, hoping this will reveal interesting patterns of similarity that will spark a question about the processes of nature that led to the patterns of similarity, or a question about what else the patterns could be related to.

Research Goal 2: Create a Hypothesis. Here we already have a good question in mind before we perform the cluster analysis. The question helps us to frame the collection of data—that is, helps us to collect the right kind of data. Then we cluster-analyze the data matrix and study the patterns of similarity in the tree; we hope that the patterns will suggest a research hypothesis. When this happens, we have a tentative answer to the question.

This use of cluster analysis is an example of a scientific method called **retroduction**. In using retroduction, scientists argue from observed facts to create a hypothetical reason—that is, a research hypothesis—that accounts for the observed facts.

Research Goal 3: Test a Hypothesis. Here we already have a good question and a good hypothesis in mind. Most likely we got them without using cluster analysis. But now we want to test the hypothesis by using cluster analysis. To do this we must be able to predict the patterns of similarity in the tree that we would expect should the hypothesis be true. That is to say, if the hypothesis is true, the tree will have certain properties that we can write down.

Then we collect the data matrix, cluster-analyze it, and obtain the actual tree. If the predicted patterns of similarity that we wrote down agree to within an acceptable tolerance with the actual patterns, then we say the hypothesis has been *confirmed*. And if they disagree, then we say the hypothesis has been *disproved*—that is, it has been shown to be false. Confirmed ideas become knowledge, whereas false ideas are discarded into a scrap heap of ideas whose initial promise failed to live up to expectations.

This use of cluster analysis is an example of a scientific method called the **hypothetico-deductive method**. Its name means that first we hypothesize an explanation of a natural or historical process and then we deduce what events ought to be true if the explanation is correct. Whereas the method of retroduction argues from observation to the creation of a hypothetical explanation, the hypothetico-deductive method argues in the reverse direction.

As you read the Research Applications throughout the rest of this book, you may find yourself wanting to learn more about scientific methods.

Hanson's (1958) book discusses retroduction. The books by Popper (1968), Searles (1968), Harvey (1969), and Chalmers (1976) discuss the method of induction and the hypothetico-deductive method, as does the article by Romesburg (1981). Polya (1954, 1957) and Hesse (1966) are among the few books that discuss the use of analogy as a scientific method.

4.1 RESEARCH GOAL 1: CREATE A QUESTION

The following two Research Applications show how cluster analysis can be used to create scientific questions.

RESEARCH APPLICATION 4.1

Grant and Hoff (1975), two anthropologists, think highly of using cluster analysis to spark new questions. The objects they cluster-analyzed were the

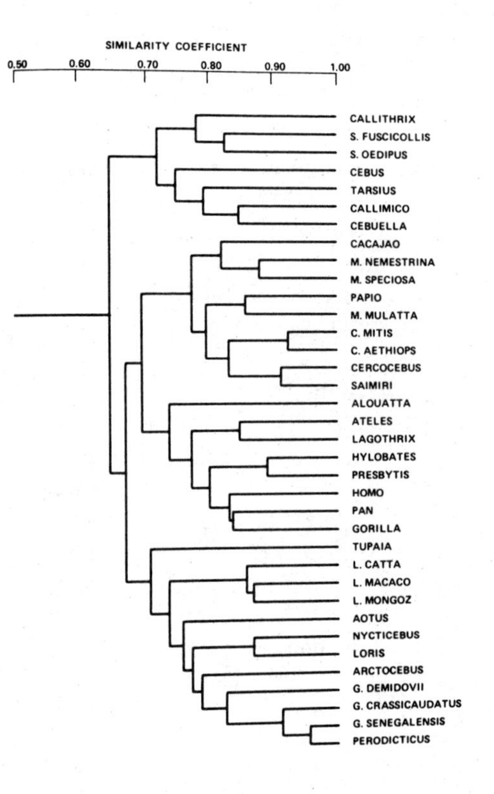

Figure 4.1. Tree of primate skins. (By permission, reproduced from Grant and Hoff (1975), Figure 3.)

skins of 36 species from several genera of primates, each skin described by 34 qualitative attributes. They hoped that seeing which skins were similar and dissimilar would raise interesting questions of "Why?" They state (p. 164) that "cluster analysis may have its greatest value in reducing a vast amount of data... to a form that is useful to investigators by allowing them to see relationships that are not clear in the untreated data. Hence, they may ask questions of the data that they were not able to formulate prior to treatment. We think that the use of numerical classification [cluster analysis] is more important [in anthropology] than its use in numerical taxonomy."

Figure 4.1 shows Grant and Hoff's tree of the primate skins. The tree, which summarizes the 36 × 34 = 1,224 values in the data matrix, brings out the similarities and dissimilarities. This made it possible for Grant and Hoff to see that the *Alouatta*, *Ateles*, and *Lagothrix* genera of primates tend to group with the hominoids. They say (p. 164), "The grouping with hominoids may be fortuitous but we find it intriguing that the highest grade of New World primates should group with the highest grade of Old World primates. This grouping is apparently related to characteristics of the dermis and epidermis and of the sweat glands. This may be worth further study."

RESEARCH APPLICATION 4.2

The objects in the data matrix shown in Figure 4.2,*a* are the dental formulas of 23 mammals. These formulas record the number of teeth of each kind on one side of the upper and lower jaws. For example, the formula of old-world monkeys is 2–1–2–3 / 2–1–2–3. The values before the slash record the upper jaw: 2 incisors, 1 canine, 2 premolars, 3 molars; after the slash they record the lower jaw. The data matrix strings the formula out as eight attributes; the first attribute is "number of incisors on one side of the upper jaw"; the second is "number of canines on one side of the upper jaw"; and so on until the eighth, "number of molars on one side of the lower jaw." (Without looking at the tree, look at the data matrix and see if you can find (1) the most similar pair of dental formulas and (2) the most dissimilar formula.)

By itself the data matrix is difficult to make sense of, and it raises few questions of interest. But this changes when it is cluster-analyzed. Figure 4.2,*b* shows the tree, and Figure 4.2,*c* shows the rearranged data matrix. Here are some of the questions the cluster analysis helps raise.

Question 1. The mole and the pig have identical dental formulas. Why? What research hypothesis could explain this? Should it be genetically or functionally oriented?

Question 2. Why is the walrus so different from the other carnivores? Why is it different from the other aquatic carnivores, for example the seals?

Object

Attribute	Old-world monkeys	New-world monkeys	Marmosets	Insectivorous bats	Frugivorous bats	Hedgehogs	Moles	Cats	Dogs	Raccoons	Hyaenas	Weasels	Eared seals	Ordinary seals	Walruses	Hippopotamuses	Pigs	Camels	Chevrotains	Hollow-horned ruminants	Tapirs	Horses	Rabbits
	2	2	2	2	2	3	3	3	3	3	3	3	3	3	1	2	3	1	0	0	3	3	2
	1	1	1	1	1	1	1	1	1	1	1	1	1	1	1	1	1	1	1	0	1	1	0
	2	3	3	3	2	3	4	3	4	4	4	4	4	4	3	4	4	3	3	3	4	3	3
	3	3	2	3	3	3	3	1	2	2	1	1	1	1	0	3	3	3	3	3	3	3	3
	2	2	2	3	2	2	3	3	3	3	3	3	2	3	0	2	3	3	3	3	3	3	1
	1	1	1	1	1	1	1	1	1	1	1	1	1	1	1	1	1	1	1	1	1	1	0
	2	3	3	3	3	2	4	2	4	4	3	4	4	4	3	4	4	2	3	3	3	3	2
	3	3	2	3	3	3	3	1	3	2	1	2	1	1	0	3	3	3	3	3	3	3	3

(a)

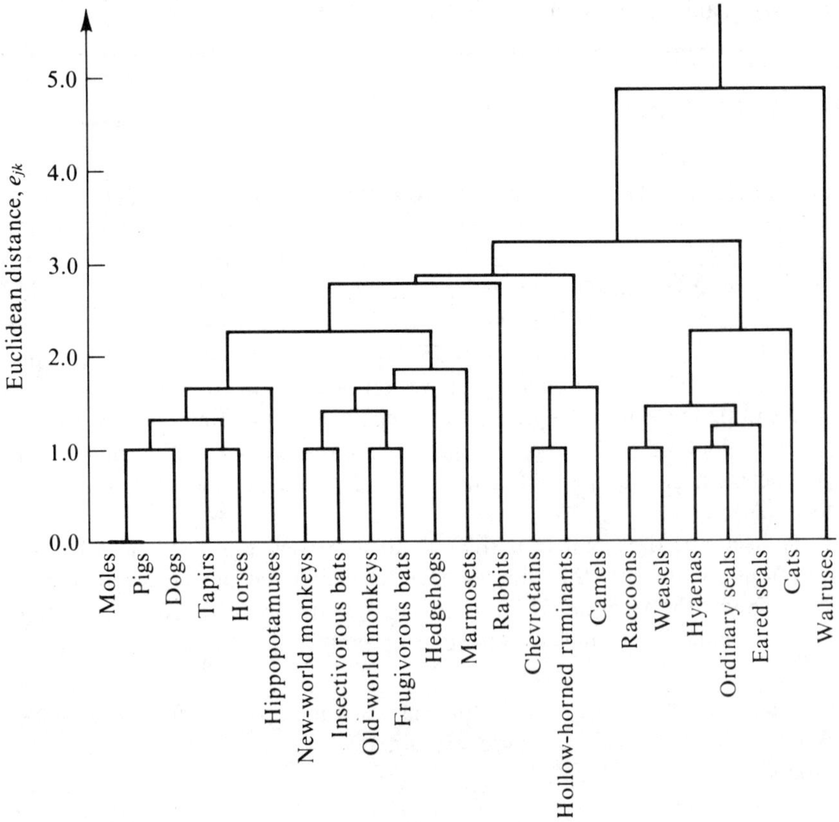

(b)

Figure 4.2. (a) Data matrix of dental formulas of 23 mammals. (b) Tree showing similarities among dental formulas. (c) Rearranged data matrix of dental formulas.

Object

Attribute	Moles	Pigs	Dogs	Tapirs	Horses	Hippopotamuses	New-world monkeys	Insectivorous bats	Old-world monkeys	Frugivorous bats	Hedgehogs	Marmosets	Rabbits	Chevrotains	Hollow-horned ruminants	Camels	Raccoons	Weasels	Hyaenas	Ordinary seals	Eared seals	Cats	Walruses
	3	3	3	3	3	2	2	2	2	2	3	2	2	0	0	1	3	3	3	3	3	3	1
	1	1	1	1	1	1	1	1	1	1	1	1	0	1	0	1	1	1	1	1	1	1	1
	4	4	4	4	3	4	3	3	2	2	3	3	3	3	3	3	4	4	4	4	4	3	3
	3	3	2	3	3	3	3	3	3	3	3	2	3	3	3	3	2	1	1	1	1	1	0
	3	3	3	3	3	2	2	3	2	2	2	2	1	3	3	3	3	3	3	3	2	3	0
	1	1	1	1	1	1	1	1	1	1	1	1	0	1	1	1	1	1	1	1	1	1	1
	4	4	4	3	3	4	3	3	2	3	2	3	2	3	3	2	4	4	3	4	4	2	3
	3	3	3	3	3	3	3	3	3	3	3	2	3	3	3	3	2	2	1	1	1	1	0

(*c*)

Figure 4.2. (*Continued*)

Walruses eat mollusks, which they scrape from the ocean floor. Has this accordingly modified their dental structure?

Question 3. Why is there no apparent relation between an animal's dental formula and its size? Why is the horse similar to the mole but dissimilar to the rabbit?

Question 4. New-world monkeys and insectivorous bats have quite similar dental formulas. Old-world monkeys and frugivorous (fruit-eating) bats also have similar formulas. Does the symmetry between these two sets of similarities mean anything?

Question 5. Why are the frugivorous bats, being the only obligatory frugivores in the set, quite similar to the insectivorous bats and other nonobligatory frugivores, for example the monkeys? Why has the specialization to obligatory fruit-eating not modified the dental formulas more?

To ask a good question requires two things. First, we must see the patterns in the data clearly. Second, we must know enough to allow the patterns to interact with what we know and raise questions. Cluster analysis helps us to see patterns. While it cannot guarantee that good questions will follow, it can increase the probability of this.

The tree and the rearranged data matrix show the power of cluster analysis for finding patterns. Few people see at first glance that the mole and the pig are identical, but the tree and rearranged data matrix make the patterns jump out. This becomes even more valued when the data matrix is larger, for then our innate abilities leave us even more unable to perceive patterns.

4.2 RESEARCH GOAL 2: CREATE A HYPOTHESIS

Suppose we have a question already in mind (it needn't have come from a cluster analysis). We want to use cluster analysis to find a research hypothesis that tentatively answers it.

To use cluster analysis for this, *it is crucial that the question can be answered by a hypothesis whose truth entails similarities and dissimilarities among objects*. In other words, there is a class of questions that can be answered by studying similarities and dissimilarities among objects. It is this class we are interested in. Scores of questions cannot be answered this way, and in this book we are not interested in them. For example, we cannot use cluster analysis to answer, What is the heartbeat rate of an excited ground squirrel? Or, Why does the grunion spawn when the moon is full? Finding answers to these questions does not start with a data matrix and its cluster analysis.

We have to answer for ourselves, Can our question be answered by studying similarities and dissimilarities among objects? Only when the answer is *yes* is cluster analysis likely to be a successful approach.

As noted earlier, arguing from the facts of a tree and rearranged data and resemblance matrices to create a hypothesis is an example of the scientific method of retroduction. Hansen (1958, p. 86) describes retroduction as the following sequence (to which we add the bracketed phrases):

1. Some surprising phenomenon *P* is observed [in our case, as some aspect of the tree or the rearranged data and resemblance matrices].
2. *P* would be explicable as a matter of course if *H* [a hypothesized reason for *P* having occurred] were true.
3. Hence there is reason to think that *H* is true.

RESEARCH APPLICATION 4.3

Sigleo (1975), an anthropologist, asked herself this question: Geographically, how wide-ranging were the trading patterns among prehistoric Indian cultures in the American southwest? At first glance there is no obvious way that similarities and dissimilarities among the columns of a data matrix can suggest an answer.

What we must do is examine the kinds of facts that are available (or could be made available) and that are in some sense framed by the question. For example, examining the past weather of the American southwest would give us facts, but these facts would not in any sense be framed by the question about trading patterns. Although available in tree ring data, these facts are oblique to the question. But in turquoise artifacts Sigleo saw a way that did relate the question to the study of similarities, and she saw how cluster analysis could be used to generate a research hypothesis that tentatively answered the question.

For the data matrix, Sigleo collected samples of turquoise from 24 prehistoric mining sites covering a wide area stretching over Nevada, California, Arizona, Colorado, and New Mexico; these were mines known to have been active during the period marked by the beginning of the Christian Era and A.D. 1200. To these she added 13 beads found at an Indian dwelling site in south-central Arizona called Snaketown.

She measured the attributes of these objects in a way that is best described by an analogy. Have you ever taken a coin, laid it under a piece of paper, and gently shaded it with a pencil? Or have you ever made a rubbing of an old tombstone? Then you know how objects often carry latent images that can be made to surface. In the same way that paper over an object can be shaded, so too can turquoise objects be imaged with the neutron flux from a nuclear reactor. The flux makes some of the elements in the turquoise radioactive, and their concentrations can be detected.

Sigleo put the samples from the mines and from Snaketown in a reactor and irradiated them. By counting radioactive decay rates, she measured the concentrations of their elements. This gave her a data matrix of turquoise samples (the objects) and 30 elemental concentrations (the attributes). Thus, each sample was described by its nuclear signature, a profile of radioactivity proportional to its composition.

Would all 13 Snaketown beads have similar composition and therefore cluster together? If so, would they cluster alone, or would they cluster with samples found in one or more of the mines? With the nearest mine? Or with mines as far away as Colorado and California?

The results were that the 13 Snaketown beads clustered together, indicating their turquoise probably had come from the same mine. As a group they clustered not with the samples from the mine nearest Snaketown, but with those from a remote mine—in Halloron Springs, California. Moreover, the cluster of Snaketown beads and Halloron Springs samples were dissimilar to all of the remaining samples from the other mines. This finding led Sigleo to a retroductively derived hypothesis: *the prehistoric Snaketown Indian culture traded over distances greatly exceeding the nearest sources of turquoise.*

In this way she used cluster analysis as a time machine to shine a light of understanding into the past.

4.3 RESEARCH GOAL 3: TEST A HYPOTHESIS

We have seen how a tree can suggest a question or suggest a research hypothesis. Now we will see how we can begin an analysis with a hypothesis in hand, make the hypothesis predict the form that a tree will take, and then compare this predicted tree to an actual tree obtained with cluster analysis. If the hypothesis is true, it ought to predict a tree that agrees with one obtained from experimental facts.

To use cluster analysis to test a hypothesis, *it is crucial that the form of a predicted tree can be deduced from the hypothesis*. Note that not all hypotheses allow this. For example, the hypothesis that governmental deficit spending increases inflation seems unrelated to sample objects and attributes. It is unclear how a tree could be deduced from this hypothesis.

On the other hand, the hypothesis that the moon is made of green cheese could be tested by using cluster analysis. If this hypothesis were true, then we predict that samples of the moon and samples of green cheese, each described by their elemental compositions, should be intermingled in the tree. But if the moon is not made of green cheese—that is, if the counter-hypothesis is true, then there should be (we predict) two well-defined clusters—one made up of green cheese samples, the other made up of moon samples.

It would remain for us to gather the samples of cheese and moon rock, code their elemental compositions, and cluster-analyze the data matrix. The actual tree will agree with one or the other of the predicted trees within an acceptable tolerance (Chapter 14 shows how to compare trees). If the samples of the moon and green cheese are actually intermingled in the tree, the hypothesis that the moon is made of green cheese will be confirmed. If they are not intermingled, the hypothesis that the moon is *not* made of green cheese will be supported.

This is the hypothetico-deductive method of testing research hypotheses. Whereas the retroductive method is used to create hypotheses that explain existing facts, the hypothetico-deductive method is used to test existing hypotheses by making them predict facts not yet collected. Science often uses the two methods back to back (Research Application 4.6 illustrates this). It uses retroduction with one kind of data to create a hypothesis, and then uses hypothetico-deduction with another kind to test the hypothesis. Two independent methods of certification make for more reliable results: false ideas that may have slipped by during retroduction have a second chance of being caught during hypothetico-deduction.

RESEARCH APPLICATION 4.4

Charlton et al. (1978) performed a cluster analysis that is superficially similar to the Snaketown study (Research Application 4.3). Both studies irradiated samples of Indian artifacts and source mine material. Both cluster-analyzed a data matrix. Both were concerned with the extent of prehistoric trading patterns. But whereas the Snaketown study had few preconceived ideas of the extent of trading patterns, and hence had no hypothesis to test, Charlton and his colleagues started with a hypothetical trading pattern that other scientists had retroductively derived by using other kinds of data. Then they used the hypothetico-deductive method to test this hypothesis. Let us see how.

They collected obsidian samples from 11 pre-Hispanic mining sites in central Mexico. To this they added 90 excavated samples from the pre-Hispanic dwelling site of Chalcatzingo. They irradiated these 101 objects, determined their elemental compositions, and cluster-analyzed the resulting data matrix.

Although the hypothesized trading pattern had not been defined in enough detail to allow a full prediction of the tree (expected similarities and dissimilarities had not been theorized for all pairs of 101 samples), sufficient detail existed to allow the hypothesis to be tested: if the hypothesis were true, then samples of obsidian from a source mine called Paredón and artifacts from Chalcatzingo should cluster in different branches of the tree; for, according to the hypothesis, the culture at Chalcatzingo did not have access to obsidian from the Paredón mine, some 150 kilometers away.

The actual tree showed that the prediction was false. Of the total artifact sample from Chalcatzingo, 32 percent was most similar to the Paredón obsidian. The scientists concluded that the "hypothesized Early Formative obsidian networks will have to be reanalyzed." This is how science is self-correcting.

The next application of the hypothetico-deductive method is from geology. It follows the same logic of predicting the form the tree would take in the event that a research hypothesis were true, and then collecting data and using cluster analysis in order to see how the actual tree compares with the predicted tree.

RESEARCH APPLICATION 4.5

Volcanic ash beds, deposited in the American West during the Pleistocene epoch, stretch from California to Nebraska. Borchardt et al. (1972) used cluster analysis to confirm the hypothesis that a particular ash bed—the Bishop ash bed—originated from volcanic eruptions of the Long Valley caldera in California.

The objects in the data matrix were 15 samples of ash: 13 samples were from sites in California, Utah, Colorado, and Nebraska; and two samples were of ash gathered at the Long Valley caldera. Some of the 13 samples were from deposits that are part of the Bishop ash bed; some were from other ash beds. Each sample was described by the percentage composition of 25 different elements.

When this data matrix was cluster-analyzed, the samples from the Long Valley caldera and the Bishop ash samples from California, Utah, Colorado, and Nebraska clustered together in the same branch of the tree. Samples from the other ash beds did not join this cluster; they were in other branches of the tree. This was as the hypothesis predicted. Thus, the hypothesis that the Long Valley caldera was the source of the Bishop ash bed was confirmed.

For a final research application, let us look at some exciting scientific research that has all the detective work of a great mystery. It very neatly combines the methods of retroduction and hypothetico-deduction, first to create a hypothesis, and then to test the hypothesis. In the process it debunks the myth that historical knowledge cannot be created by using the scientific methods customarily used in the "hard sciences."

RESEARCH APPLICATION 4.6

The research of Harris et al. (1978) begins with a question: Who was the Egyptian mummy of unknown identity found in the tomb of Amenhotep II? At the time of the research, this mummy was known only by the name of the "Elder Lady" given to her by Elliot Smith, the famed Egyptologist.

Attempting to answer the question, these researchers retroductively proposed two alternative hypotheses:

H_1: The Elder Lady is the mummy of Queen Tiye.

H_2: The Elder Lady is the mummy of Queen Hatshepsut.

The retroductive support for these hypotheses was this: (1) the Elder Lady's arms were crossed, indicating she was someone of great importance; (2) both queens Tiye and Hatshepsut commanded respect sufficient to warrant a crossed-arms burial; (3) their mummies had never been found.

Both of these queens were known to have lived at a time that could have resulted in their mummies being brought into Amenhotep II's tomb. Let us read the evidence these researchers cited when describing these hypotheses in *Science*:

> One of us (E.F.W.) has suggested that the unique position of the arms of this lady [the Elder Lady] indicated that she was of great importance and was perhaps Queen Hatshepsut or Queen Tiye.

> The former queen had indeed ruled Egypt as a pharaoh and had assumed the titles and raiment of a king as depicted in temple scenes and in statuary after the death of her husband Thutmosis II and during her coregency with her stepson Thutmosis III, who had been born to Thutmosis II by a minor queen. Queen Tiye was the beloved wife of Amenhotep III and mother of the heretic pharaoh Amenhotep IV who changed his name to Akhenaton. Her importance during her husband's reign and the reign of her son Akhenaton was reflected in the statuary, inscriptions, and reliefs, in temples and tomb chapels. She was even involved in diplomatic correspondence with heads of foreign states. Representations of a queen illustrate a similar arm position to that of the mummy [the Elder Lady].

The researchers used cluster analysis to hypothetico-deductively test the two hypotheses in the following way. The Egyptian Museum held the mummies of ten other queens whose identities were known. Their cephalograms (lateral head radio-grams) were available. The Elder Lady's cephalogram was added, making 11 objects, several of which are shown in Figure 4.3,*a*. For attributes, measurements describing the sizes and shapes of the cephalograms were used. One of the mummies was that of Queen Thuya, the mother of Queen Tiye. This made possible the test prediction P, deduced from H_1:

> P: There should be greater similarity between the cephalograms of the Elder Lady and Queen Thuya than between the cephalograms of the Elder Lady and the remaining nine mummies.

Figure 4.3,*b* shows the tree resulting from the cluster analysis. The close similarity of the cephalograms of the Elder Lady and Queen Thuya agrees with the prediction. P is true, thus confirming H_1.

Because the deduction "If H_1, then P" could have possibly been in error, which would have made the confirmation of H_1 suspect, further tests were needed. That is, the deduction of P from H_1, being based on the probabilities of genetics, would tend to be true on average, but on any one trial it might be in error; daughters do not always inherit the morphologic characteristics of their mothers.

When these researchers presented their findings to the Egyptian Department of Antiquities, they convinced the officials to permit a second, independent test of H_1. Based on the matching of hair samples, the second test used a prediction P deduced from H_1:

> P: A sample of the Elder Lady's hair should match a sample of hair taken from a wood sarcophagus found in King Tutankhamon's tomb bearing the name of Queen Tiye.

The hair samples did match. H_1 was reconfirmed. Thus the hypothesis that the Elder Lady and Queen Tiye are one and the same was jointly

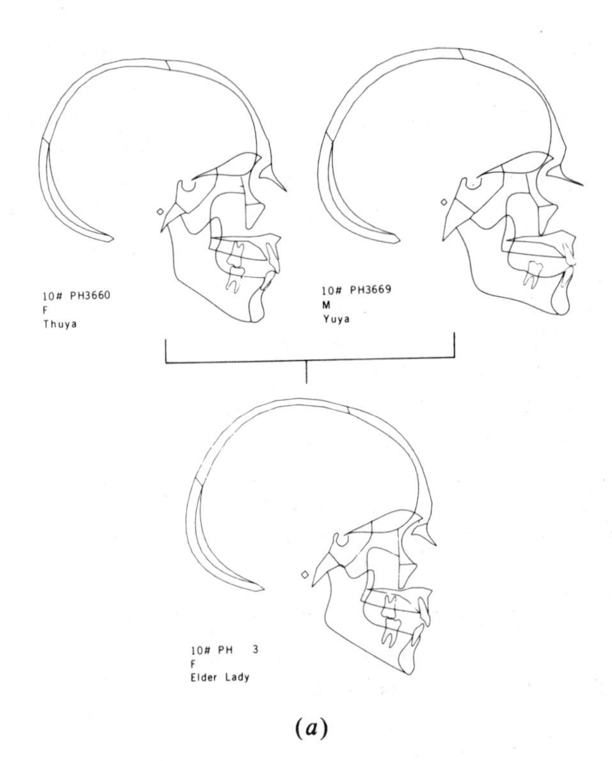

(*a*)

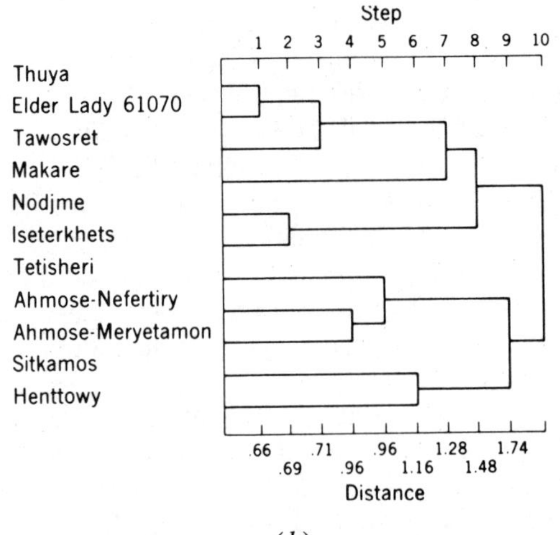

(*b*)

Figure 4.3. (*a*) Cephalograms of Egyptian mummies that were coded as data and cluster-analyzed. (*b*) Tree of mummy cephalograms. (Reproduced from Harris et al. (1978), Figures 3 and 4; copyright 1978 by the American Association for the Advancement of Science.)

supported by three kinds of evidence: (1) The "arm-crossing" evidence used in the retroduction, (2) the hypothetico-deductive test done through cluster analysis, and (3) the hypothetico-deductive test done with hair samples. Three independent kinds of evidence lend a strong belief that the hypothesis is true.

Notice that the objects in this Research Application are mummy cephalograms. It is worth reflecting on how such objects might be measured. When objects like cephalograms are superimposed on a two-dimensional system of Cartesian coordinates, there are at least two approaches to coding the attributes that describe their shapes. One is to let the attributes be the important measures of linear dimension—in this Research Application, length of a mummy's nose. Another is to let the attributes be coordinate positions on the grid. The shapes of the objects can then be described mathematically with a Fourier series, and the values of the coefficients of the series can be used as attributes. In this approach, described by Younker and Ehrlich (1977), the objects are replaced by coefficients that define their mathematical equation—a vector of fitted coefficients—and the equations are cluster-analyzed. Vasicek and Jicin (1971) present a comparable method, based not upon Fourier series, but upon trigonometric functions.

4.4 THE IMPORTANCE OF BACKGROUND KNOWLEDGE

There are fine distinctions between using cluster analysis (or any other technique) to (1) create a question, (2) create a hypothesis, (3) test a hypothesis. Essentially, which of the three objectives should be pursued depends on how much background knowledge is available. To show this, let us consider three scenarios for studying the data matrix of dental formulas presented in Research Application 4.2.

Scenario 1. Suppose the scientist is an expert in dental structures but not in animal diets. He has access to the dental formulas, so it is easy for him to cluster them. This he does out of curiosity. As he looks at the tree he naturally relates it to other fields of study. One of these fields concerns animal diets. But he does not know enough about this to propose a specific relation between dental structure and diet. Thus, he is only in a position to ask, not to answer. He asks, Do animals with similar dental formulas have similar diets? That is, are their dental formulas a reflection of their diets?

Scenario 2. Suppose the scientist is an expert in dental structures and knows a little about animal diets. He questions the possible relation between the two. He cluster-analyzes the data matrix of dental formulas. He looks at the tree and notes that the carnivores (such as cats, dogs, or hyaenas), herbivores (horses, rabbits, chevrotains, and so on), and insectivores (insectivorous bats, or moles) do not group themselves in separated branches of the tree. From this he forms the following research hypothesis: what an animal eats and its dental structure are not, on the whole, related.

Scenario 3. Suppose the scientist knows a great deal about dental formulas and diets. He also holds the following hypothesis, which he believes is supported by the theory of evolution: the diets of animals are more related to their tooth morphology—that is, to the shapes of their teeth—than they are to their dental formulas.

If this hypothesis is true, then the following prediction should hold. Suppose there are three data matrices, each having the same animals for its objects. The first data matrix has dietary attributes—the different foods the animals eat. The cells of this data matrix give the percentage of each food an animal eats, summing to 100 percent down each column. The second data matrix has attributes that describe tooth morphology. The third data matrix contains the dental formulas. When each data matrix is cluster-analyzed, the tree based on diets should be more similar to the tree based on dental morphology than it is to the tree based on dental formulas.

The scientist performs the three cluster analyses and decides whether the prediction holds. If it does hold, he declares the hypothesis to be confirmed. If not, he declares it to be false.

In summary, the scientific use of data depends upon how much the scientist already knows and on how much other data he has access to. To one scientist, a given set of data may raise only questions. To another, it may let him create a research hypothesis. To another, it may let him test a hypothesis.

SUMMARY

One of the uses of cluster analysis is to discover knowledge about natural or historical circumstances. There are three phases to this discovery process: (1) create a scientific question; (2) create a research hypothesis; (3) test a research hypothesis. The Research Applications in this chapter show how cluster analysis has been used in each of the three phases; in each Research Application similarities and dissimilarities among objects were important inputs to the discovery process. Hence, cluster analysis was an appropriate multivariate method.

Background knowledge and what other data are available partly determine which phase of the discovery process a scientist will pursue. Depending on the background information, a given data matrix may only be suitable for helping to raise scientific questions, or it may lead beyond this to the generation or testing of a research hypothesis.

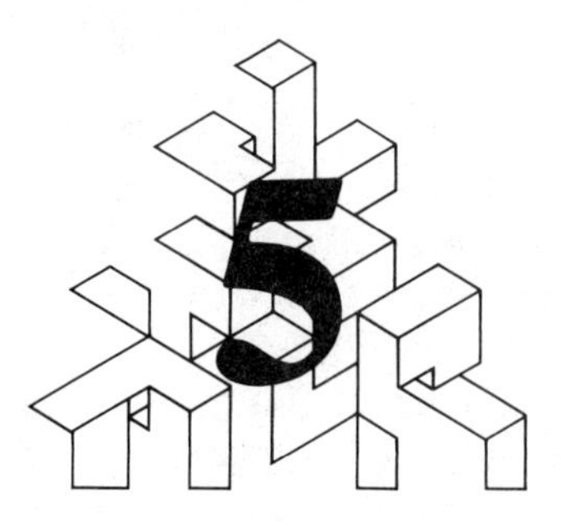

Applications of Cluster Analysis in Making Classifications

OBJECTIVES

This chapter examines two more research goals you can address with cluster analysis:

- How to make a general-purpose classification.
- How to make a specific-purpose classification.

These two goals will extend the sequence of Research Goals 1, 2, and 3 given in Chapter 4.

The two objectives listed on page 53 are as follows:

Research Goal 4: Make a General-Purpose Classification. This type of classification serves general scientific purposes, as a basic catalog of the entities that scientists study, and as a building block that scientists can incorporate in their theories.

We distinguish between two types of general-purpose classifications in this chapter: (1) biological classifications (for example, a classification of species); (2) nonbiological classifications (for example, a classification of soils).

Research Goal 5: Make a Specific-Purpose Classification. This is the antithesis of a general-purpose classification. It differs in that at the time we make the classification we have a specific use for it in mind—namely, relating it to specific qualitative or quantitative variables.

Although both the research goals of making a general-purpose classification and a specific-purpose classification share much in common, it is important to differentiate them. The reason is fully explained in Chapter 22. There it is shown that the cluster analyses for these two research goals must be framed and validated differently. To have an idea of what this means, consider that a numerical taxonomist whose aim is to make a general-purpose classification of species of oak trees chooses different attributes to describe the trees than does a scientist whose aim is to make a specific-purpose classification of oak trees with the goal of relating the oaks to the properties of the soils they are growing in. There may also be differences in how the two scientists standardize their data matrices, and in the resemblance coefficients each chooses. Furthermore, each will use a different criterion of validity for judging the resulting classification.

Let us see how a specific-purpose classification can be related to a qualitative variable—a nominal-scaled binary or multistate variable, let us say.

Suppose we have cluster-analyzed a data matrix having people as its objects and measures of their socioeconomic status as its attributes. Cutting the tree gives us a classification of people according to their socioeconomic status. (This classification is itself a qualitative variable, one which we have invented.) Our next step is to examine the relation of the classification's classes to a person's race, an existing qualitative variable having classes of (1) white, (2) black, (3) yellow, and so on. In doing so, we might find that the white people we have measured tend to fall in one socioeconomic class while the yellow people tend to fall in another.

We could then construct a contingency table to test the hypothesized relation. Its rows would be the classes of socioeconomic status that we derived by using cluster analysis. Its columns would be the states of the qualitative variable of race. We would survey another group of people, identify them with respect to the rows and columns of the table, and perform a contingency table analysis (as is explained in almost any introductory statistics text). This would tell us whether there is a statistically significant relation between the classification and the qualitative variable.

As another example of the use of a specific-purpose classification, we could use cluster analysis to produce two classifications, and use a contingency table analysis to see whether they are related. One classification might be of forested plots, based on the properties of the vegetation growing on them; the second classification might be based on their soil properties. A contingency table analysis of forested plots, cross-classified according to vegetation and soil classes, would tell us whether the two classifications are statistically independent or not.

A third kind of specific-purpose classification is one that is related to a quantitative variable (an ordinal-, interval-, or ratio-scaled variable). For example, we could use cluster analysis to form a socioeconomic classification of people and then use regression analysis to examine the relation between the ages of people (the quantitative dependent variable) and their socioeconomic classes (the independent variable).

Having set the stage, we are now ready to examine some applications of cluster analysis that illustrate the two research goals of making general-purpose and specific-purpose classifications.

5.1 USING CLUSTER ANALYSIS TO MAKE GENERAL-PURPOSE CLASSIFICATIONS

This section describes both biological and nonbiological general-purpose classifications, and demonstrates these with examples.

Biological Classifications

The modern system of biological classification started in the 18th century and is attributed to the Swedish scientist Carolus Linnaeus. This system can be visualized as a stepped pyramid. Imagine that we are standing at this pyramid's top, in the manner of a Noah surveying prospective passengers for his ark. We take the first step down and come to the classification called **kingdoms** with its two classes, plants and animals. We take another step down and see that each kingdom is divided into classes termed **phyla**. Then another step down and each phylum is divided into classes that are literally called **classes**. And so on down—from kingdoms, phyla, and classes, to **orders, families, genera**, and finally at the pyramid's base, at the edge of ordinary experience, **species**. For two centuries taxonomists have been swarming antlike downward over this pyramid trying to locate its bottom.

Most current work in numerical taxonomy involves taking specimens from a family and clustering them into a tree that is cut to define a classification of genera. Or it involves taking specimens from an accepted genus and using them to define a classification of species. Some of the work is new, and some is a reworking and correcting of older work.

RESEARCH APPLICATION 5.1

Larson and Tanner (1974) used cluster analysis to make a classification of the lizard genus *Sceloporous*. For objects they selected specimens within this genus. They described the skull of each specimen with 80 attribute measurements, some of which are shown in Figure 5.1. Most of these attributes were expressed as ratios of dimensions to standardize them for differences in skull size. After cluster-analyzing the data matrix, Larson and Tanner cut the tree into three clusters of specimens. They argued that these represented three previously unrecognized genera and recommended that *Sceloporous*, then considered a single genus, be replaced by three new unnamed genera.

In a similar study, Robins and Schnell (1971) cluster-analyzed 12 species from six sparrow genera. The objects were 12 species of sparrows; the attributes were 48 skeletal measurements. The data matrix thus contained

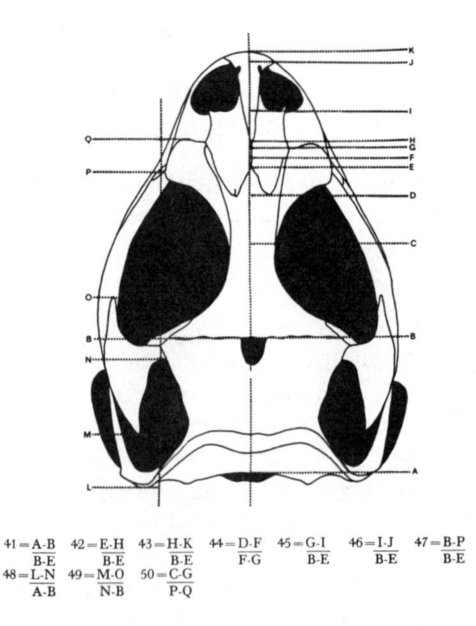

Figure 5.1. Some features of skulls of specimens of the lizard genus *Sceloporous* used as attributes to make a general-purpose classification. (By permission, reproduced from Larson and Tanner (1974), Figure 2.)

$12 \times 48 = 576$ values. Each species was not represented by a single specimen. Instead, each was represented by a sample of between five and ten skeletons of that species, and their measurements were averaged. The researchers cut the tree from the cluster analysis into two clusters. They argued that instead of six genera, as previously believed, there were really only two.

For a classification of ticks, Herrin and Oliver (1974) clustered 117 specimens of ticks, sampled from several known species from within one genus. For attributes they used measurements of tick body-parts. When they cut the tree into clusters, they found that the specimens within clusters were of the same known species. This reconfirmed the classical taxonomic classification of the genus.

Note that the lizard study used cluster analysis to partition an existing genus into three new ones, while the sparrow study collapsed six existing genera into two new ones. The tick study changed nothing but did reconfirm the existing classification.

The power of cluster analysis increases when the objects being classified are unfamiliar in common experience. Lizards, sparrows, and ticks can be held and touched. Their familiarity makes their traditional classifications fairly obvious. But in worlds with sizes beyond ordinary experience—the tiny world of bacteria, for example—taxonomists find cluster analysis an increasingly dependable method.

RESEARCH APPLICATION 5.2

Bacteria of the genus *Bacillus* are of commercial interest because of their abilities to form useful chemical substances such as antibiotics, enzymes for tenderizing meats, and detergents. Bonde (1975) cluster-analyzed 460 strains of *Bacillus*. He did not describe each strain by **descriptive attributes** (attributes of size and shape, or physical and chemical properties). Instead he described each by **functional attributes** (attributes that revealed the effects of strains acting on other substances). For example, some of the functional attributes he used were the effects of strains on the formation of pigments, their effects on the fermentation of sugars, and so forth. After the cluster analysis, Bonde cut the tree's 460 branches into seven clusters. Based on this, he recommended that the descriptions of some existing genera and species be amended.

Tsukamura et al. (1979) performed a similar study, clustering 369 strains of bacteria of the genera *Mycobacterium*, *Rhodococcus*, *Nocardia*, and *Corynebacterium*. The 88 attributes were a mixture of the descriptive and the functional. Several of the functional attributes used were tolerances of the strains to different chemicals. The strains separated according to the existing classification, which was based on traditional taxonomic methods, thus reconfirming it.

In the preceding Research Application, note that a functional attribute can be either the effect of an object on a material, or the tolerance of an object to a material. If descriptive attributes are used, the objects will yield measurements of themselves, but if functional attributes are used, the objects will not yield measurements unless brought into contact with another material.

An even smaller microscopic world than that of bacteria is that of the cell. Cells are not classified according to species, or strains, but according to **lines**. For example, Olsen et al. (1980) used cluster analysis to make a classification of mammalian cells. The lines agreed with the lines derived from a traditional classification.

Nonbiological Classifications

Earlier in this chapter we mentioned a classification of soils as an example of a nonbiological classification. As an example, Gaikawad et al. (1977) used cluster analysis to make such a classification. It agreed with and broadly confirmed an earlier conventional soil classification. In addition, the use of cluster analysis and other multivariate methods to make soil classifications is described in a book by Webster (1977). We now look at two more examples of how researchers have used cluster analysis to make general-purpose, nonbiological classifications—one is from economics and the other is from meteorology.

RESEARCH APPLICATION 5.3

Classifications of industries, of wealth, and of raw materials are central to economics. Most of these classifications have been built with conventional methods whereby economists study data compiled in reports and computer outputs and subjectively group together what appear to be similar objects.

Blin and Cohen (1977) have shown the potential of cluster analysis for aiding the conventional methods. They cluster-analyzed a set of 83 American industries. They described each industry by its production function, a vector of attributes giving the amounts of different products it produces. Upon cutting the tree into 29 clusters of industries, they concluded that the classification agreed with the conventional one and reconfirmed it.

RESEARCH APPLICATION 5.4

McCutchen (1980) used cluster analysis in a meteorological application to make a classification of weather types for southern California. The objects were 603 days dated over the period of May–October, 1973–1975. He described each with five attributes: (1) temperature, (2) wind speed, (3) dew point, (4) atmospheric pressure gradient, and (5) temperature gradient. He did not measure the data for these attributes directly, but instead used 12-hour weather predictions made by a mathematical model.

McCutchen cluster-analyzed the data matrix—603 objects and five attributes—by using a nonhierarchical method called ISODATA (described by Anderberg, 1973, pp. 170–173). This produced five clusters of days whose weather types had the following defining properties: Type 1—moist marine air and very hot afternoon; Type 2—relatively dry forenoon air and hot afternoon; Type 3—moist marine air and warm afternoon; Type 4—cool and cloudy, often rainy throughout the day; Type 5—hot and dry throughout the day (the famed Santa Ana wind).

5.2 USING CLUSTER ANALYSIS TO MAKE SPECIFIC-PURPOSE CLASSIFICATIONS

Scientists often make classifications with the specific purpose in mind of investigating their relation to other variables. The variables may be (1) qualitative or (2) quantitative.

Classifications Related to Qualitative Variables

The first example in this section is from medicine, and the second is from oceanography.

RESEARCH APPLICATION 5.5

Through a cluster analysis using chemical results from a medical laboratory, Solberg et al. (1976) made a classification of patients with liver diseases. They found a strong tendency towards correspondence between this classification, based on functional criteria, and previous diagnoses, which were based on liver biopsy findings.

The objects in the cluster analysis were 39 patients having liver diseases. Twenty clinical laboratory tests were performed on their blood serum or plasma. The result of each test was used as the value of an attribute. The data matrix, consisting of 39 patients described by 20 attributes, was then cluster-analyzed and the tree was cut into four clusters labeled A, B, C, and D.

A previous classification had been based on morphological criteria obtained from liver biopsy tests. Solberg et al. used a contingency table to cross-classify the patients according to the two classifications; a contingency table analysis showed that the classification derived through cluster analysis and the previous classification were quite consistent. The two systems of measurement—one based mainly on morphological data and the other on chemical data—were thereby shown to be related.

Where more than two classifications are related, the methods of multidimensional contingency table analysis are appropriate (for example, see Everitt, 1977*a*; Fienberg, 1977; Bishop et al., 1975).

RESEARCH APPLICATION 5.6

There is a natural barrier to small ocean organisms that is known as the 10° S Hydrochemical Front. Running from Asia on the east to Africa on the west, the Front is a band of low-salinity water that separates the high-salinity water masses of the northern and southern Indian Ocean. Hecht et al. (1976) used cluster analysis to test the following research hypothesis about the Front:

H: The 10° S Hydrochemical Front has been in the same location for at least the last 18,000 years.

The test prediction that they deduced from the hypothesis was this:

P: The forms of the shells in samples of 18,000-year-old fossil *Orbulina universa* collected from sediments will be different according to the present location of the Front.

Shortly we will see that *P* involves a relation between a classification and a qualitative variable.

As an indicator of the age of the Front, the researchers used *O. universa*, a tiny shelled organism that exists in sediments as 18,000-year-old fossils and also continues to live today. The Front is a barrier to this organism. To live it must have high-salinity water; the low-salinity water of the Front will not sustain it.

The researchers collected two data matrices. The first contained 18 objects, samples of fossil *O. universa* collected from sediments at the locations shown in Figure 5.2, *b*. The second contained 11 objects, samples of living *O. universa* collected from plankton at the locations shown in Figure 5.2, *b*. Both data matrices used the same eight attributes: average concentrations of (1) small, (2) large, and (3) total number of pores in the specimen shells; average porosities of (4) small, (5) large, and (6) total pores; (7) average shell diameter, and (8) average shell thickness.

Figure 5.2, *a* shows the trees from the two cluster analyses, each cut into two clusters (sediment-sample *H* was considered anomalous and disregarded). This gave two classifications of shells, C_1 (fossil) and C_2 (living).

Now consider a qualitative spatial variable *Y* that describes the locations of samples according to the map shown in Figure 5.2, *b*. The two binary classes of *Y* are: (1) south of the Front, samples the scientists called the "central group"; and (2) north of the Front, samples called the "equatorial group." Clearly there is a relation between the classification of the fossils, C_1, and their position *Y*, which we note as $C_1 \leftrightarrow Y$. There is also a relation

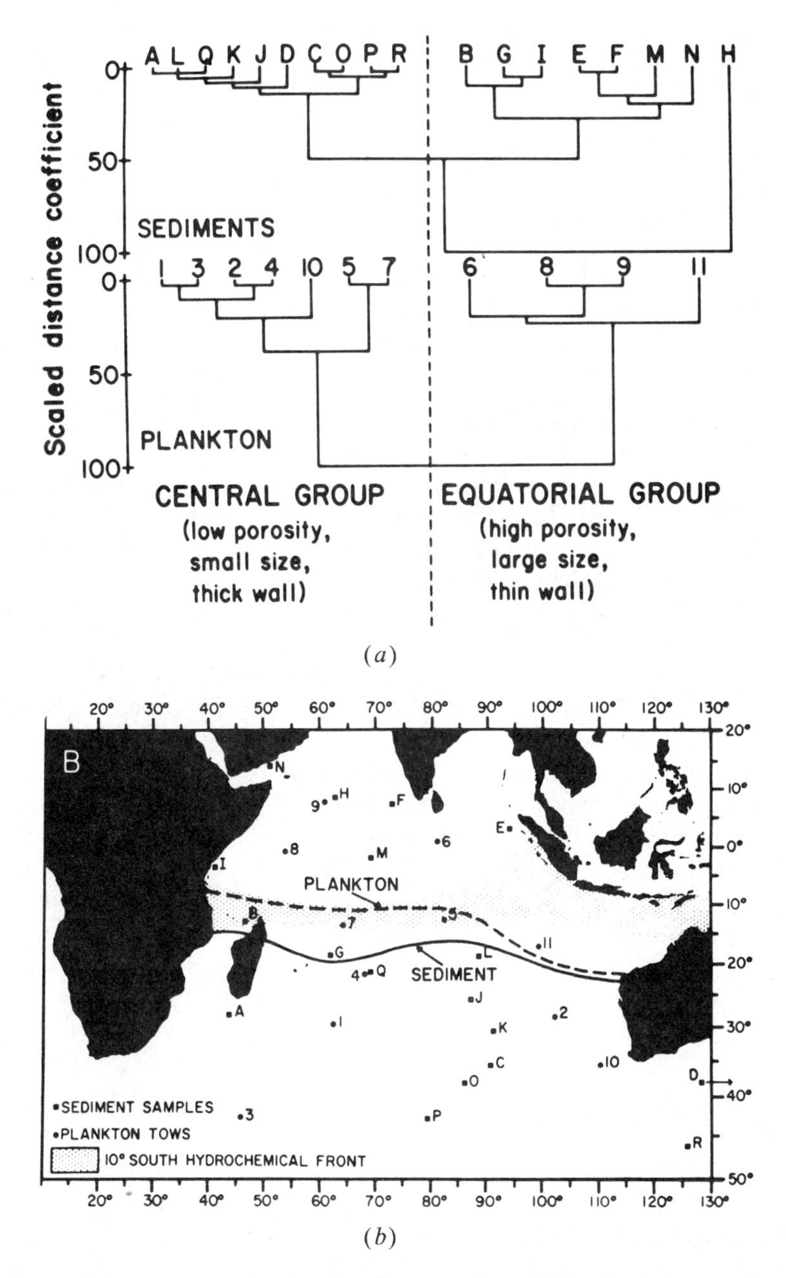

Figure 5.2. (*a*) trees showing the cluster analyses of fossil specimens of *O. universa* collected from sediments, and living specimens of *O. universa* collected from plankton. (*b*) location of samples of *O. universa* used in the cluster analysis. (Reproduced from Hecht (1976), Figure 1; copyright 1976 by the American Association for the Advancement of Science.)

between the classification of living organisms C_2 and their position Y, $C_2 \leftrightarrow Y$. Moreover, the two relations are themselves similar, as can be diagrammed:

$$\begin{matrix} C_1 & & C_2 \\ \updownarrow & \leftarrow \text{Relation of similarity} \rightarrow & \updownarrow \\ Y & & Y \end{matrix}$$

Thus the hypothesis that the Front's position has remained largely unchanged over the past 18,000 years was confirmed.

In the foregoing application, cluster analysis was used twice to make a classification that was then related to a qualitative variable. But this was not an end in itself. The grander end was to test a hypothesis. In the examples of Section 4.3, in which trees were used to test hypotheses, the test prediction P was only about the form the tree should take. Here, the prediction is more complex. P is about the forms that the *relation* between two classifications and a qualitative variable will take. The relation of similarity between the two relations $C_1 \leftrightarrow Y$ and $C_2 \leftrightarrow Y$ is an example of a second-order pattern, which is discussed in Chapter 23.

There are numerous other marine research applications in which a classification produced with cluster analysis has been related to a qualitative spatial variable. Kaesler et al. (1977) found that benthic samples (bottom-dwelling ocean organisms) from the American Quadrant of the Antarctic Ocean produced two clusters of samples that spatially corresponded with the water masses on each side of the Antarctic Convergence. Holland and Claflin (1975) found that samples of planktonic diatoms from different locations in Green Bay (Lake Michigan), formed clusters related to sampling location. Ali et al. (1975) found that samples of marine sediments collected from the New York Bight, a region in the Atlantic Ocean near New York City where solid wastes from the city are dumped, formed clusters related to sampling location.

Classifications Related to Quantitative Variables

For this type of a specific-purpose classification, cluster analysis is used to make a classification that is then related to a quantitative variable. Frequently the quantitative variable is a measure of time or space. The following examples illustrate this in several contexts.

RESEARCH APPLICATION 5.7

West (1966) placed a twenty-mile transect through a forest on the east flank of the central Oregon Cascade Mountains. The transect was oriented from west to east. West's question was, Is the classification of forest vegetation, formed by applying cluster analysis to inventories of the forested plots, related to position along the transect?

For the objects of the data matrix, West chose 40 forested plots, spaced at two-mile intervals along the transect. For attributes he chose the relative frequencies of 62 trees, herbs, and shrubs. After performing the cluster analysis, he cut the tree to form a classification of plots. The classes of plots tended to be located adjacent to each other along the transect: most "type A's" were located together, most "type B's" were together, and so forth. West retroductively postulated that the gradient of annual precipitation was the likely cause of the clustering. He found that this was so; the transect corresponded to the decreasing gradient.

RESEARCH APPLICATION 5.8

Transects need not be on land or be horizontal. They can be extended through a body of water vertically, from the surface to the bottom.

Lighthart (1975) did just this. He located a "water column," a vertical transect, in Green Lake, Washington. For objects he took samples of water from different depths in the column. These he described with 51 attributes, the results of laboratory tests of the water's bacterial composition. After cluster-analyzing the data matrix, he cut the tree into clusters, forming a classification of samples. He then showed that the classes were spatially dependent: the bacterial composition of the classes, which differed from class to class, was related to the depths from which the samples were taken.

One way used in juvenile detective stories to catch a crook is to tie a bag of flour with a hole in it underneath the crook's car. The telltale path of flour is then traced to the crook's hideaway. As the next two applications illustrate, there are many ways in nature in which "bags of flour" can become tied to objects to make it possible to trace their paths through time. Both applications required that cluster analysis be used to make specific-purpose classifications.

RESEARCH APPLICATION 5.9

Hodson et al. (1966) studied the changes in the designs of decorative brooches worn as jewelry during the Iron Age. These researchers obtained a sample of brooches from an Iron Age cemetery in Switzerland and used them for the objects in a cluster analysis. They described the brooches with attributes of form—type of decoration, presence or absence of characteristic features, and so forth. Then they cluster-analyzed the data matrix, cut the tree, and made a classification of brooches.

They inferred the relative age of the brooches by using an analogy to the flour-bag trick. In Iron Age societies, graves were usually dug in rows that progressed across a field linearly. Thus a grave's position corresponds to its relative burial date, and the brooches can be ranked according to their presumed order of manufacture.

The researchers found a relation between the classification of brooches, based on their design, and their order of manufacture.

RESEARCH APPLICATION 5.10

Plants leave traces of pollen that are fossilized in the earth's crust. Cores drilled into the crust give a record of the diversity of pollen types, and hence of the diversity of past plant life.

Using cluster analysis, Adam (1974) studied changes in the diversity of pollen types over time. First he obtained a core sample of the earth's crust. He then chopped the core, 440 cm long, into 46 objects, each about 10 cm long. He described these samples by the counts of 42 pollen types he observed in them. He found that a classification of the samples was related to their positions in the core. He concluded that the earth's vegetation has changed over the period signified by the core.

A transect is a one-dimensional map. It is natural to wonder whether the needs to scientists to investigate the relations between classifications and one-dimensional maps generalizes to two-dimensional maps. The next example shows this to be the case.

RESEARCH APPLICATION 5.11

For a geographic study, Fesenmaier et al. (1979) used a classification created by cluster analysis to investigate the zone of transition between built-up urban areas and the surrounding countryside—the zone being what geographers term the "rural-urban fringe." To do this, they used sections of land adjoining London, Ontario, as the objects of a cluster analysis, with attributes that measured socioeconomic properties. They cluster-analyzed the data matrix, obtained a classification, and mapped the land according to the land-use classes. Their results showed a gradual gradient of classes radiating outward from London's center.

From accepted theory, these geographers held a hypothesis that predicted that a metropolitan map would show an abrupt urban-to-rural transition, as if urban development pushed outward like the leading wall of a lava flow. The land-use map of London showed a continuous transition without well-defined boundaries, thus falsifying the hypothesis (as it applies to London, Ontario, in its present stage of growth).

Other examples involving maps are the following. Kendig (1976) used cluster analysis to make a classification of neighborhood types based on socioeconomic attributes for the city of Los Angeles. Using this, he drew a map of the city shaded to indicate the classes of neighborhood types and investigated their contiguities. In another application, Ashley and Lloyd (1978) surveyed a coordinate grid over a region of Chile and then made a classification of samples of ground water collected from locations throughout the grid. The samples became the objects of the data matrix and were described by their hydrochemical properties. After making their cluster analysis, the researchers cut the tree into clusters of samples and mapped the region according to the classes.

SUMMARY

Cluster analysis can be used to make general-purpose classifications of biological or nonbiological objects. The classifications are called general-purpose because they are designed to serve a variety of general functions in unspecified scientific theories. In contrast is the use of cluster analysis to make specific-purpose classifications; these are designed to be related to specific qualitative or quantitative variables.

It is important to distinguish between general-purpose and specific-purpose classifications. The choices of objects and attributes, as well as of scales of measurement and resemblance coefficients, depend on the purpose of the classification. In other words, as is shown in Chapter 22, different ends dictate different means.

The research goals in Chapters 4 and 5 are not always mutually exclusive. For example, one can make a general-purpose classification, or a specific-purpose classification, as a research goal in itself. But these can also be merely steps toward the ultimate research goals of (1) creating a question, (2) creating a hypothesis, or (3) testing a hypothesis, as the applications in this chapter show.

Applications of Cluster Analysis in Planning and Engineering

OBJECTIVES

In this chapter we shall describe some applications of cluster analysis in

- Planning.
- Engineering.

It is important to recognize that

- These applications differ from applications in science.

The first section in the chapter explains in what way this is so.

6.1 DIFFERENCES BETWEEN SCIENCE, PLANNING, AND ENGINEERING APPLICATIONS

The goal of science is to build a mental picture of reality—a picture we call knowledge and understanding—that will be consistent with facts determined through experiment whenever the picture is put to test. The knowledge can be of natural or human processes, or of relations among events.

Scientists differ from other professionals in how they are motivated and in how they validate their work. First, scientists are primarily motivated by a curiosity to discover how nature works, rather than by a need to apply the knowledge they discover for the benefit of society. For example, the scientists who tested the hypothesis described in Research Application 4.6, about the relationships among Egyptian mummies, were satisfying their intellectual curiosities, and it is difficult to see how their work might materially and directly benefit society. Second, the validity of scientific knowledge rests with how closely and broadly it agrees with experimental facts.

In contrast, the goal of planning is to make the world materially better. Planning is used in this book generally to mean land-use planning, or market research planning, or national defense planning—that is, essentially all kinds of planning that take place in business or in other organizations. Planning is the part of management concerned with the best way to accomplish a goal (the other part of management is concerned with control—with implementing plans).

Planners differ from scientists: first, planners are primarily motivated by a need to change the world for the better; second, the validity of their plans depends on how well the implemented plans improve the human condition.

Since the values and goals of science and planning are distinct, it follows that each has its distinct standards for framing and validating applications. This is the reason we separate the two disciplines in this book. Romesburg (1981) describes the distinction in fuller detail, and the necessity for the distinction is made apparent in Part IV, which presents details of how applications of cluster analysis are framed and validated.

Superficially, science and planning applications share so much in common that their differences are easily masked. For example, both activities use data, use largely the same quantitative methods, use computers, and sometimes publish their results in the same journals. Both even call their practitioners by the same term—researchers. And scientists and planners occasionally switch roles in the course of doing their main work—for example, a scientist acts as a planner when designing experiments and a planner commissions applied scientific studies to obtain the specific knowledge needed for formulating a plan.

Yet an applied scientific study done by a planner will usually be done according to tolerances of validity typically used in planning, not according to the tolerances used in pure science. Thus, a drug manufacturer may run many more trials of a drug to demonstrate all of its effects than would a scientist who wanted to determine if the drug merely does have a measurable effect. And in other cases, such as the planning of national defense, much of the data planners use are based on opinions, and the quality of such data would be intolerable for use in purely scientific research. To be successful in achieving the goal of national defense, the defense planner selects the best available data, using the opinions of experts in defense, even if the resulting data are not of the quality a scientist would require.

Engineering is a special branch of planning, and applications of cluster analysis in engineering are singled out for special discussion. However, because the planning and engineering applications described in this chapter come in diverse forms they are all grouped under the one heading of "Research Goal 6: Facilitate Planning and Management."

6.2 USES OF CLUSTER ANALYSIS IN PLANNING

Let us begin by examining some of the many forms that planning applications can take. To start, alternative decisions or plans can be used as the objects of a cluster analysis. The attributes can either describe the features of the alternatives or their expected outcomes. Finding clusters of similar alternatives reduces the decision problem to two stages: (1) select the cluster that can best achieve the planning objective, and (2) select the best alternative within the best cluster. Consumers often shop for automobiles in this way, without realizing they are making a primitive application of cluster analysis. First they narrow their choice to a cluster of cars, any of which they would not be displeased to own. Then they make their final selection from the preferred cluster.

Environmental impact analysis lends itself to this form of cluster analysis. Suppose a recreational area planner is considering developing a campground at one of t sites along a river. These can be described by their biological and human-interest attributes. A problem in environmental analysis, which Leopold and Marchand (1968) describe, is to find which sites are unique and which are not. The planner will want to protect a unique site from being developed while narrowing his choice of sites for development to those that are almost twins of each other. A variation on this is to make the data matrix contain the t sites as they are, plus the t sites as they would be expected to be if each were developed. A cluster analysis of the data matrix of the $2t$ sites will identify the sites expected to be most seriously impacted. Sites not expected to be seriously impacted by development will show present and future forms that are quite similar. Conversely, those expected to be seriously impacted will show present and future forms that are quite dissimilar.

Another form that a planning application may take is a cluster analysis to find classes of objects that should be treated or addressed in the same way. An example of this in the area of investment management is Jensen's (1971) cluster analysis of New York Stock Exchange companies based on their dividends and earnings data. It gave a classification of stocks that would enable a portfolio manager to find similar stocks to substitute for one another in narrowly focussed portfolios, or dissimilar stocks for diversified portfolios. Jensen also identified certain classes of stocks that had performed better than the stock market averages for certain periods, thus suggesting that this use of cluster analysis could also help portfolio managers to improve profits for their clients.

In a more complex investment application, Frankfurter and Phillips (1980) used cluster analysis to arrange securities into similar groups. For data they used "orthogonalized score components," which they obtained by means of a factor analysis performed on financial data describing each security. They claim that their approach using cluster analysis can lead to the design of security portfolios that have certain advantages over portfolios designed by using the well known Markowitz method—a method that uses mathematical optimization techniques.

Marketing and advertising planners have also used cluster analysis to form groups of similar objects. Lessig and Tollefson (1971) cluster-analyzed a sample of coffee drinkers that they described with such attributes as income, race, and occupation, which gave a classification of market segments. And in the same vein, Sethi and Holton (1973) used cluster analysis to form market segments of nations that multinational companies could address with different marketing strategies. In concluding their analysis, they stated that their formation of clusters of countries "leads to some rather strange bedfellows. Markets which intuitively might have seemed quite different from each other prove to have more similarities than one would guess at first glance. In other cases markets which might be considered very much alike, judging from casual observation, prove to fall into different clusters."

Although planning and management applications often generate knowledge as a byproduct, the primary purpose of the ones mentioned so far was to aid decision-making. Keep this in mind as you read through the following detailed research applications.

RESEARCH APPLICATION 6.1

McCormick et al. (1972) used cluster analysis to find which air force transport aircraft could be used interchangeably. They cluster-analyzed 49 types of transport aircraft described by 37 functional attributes: ability to function as an aerial refueler, as an assault transport, as an aeromedical transport, as a personnel transport, as a utility transport, and so forth. They scored the aircraft on an ordinal scale of 0 to 2 for their ability to perform each function. Thus their data matrix contained 49 objects and 37 attributes, and each cell contained a score of 0, 1, or 2.

The tree resulting from the cluster analysis showed those aircraft having similar overall capabilities. They would be candidates for interchangeable use, which is important information when deciding what mix of aircraft should be stocked at an airbase.

Mobley and Ramsay (1973), and Lissitz et al. (1979), in similar applications, used cluster analysis to group jobs in an organization; workers in similar jobs can be more easily substituted for each other. Likewise, Chu (1974) suggested using cluster analysis to identify drugs that are candidates for substitution. The objects Chu used were drugs, described by functional attributes giving their effects on animals and people.

RESEARCH APPLICATION 6.2

Consider how a new drug molecule is selected for manufacture. A drug molecule's beneficial and toxic properties are determined by its substituents, atoms or groups of atoms that can replace atoms of the basic molecule, thereby forming a derivative. In essence the molecule is like a stripped-down automobile and the substituents are like accessories; different ways of equipping the molecule result in different derivatives.

The problem a drug designer faces is that the number of possible derivatives of a molecule is huge. Many will have similar shapes and similar effects. Only a few can be made and tested because the cost of doing this is great. How should a drug designer select derivatives to test from among the huge smorgasbord of possibilities?

Hansch and Unger (1973) suggested that cluster analysis be used to obtain a systematic sample of diverse derivatives. Working with the premise that molecules with similar chemical structures will have similar effects, they clustered a sample of substituents and selected one or two from each of the different branches of the tree, thus obtaining a diverse sample presumably covering the range of possible drug effects. From this a selected set of derivatives could be made and tested.

RESEARCH APPLICATION 6.3

In the method of statistical inference termed **stratified random sampling**, a finite population of sample units (plots of land, for example, or people) must be segregated by gathering similar units into separate sets called **strata**, before a random sample of units is drawn (Mendenhall et al., 1971; Levy and Lemeshow, 1980). Cluster analysis can be used to group similar sample units into the requisite strata.

In an illustrative study, Derr and Nagle (1977) used cluster analysis to stratify townships in New Jersey. In all, 107 townships were used as the population of sample units to be cluster-analyzed. The attributes described their agricultural, commercial, and demographic features. After cutting the tree into ten clusters, each a stratum containing similar townships, these researchers drew a stratified random sample from the strata. From the random sample, they estimated economic parameters of the whole population of townships.

6.3 USES OF CLUSTER ANALYSIS IN ENGINEERING

We shall treat engineering as a field of planning, inasmuch as engineering almost always entails transformation of a plan into a finished system or construction capable of achieving a predetermined goal.

Ordinarily when we encounter the term engineering we think of traditional core disciplines such as mechanical and electrical engineering. Other, less well known engineering disciplines, such as environmental engineering, mining engineering, and operations research have generated recent, interesting applications of cluster analysis, of which the following are examples.

Environmental Engineering. Crecellus et al. (1980) cluster-analyzed atmospheric aerosols collected from air filter samples near Colstrip, Montana. Their purpose was to establish baseline data from which manmade air pollution could be gauged and ultimately be controlled through engineering. This type of environmental engineering study has a counterpart in water pollution studies.

In an application to assess the amount of pollution caused by an oil rig, Lytle and Lytle (1979) collected samples of water at a site in the Gulf of Mexico. Some of the samples were taken before an oil rig was put in place; others were taken after. By means of gas chromatography, the samples were analyzed for their hydrocarbon content and the resulting measurements were made the attributes of a cluster analysis. The distribution of water samples in the resulting tree was unrelated to whether the samples had been gathered before or after installation of the rig, suggesting the rig had caused no measurable pollution.

Mining Engineering. Neogy and Rao (1981) used cluster analysis to delineate the grades of mineable iron ore in the Kiriburu district of India. For objects they used cores taken from bore holes made in an area of the district, and they described the cores by their chemical composition. The cluster analysis gave a classification that was correlated with grades of iron ore lumps, also found in the district, thereby making it possible to predict what grade of the ore might be found from the chemistry of sample cores. Neogy and Rao concluded that this kind of analysis could provide information that could reduce the costs of mineral exploration.

Remote Sensing. Airborne and satellite-borne multispectral scanners are used as remote sensors of the earth's environment. The output of a multispectral scanner is a set of *n* measurements, one for each channel of the scanner. In this way, each scanned area of the earth's surface signs its name with its spectral signature. As Swain (1978) shows, these signatures can be cluster-analyzed to give a classification of surface types. Once the classification has been made, it can be used for identification of unclassified areas: the spectral signatures from unmapped surfaces can be obtained, their surface types identified through comparison with the classification, and a map of the earth drawn according to the types of surface present. This use of cluster analysis can provide information to agricultural engineers, irrigation engineers, environmental engineers, mining engineers, and others. (For a specific cluster-analysis method for analyzing remotely sensed data, see Wharton and Turner (1981).)

Many scientific studies have recorded time series of data and then cluster-analyzed the time series to find out which ones were similar. In a study to determine whether the singing behavior of birds varied with the location of bird populations, Martin (1979) recorded the sonographs of song types of birds and cluster-analyzed them; Adams et al. (1978) recorded the respiratory time series of experimental animals and cluster-analyzed them; two recent studies have analyzed electrocardiogram (ECG) data: Rompelman et al. (1980) compared the ECG's of heart patients, and Battig et al. (1980) compared those of psychiatric and normal patients. And Otter et al. (1980) cluster-analyzed time series data obtained from strip-chart recordings of the electrophysiological responses of the cabbage white butterfly to different olfactory stimuli. Engineering offers many potential applications of cluster-analyzing times series data, yet the literature of engineering has not reported many applications in this area.

SUMMARY

Whereas research goals in science are concerned with the creation of knowledge, in contrast, research goals in planning and management are concerned with deciding how to change the world in certain predetermined ways. Both kinds of research goals demand specific information, the quality of which is determined by tolerances that may differ between disciplines.

Science is often interwoven with planning and management: planning and managing are necessary in any scientific project, and scientific knowledge is an important ingredient in many forms of planning and management. A scientist becomes a planner when deciding, for example, how to form strata for a stratified random sampling problem. And a planner, for example of national defense, must be able to evaluate certain kinds of scientific information in order to plan applications of it.

Engineering is a special kind of planning, the aim of which is to build a device or control an environment to achieve a well-defined goal. Information for planning and engineering can be obtained by using cluster analysis. Whereas Chapters 4 and 5 discussed applications of basic scientific research —knowledge for the sake of knowledge, essentially—the goals of engineering studies such as those dealt with in this chapter prescribe certain kinds of applied science research—knowledge for the sake of a particular use.

Applications in science are fundamentally different from those made in planning and management, from the viewpoint of how a cluster analysis should be framed. This distinction controls how the objects and attributes should be chosen, which resemblance coefficient should be used, and other considerations. This in turn controls how the analysis should be validated. These philosophical concerns are discussed in Part IV.

EXERCISE

6.1. Now that you have read Chapters 4, 5 and 6, you should begin to search for applications articles that use cluster analysis. Read from a wide variety of fields. Develop the practice of identifying which of the six research goals a given application fits. Begin to feel how the methods of cluster analysis are subservient to these research goals. Cultivate the use of your library's reference section to find applications that interest you. One reference book that is a good place to start is *Science Citation Index*.

At first, unfamiliar terms and authors' failures to outline their research goals may present roadblocks. Over time, as you become adept at reviewing the literature, you will learn ways to become more creative in the use of mathematics.

PART II

How to Do Cluster Analysis in Depth

If the only tool you have is a hammer, it is tempting to treat everything as if it were a nail.

ABRAHAM H. MASLOW
The Psychology of Science

To understand cluster analysis well enough to use it in your research, you must build upon the basic steps described in Part I. How to do this is the subject of Part II.

Chapter 7 gives alternative ways of standardizing the data matrix.

Chapters 8 and 10 through 12 treat alternative resemblance coefficients: those for data matrices based on quantitative attributes are described in Chapter 8; Chapter 10 describes those for data matrices based on qualitative attributes; Chapter 11 covers special resemblance coefficients for ordinal-scaled attributes; and Chapter 12 shows several ways of computing resemblance coefficients for data matrices comprising mixtures of quantitative and qualitative attributes.

Chapter 9 explains clustering methods that are alternatives to the UPGMA clustering method treated in Chapter 2.

Chapter 13 shows how, in some instances, you can collect data directly in the form of the resemblance matrix, thereby bypassing the data matrix.

Chapter 14 describes how to use matrix correlation methods in cluster analysis. These methods go beyond the correlation between a resemblance matrix and a cophenetic matrix, dealt with in Chapter 2, to the correlation between two resemblance matrices, or between two cophenetic matrices.

Finally, Chapter 15 explains how to present the results of your cluster analysis to others so that no one mistakes the alternatives you considered, and so that everyone understands what you did.

Standardizing the Data Matrix

OBJECTIVES

This chapter explains

- The reasons for standardizing the data matrix.
- Alternative methods for standardizing the data matrix.

7.1 REASONS FOR STANDARDIZATION

There are two main reasons for standardizing a data matrix. First, the units you choose for measuring attributes can arbitrarily affect the similarities among objects. By standardizing the attributes and recasting them in dimensionless units, you remove the arbitrary affects. Second, standardizing makes attributes contribute more equally to the similarities among objects. For example, if the range of values on the first attribute is much greater than the range of values on the second, then the first attribute will carry more weight in determining the similarities among objects. When this is undesirable in the context of your research goal, you can compensate for the effect by standardization.

7.2 STANDARDIZING FUNCTIONS

To standardize a data matrix, you must first choose an equation called a **standardizing function** and then apply it to the data matrix. This section describes various standardizing functions.

To understand these equations, you need to understand the symbolic notation of the data matrix and the standardized data matrix, shown in Figure 7.1. Each matrix has n attributes that are subscripted $i = 1, 2, \ldots, n$; and t objects subscripted $j = 1, 2, \ldots, t$. In the data matrix, the value of data for any ith attribute and jth object is denoted as X_{ij}. The corresponding value in the standardized data matrix is denoted as Z_{ij}.

The standardizing function most commonly used in practice is

$$Z_{ij} = \frac{X_{ij} - \overline{X}_i}{S_i}, \tag{7.1}$$

where

$$\overline{X}_i = \frac{\sum_{j=1}^{t} X_{ij}}{t},$$

and

$$S_i = \left(\frac{\sum_{j=1}^{t} \left(X_{ij} - \overline{X}_i \right)^2}{t - 1} \right)^{1/2}$$

In words, to calculate the standardized value Z_{ij} for any ith attribute and jth object, we take the corresponding value X_{ij} in the data matrix,

Data matrix

Attribute	Object 1	2	$\cdots$	j	$\cdots$	t
1	X_{11}	X_{12}	$\cdots$	X_{1j}	$\cdots$	X_{1t}
2	X_{21}	X_{22}	$\cdots$	X_{2j}	$\cdots$	X_{2t}
$\vdots$	$\vdots$	$\vdots$		$\vdots$		$\vdots$
i	X_{i1}	X_{i2}	$\cdots$	X_{ij}	$\cdots$	X_{it}
$\vdots$	$\vdots$	$\vdots$		$\vdots$		$\vdots$
n	X_{n1}	X_{n2}	$\cdots$	X_{nj}	$\cdots$	X_{nt}

Standardized data matrix

Attribute	Object 1	2	$\cdots$	j	$\cdots$	t
1	Z_{11}	Z_{12}	$\cdots$	Z_{1j}	$\cdots$	Z_{1t}
2	Z_{21}	Z_{22}	$\cdots$	Z_{2j}	$\cdots$	Z_{2t}
$\vdots$	$\vdots$	$\vdots$		$\vdots$		$\vdots$
i	Z_{i1}	Z_{i2}	$\cdots$	Z_{ij}	$\cdots$	Z_{it}
$\vdots$	$\vdots$	$\vdots$		$\vdots$		$\vdots$
n	Z_{n1}	Z_{n2}	$\cdots$	Z_{nj}	$\cdots$	Z_{nt}

Figure 7.1. Canonical forms of the data matrix and the standardized data matrix.

subtract from it the mean $\overline{X}_i$ of the values of the ith attribute, and divide the result by the standard deviation S_i of the values of the ith attribute.

Note that the standardized attributes will be unitless. To illustrate, suppose the units of X_{ij} are centimeters. Then $X_{ij} - \overline{X}_i$, which is the numerator of equation 7.1, will also be in centimeters. When this is divided by S_i, also measured in centimeters, the value Z_{ij} will be unitless.

Note also that the mean $\overline{Z}_i$ of each ith row of the standardized data matrix will be $\overline{Z}_i = 0.0$, and the standard deviation will be $S_i' = 1.0$. Moreover, all of the values of Z_{ij} in each ith row of the standardized data matrix are likely to be in the range of $-2.0 \leq Z_{ij} \leq +2.0$.

Let us make these points clear with an example, in which we compute a standardized data matrix by applying equation 7.1 to a data matrix like the one used in Chapter 2 to illustrate the basic steps of cluster analysis. The data matrix is shown in Figure 7.2,*a*.

The calculations that transform the data matrix into the standardized data matrix shown in Figure 7.2,*b* are as follows. First we calculate $\overline{X}_i$ and S_i for each row, as Figure 7.2,*a* shows. Using equation 7.1, we then calculate Z_{11}—that is, the standardized value where the first row and the first column meet. It is $Z_{11} = (X_{11} - \overline{X}_1)/S_1 = (20.0 - 21.4)/2.0736 = -0.68$. Then we calculate $Z_{12} = (X_{12} - \overline{X}_1)/S_1 = (24.0 - 21.4)/2.0736 = 1.25$. Continuing, we obtain all of the standardized values for the first attribute: Z_{11}, Z_{12}, Z_{13}, Z_{14}, and Z_{15}. Moving to the second row, $Z_{21} = (X_{21} - \overline{X}_2)/S_2 = (19.0 - 15.8)/6.9426 = 0.46$. And so on. As we noted, the mean of each row of the standardized data matrix, $\overline{Z}_i$, should be 0.0, and the standard deviation, S_i',

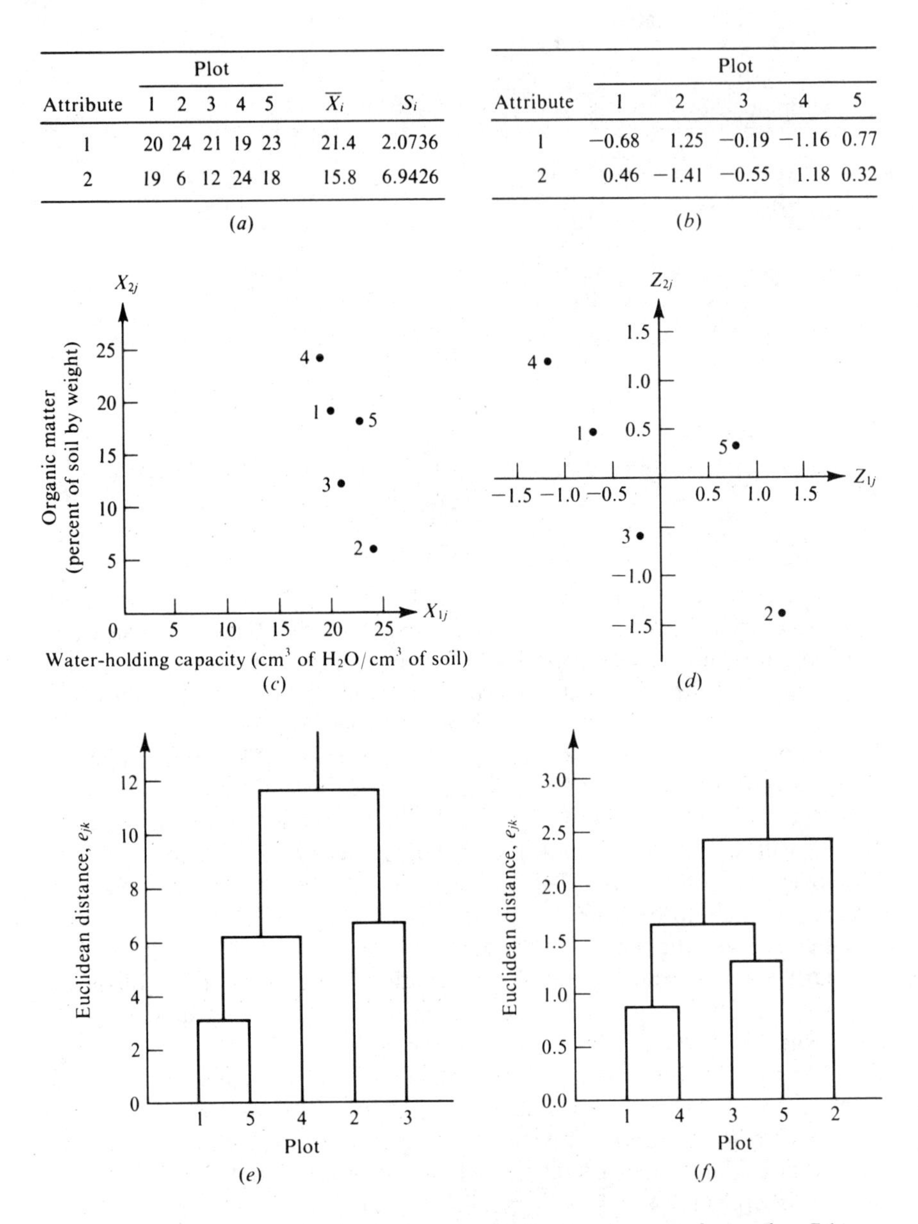

Attribute	Plot 1	2	3	4	5	$\overline{X}_i$	S_i
1	20	24	21	19	23	21.4	2.0736
2	19	6	12	24	18	15.8	6.9426

(*a*)

Attribute	Plot 1	2	3	4	5
1	−0.68	1.25	−0.19	−1.16	0.77
2	0.46	−1.41	−0.55	1.18	0.32

(*b*)

Figure 7.2. Example illustrating standardization by use of equation 7.1. (*a*) Initial data matrix; (*b*) standardized data matrix; (*c*) attribute space plot of the data matrix; (*d*) attribute space plot of the standardized data matrix; (*e*) tree produced by cluster analysis of the data matrix, using the Euclidean distance coefficient e_{jk} and the UPGMA clustering method; (*f*) corresponding tree produced by cluster analysis of the standardized data matrix.

should be 1.0. A check for the case of $i = 1$ shows this is true:

$$\bar{Z}_1 = [-0.68 + 1.25 + (-0.19) + (-1.16) + 0.77]/5$$
$$= 0.0;$$
$$S'_1 = \left[\left((-0.68)^2 + (1.25)^2 + (-0.19)^2 + (-1.16)^2 + (0.77)^2\right)/(5-1)\right]^{1/2}$$
$$= 1.00.$$

A similar procedure holds for $i = 2$.

For a resemblance coefficient like the Euclidean distance coefficient, e_{jk}, the second attribute of this data matrix contributes the most weight. As an illustration, consider e_{12}, calculated by the method given in Section 2.3. Using the values in the data matrix, it is

$$e_{12} = \left[(20 - 24)^2 + (19 - 6)^2\right]^{1/2} = (16 + 169)^{1/2}.$$
$$= 13.6.$$

The second term in the sum contributes the most. But note what happens when e_{12} is calculated by using the standardized data matrix:

$$e_{12} = \left[(-0.68 - 1.25)^2 + (0.46 - (-1.41))^2\right]^{1/2}$$
$$= (3.72 + 3.50)^{1/2} = 2.69.$$

The contributions of both attributes to e_{12} are more nearly equal.

The attribute space plots shown in Figures 7.2,*c* and *d* graphically show how standardizing equalizes the contribution of each attribute to overall resemblance. Whereas interobject distances computed by using the original data matrix are dominated by the second attribute, the distances computed by using the standardized data matrix are not dominated by either attribute. The objects in the attribute space of the standardized data matrix are spread equally over both standardized attributes.

When we cluster-analyze the original data matrix and the standardized data matrix, the respective trees shown in Figures 7.2,*e* and *f* are quite different. This simply mirrors the fact that attribute 2 carries the most weight in the original data matrix, whereas both attributes carry nearly equal weight in the standardized data matrix.

To better understand equation 7.1, look at Figure 7.3, where it is graphically portrayed. To obtain the figure, we have recast the equation as

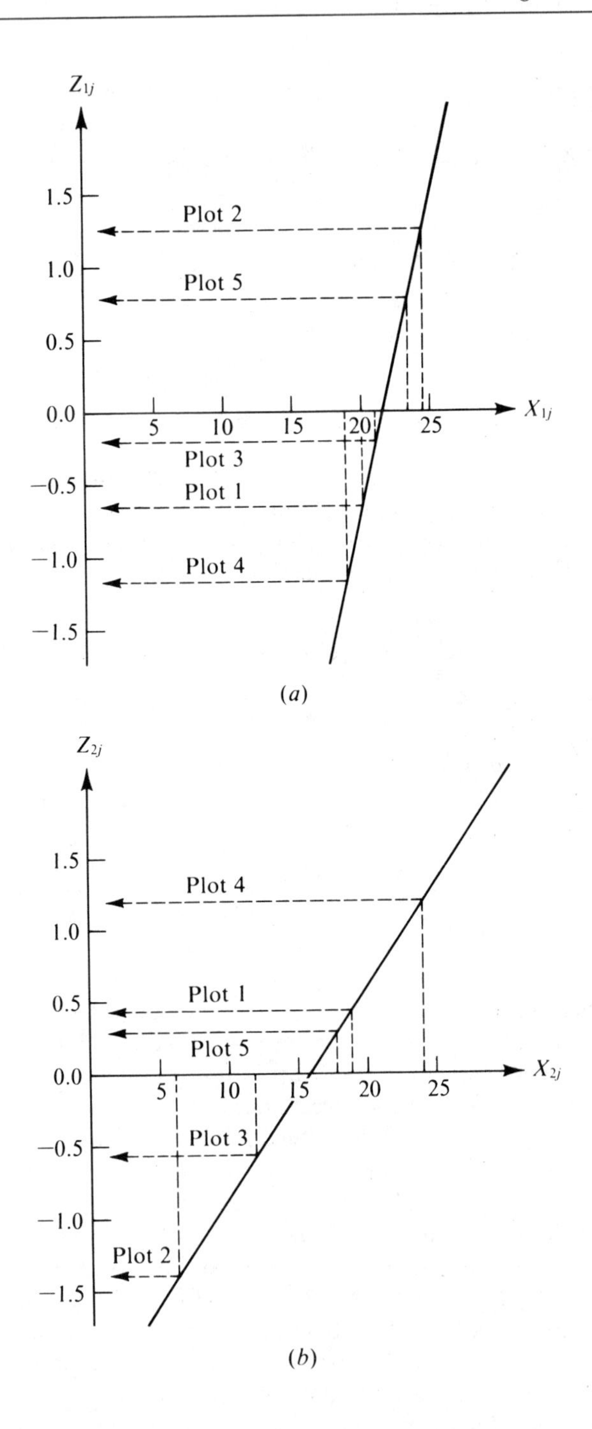

Figure 7.3. Graphic portrayal of how the data matrix of Figure 7.2,*a* is standardized through use of equation 7.1. (*a*) Attribute 1; (*b*) attribute 2.

$Z_{ij} = (-\bar{X}_i/S_i) + (1/S_i)X_{ij}$, which fits the general form of the equation of a straight line, $y = a + bx$, in which the value $a = -\bar{X}_i/S_i$ is the y-axis intercept, and $b = (1/S_i)$ is the slope of the line. Part a of the figure shows attribute 1; part b shows attribute 2. For each ith attribute, the values of $\bar{X}_i$ and S_i fix the position of the line so that the standardized data are centered about $\bar{Z}_i = 0.0$ and cover approximately the same range. For example, because the data X_{ij} in our example are less dispersed on attribute 1 than on attribute 2 (S_1 is less than S_2), the slope of the line is greater for attribute 1 than for attribute 2 ($1/S_1$ is greater than $1/S_2$).

Equation 7.1 is by far the most-often-used standardizing function in applications of cluster analysis. The following are lesser-used alternatives.

Making The Standardized Data Into Proportions.

When all of the X_{ij} values of the data matrix are nonnegative, and no row contains all zeros, we can make the standardized values into proportions such that $0.0 \leq Z_{ij} \leq 1.0$. There are three ways of doing this. The first is

$$Z_{ij} = \frac{X_{ij}}{\mathrm{RMAX}_i}. \tag{7.2}$$

The value RMAX_i is the maximum value in the ith row of the data matrix. To compute the standardized value of any ith attribute and jth object we take the corresponding unstandardized value and divide it by the maximum value in the ith row. As a result, every row of the standardized data matrix will contain at least one value of $Z_{ij} = 1.0$, which will occur when $X_{ij} = \mathrm{RMAX}_i$.

The second standardizing function for making the data into proportions is

$$Z_{ij} = \frac{X_{ij} - \mathrm{RMIN}_i}{\mathrm{RMAX}_i - \mathrm{RMIN}_i}. \tag{7.3}$$

The value RMIN_i is the minimum value in the ith row of the data matrix. To compute the standardized value of any ith attribute and jth object we first take the corresponding unstandardized value and subtract from it the minimum value in the ith row. Then we divide this result by the range of values, $\mathrm{RMAX}_i - \mathrm{RMIN}_i$, in the ith row. As a result, each row of the standardized data matrix will contain at least one value of $Z_{ij} = 0.0$, which will occur when $X_{ij} = \mathrm{RMIN}_i$. And each row will contain at least one value $Z_{ij} = 1.0$, which will occur when $X_{ij} = \mathrm{RMAX}_i$.

The third proportional function is

$$Z_{ij} = \frac{X_{ij}}{\sum_{i=1}^{t} X_{ij}}. \tag{7.4}$$

To compute the standardized value of any ith attribute and jth object we divide the corresponding unstandardized value by the sum of all the values of the ith attribute over all t objects. This normalizes the standardized values in the ith row so that they sum to 1.0.

Equations 7.1 through 7.4 are **row functions**, so named because the values in each ith row of the standardized data matrix depend only on values in the ith row of the original data matrix. For each of these equations there is a corresponding **column function** that follows the same logic. When a data matrix is standardized with a column function, the values in each jth column of the standardized data matrix depend only on other values in the jth column of the data matrix.

As a column function, equation 7.1 is written as

$$Z_{ij} = \frac{X_{ij} - \overline{X}_j}{S_j}, \tag{7.5}$$

where

$$\overline{X}_j = \frac{\sum_{i=1}^{n} X_{ij}}{n}$$

and

$$S_j = \left(\frac{\sum_{i=1}^{n} \left(X_{ij} - \overline{X}_j \right)^2}{n - 1} \right)^{1/2}.$$

Here, $\overline{X}_j$ and S_j are the mean and standard deviation of the jth column of the data matrix. When standardized, the mean $\overline{Z}_j$ of each jth column of the standardized data matrix will be $\overline{Z}_j = 0.0$, and the standard deviation will be $S_j' = 1.0$.

As a column function, equation 7.2 is written as

$$Z_{ij} = \frac{X_{ij}}{\text{CMAX}_j}. \tag{7.6}$$

Here, CMAX_j is the maximum value in the jth column of the data matrix.
As a column function, equation 7.3 is written as

$$Z_{ij} = \frac{X_{ij} - \text{CMIN}_j}{\text{CMAX}_j - \text{CMIN}_j}. \tag{7.7}$$

Here, CMIN_j is the minimum value in the jth column of the data matrix.
Finally, as a column function equation 7.4 is written as

$$Z_{ij} = \frac{X_{ij}}{\sum_{i=1}^{n} X_{ij}}. \tag{7.8}$$

In words, this equation says that to compute the standardized value of any jth object and ith attribute we take the unstandardized value and divide it by the sum of all the values in the jth column. As a result, the standardized values in each jth column will be normalized to sum to 1.0.

The following application, an example of Research Goal 6 ("Facilitate Planning and Management"), illustrates the striking effects that standardizing can have on research conclusions.

RESEARCH APPLICATION 7.1

Consider the 17 objects of the data matrix shown in Table 7.1. The first 16 objects are forecasts of the 1981 Canadian economy made prior to 1981 by various banks and investment houses. Object 17 is the actual data for 1981. The attributes of the data matrix are six economic indicies that were forecast: (1) percent change in real gross national product, (2) percent change in consumer price index, (3) percent change in the jobless rate, (4) percent change in corporate pretax profits, (5) the number of housing starts, and (6) the exchange rate with the U.S. dollar (U.S. cents).

A cluster analysis of the forecasts will show which forecasters are similar and dissimilar to each other, as well as to the actual data. If replicated over a period of years, an analysis of this kind could help find a diverse set of forecasters covering a spectrum of viewpoints.

Figure 7.4,*a* shows a computer-generated tree resulting from cluster analysis of the data matrix. Notice that forecasts are identified along the left of the tree with triplet numbers (004, 007, and so on). This tree utterly fails its purpose because attribute 5, the number of housing starts, dominates the clustering. We can see that forecast 8 is almost perfectly similar to the actual data (008 and 017 respectively in the tree) because the two agree on the number of housing starts, even though they disagree on the five other economic indicies. A cluster analysis of a data matrix containing the single attribute of housing starts would essentially give the same result. Because

Table 7.1 Data matrix in Which the Objects are Forecasts of the 1981 Canadian Economy and the Attributes are Economic Indices.

	Forecasts								
	Conference Board	*Brault, Guy*	*Data Resources*	*Burns, Fry*	*McLeod Young Weir*	*The Permanent*	*Bank of Montreal*	*Pitfield Mackay*	*Research Securities*
Economic Index	1	2	3	4	5	6	7	8	9
1. Real G.N.P.(%)	1.0	0.0	1.3	0.5	1.8	1.5	0.2	1.7	1.3
2. C.P.I.(%)	11.7	11.0	10.8	10.7	11.0	11.6	11.7	12.0	10.7
3. Jobless rate(%)	7.8	8.2	8.2	7.6	8.0	8.1	8.3	7.7	8.4
4. Pretax profits(%)	−3.8	−8.0	−2.6	−3.0	3.8	2.4	0.0	0.0	0.0
5. Housing starts	181,000	140,000	187,000	175,000	180,000	183,000	175,000	178,000	200,000
6. Exchange rate (U.S.¢)	84.7	83.5	86.9	83.5	87.0	85.0	84.0	83.5	88.0

	Forecasts							
	T-D Bank	*Wood Gundy*	*Greenshields Inc.*	*Merrill Lynch*	*Dominion Secs.*	*Informetrica Ltd.*	*Deacon, Hodgson*	*Actual Data, 1981, Statistics Canada*
Economic Index	10	11	12	13	14	15	16	17
1. Real G.N.P.(%)	1.0	0.9	1.8	1.0	−0.8	1.3	0.1	3.1
2. C.P.I.(%)	11.3	11.4	10.8	12.0	11.0	11.0	12.0	12.5
3. Jobless rate(%)	8.0	8.1	7.8	7.7	8.8	8.2	7.8	7.6
4. Pretax profits(%)	1.9	3.4	3.0	5.6	−5.0	6.0	4.4	−11.5
5. Housing starts	185,000	192,000	180,000	168,000	165,000	190,000	163,000	178,000
6. Exchange rate (U.S.¢)	85.0	85.5	85.0	84.5	85.3	87.0	85.5	83.4

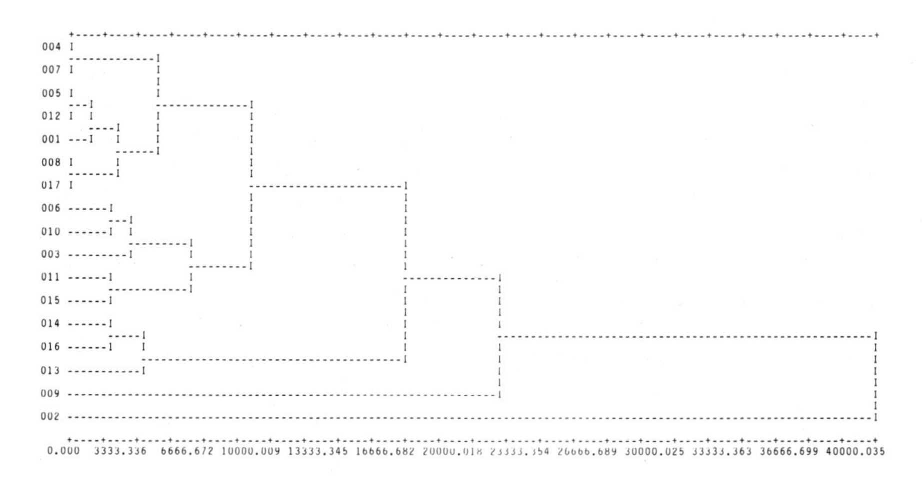

(*a*)

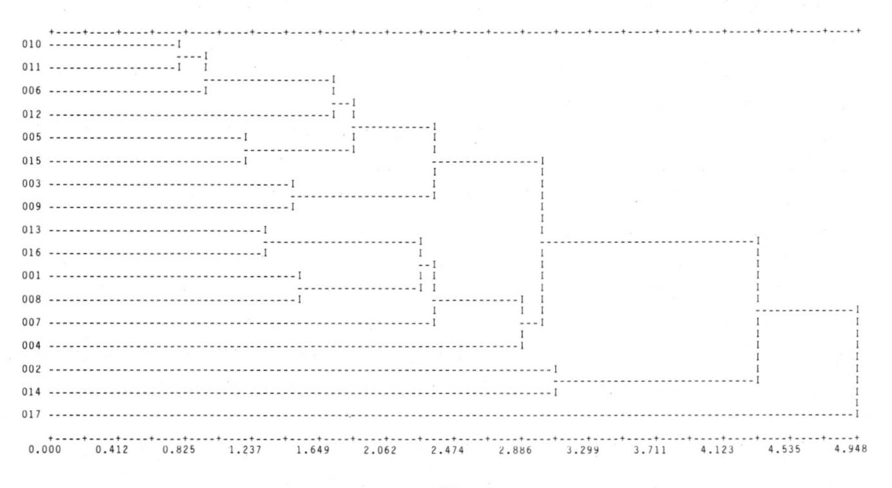

(*b*)

Figure 7.4. (*a*) Tree of forecasts that results from cluster-analyzing the data matrix of Table 7.1, using the Euclidean distance coefficient e_{jk} and the UPGMA clustering method; (*b*) corresponding tree produced by cluster-analyzing the standardized data matrix. Both trees were drawn on a line printer by the CLUSTAR program described in Appendix 2, and are reduced in size for display here.

the range of housing starts is so much greater than the range of values on the other indicies, it swamps their contributions to the similarities between objects.

When the data matrix is standardized by using equation 7.1, and then cluster-analyzed, we get the tree shown in Figure 7.4,*b*. It shows which forecasts are similar and dissimilar in relation to the actual economic performance. In this instance, the forecasts are more similar to each other than they are to the actual performance of the economy.

7.3 STANDARDIZING FOR Q-ANALYSIS AND R-ANALYSIS

Recall from Chapter 3 that a Q-analysis is a cluster analysis of objects whereas an R-analysis is one of attributes. That both Q-analyses and R-analyses are commonly used requires that we consider both row- and column-standardizing functions.

The first thing to note is that most Q-analyses based on standardized data matrices are standardized with row functions such as equations 7.1, 7.2, and 7.3. Conversely, most R-analyses are standardized with column functions such as equations 7.5, 7.6, and 7.7.

In other words, if we want to compare the similarities among objects, we standardize the attributes with a row-standardizing function. Conversely, if we want to compare the similarities among attributes, we standardize the objects with a column-standardizing function.

There is, however, an exception to this that occurs in the Q-analysis of histograms. In this case, the objects to be clustered are histograms, and the attributes are the class intervals of the histograms. Researchers usually standardize the resulting data matrix with the column function given by equation 7.8. This normalizes the histograms and makes them sum to 1.0 down the columns.

The comparison of histograms is an important use of cluster analysis, and Langohr et al. (1976) give a nice illustration. They cluster-analyzed histograms of grain size distributions from different soil samples collected in the Argentinian Pampa. Such applications, be they histograms of soil grain size, or of people's income, could conceivably fit several research goals: (1) they could lead to a question, "Why did these histograms, constructed from data collected in different places, cluster as they did?"; (2) they could lead to a research hypothesis, "Ah ha, here is a reason they clustered as they did"; (3) they could lead to testing a research hypothesis, "If my research hypothesis is true, then I would expect the histograms to cluster in the following way. . . ."

Chapter 8 explains another reason for column-standardizing a Q-analysis. There, we see that column-standardizing can be used to compensate for size displacements between the data profiles of objects. In instances where we don't want these size displacements to contribute to the similarity among objects, we can "standardize them away."

Sometimes a research goal makes it necessary to do a Q-analysis and an R-analysis on the same data matrix. When standardizing is warranted, usually the data matrix should be standardized separately for each analysis, usually by using a row function for the Q-analysis and a column function for the R-analysis. Another possibility is that the Q-analysis requires standardizing but the R-analysis does not, or vice versa.

At any rate, the data matrix should not be "double standardized" by using a row function and a column function in succession. The reason is that standardizing by rows first and by columns second generally gives a different standardized data matrix than is obtained by standardizing by columns first and rows second. The order matters.

As to whether we should standardize at all, or as to which is the best standardizing function to use, there is no categorical answer. In the first place, the correct decision depends on the research goal we are addressing, and on its context. And in the second place, the best standardizing function is dependent upon the resemblance coefficient we use, for some resemblance coefficients have built in to them self-standardizing features.

Right now we are more concerned with describing the available methods for standardizing than discussing how to select the best method. Selecting the best method entails subjective judgment, and the nature of this is discussed in Chapters 21 and 22.

7.4 DATA TRANSFORMATIONS AND OUTLIERS

Standardizing is a way of changing the original data. There are two other ways of changing the data: (1) transforming the data, and (2) identifying and removing freak values of data called outliers. Most books about multivariate data analysis that discuss cluster analysis also discuss how to transform the data and deal with outliers—for examples, see the books by Clifford and Stephenson (1975), Gnanadesikan (1977), and Chatfield and Collins (1980). We choose not to cover these topics in depth. But before seeing why, let us find out how transforming the data works and learn more about what outliers are.

To transform data, we must choose and use a **transforming function**. It differs from a standardizing function in the following way. Although both kinds of functions change the values of the attributes in the data matrix and make the values more nearly fall within the same range, a standardizing function has parameters—for examples, $\overline{X}_i$ and S_i in equation 7.1—which are determined by the data, whereas a transforming function does not.

Typical examples of transforming functions are $Z_{ij} = \log(X_{ij})$ and $Z_{ij} = \sqrt{X_{ij}}$. For instance, suppose that on attribute 1 the data vary from 5 to 100 units, and on attribute 2 they vary from 10 to 300 units. See how the range is reduced by using the transforming function $Z_{ij} = \log(X_{ij})$: the variation on attribute 1 will be from 0.70 to 2.00 (that is, log(5) = 0.70 and log(100) = 2.0), and on attribute 2 it will be from 1.0 to 2.48. The variation of the data on the transformed scale is more nearly equal, making the transformed attributes contribute more equally to the overall resemblance between objects.

Now, for what an outlier is. An **outlier** is any value of data that is atypical with respect to other values in the data set. In other words, the values in a data set establish a norm, and any values that are quite deviant—say three standard deviations or more—are labeled outliers. Because outliers are so deviant, they exert a much stronger influence than nonoutliers do in determining the parameters of a standardizing function and ultimately in determining the resemblance among objects.

The books mentioned at the beginning of this section present statistical methods for identifying outliers. Once they are identified, you must decide if they are due to errors in measuring and coding the data, in which case you should correct the errors (if their correct values are unavailable, then you should remove the erroneous outliers from the data set). Alternatively, you must decide if they are genuine, natural abnormalities, in which case you should let them remain.

Having said this, let us now say why we prefer not to cover these topics in depth. The reason is that a researcher's purpose in using cluster analysis differs fundamentally from what it is in using almost any other multivariate method—multivariate analysis of variance, principal component analysis, and so on.

Cluster analysis is a method for describing the similarities among objects in a *sample*. It is a mathematical microscope for looking at the relations of similarity among a *given set* of objects. It cannot be used for making statistical inferences about these relations to a larger population. Any inferences a researcher makes by studying the tree are made by using reasoned analogy rather than by using formal statistical methods.

Nearly all of the research applications in this book use nonrandom samples of objects. The researchers collected them partly out of interest and partly out of convenience. Furthermore, the success of much of their research depended on the inclusion of genuine outliers in the data set. For example, Sigleo's (1975) clustering of nonrandomly collected turquoise artifacts, which we described in Research Application 4.3, led her to conclusions about prehistoric Indian trading patterns precisely because two samples that were outliers from the rest clustered together. Had she removed these at the start because they were atypical to the rest, her research would have led her to a different conclusion.

In contrast to the descriptive purpose of cluster analysis, methods like multivariate analysis of variance and principal component analysis, besides being descriptive, also make inferences. To make valid inferences, they require random samples of objects, and the data must be normally distributed. Because outliers destroy normality, and because transforming the data is a way of compressing a skewed nonnormal distribution of data so that it becomes more normal, books on multivariate data analysis are rightly concerned with these topics.

The literature reviewed for this book found that researchers standardize their data matrices much more often than they transform them—applications ran over 100 to 1 in favor of standardizing. And in not one application did a researcher report use of formal statistical methods for detecting outliers, although we can assume that they used their eyes and common sense to screen their data for abnormal values.

You shouldn't categorically dismiss transformation functions or identification and resolution of outlier data as unimportant topics because you will most likely encounter a research problem in which these topics are important. Hence, you should eventually learn about them, perhaps by starting with the references mentioned at the beginning of the section. But in the context of this book, they are minor topics.

SUMMARY

Standardizing the data matrix makes each attribute contribute more equally to overall resemblance. To standardize, select a standardizing function from one of several alternatives. For each situation, the best standardizing function depends on the context of the research problem and on its research goal.

A Q-analysis is usually standardized differently than an R-analysis, even when the same data matrix is used for each.

Transforming the data can be done in conjunction with standardizing the data, or done by itself. Identifying and removing outliers can be done either by looking at the data or by using statistical methods.

Because cluster analysis is a method of description, transforming data and identifying outliers are lesser concerns than with such methods as multivariate analysis of variance, which are methods of inference whose validity depends on normality within the data.

EXERCISES

(Answers to exercises are given in the back of the book.)

7.1. To demonstrate that you can mechanically use the different standardizing functions, standardize the following data matrix using (a) equation 7.1 and (b) equation 7.3.

	Object			
Attribute	1	2	3	4
1	10	14	10	10
2	66	20	10	80
3	−30	0	−10	−20

7.2. Rework exercise 7.1, standardizing the data matrix by columns, using equation 7.5.

7.3. Which standardizing functions presented in this chapter *cannot* be used on the following data matrix?

	Object			
Attribute	1	2	3	4
1	10	2	7	4
2	6	6	6	6
3	3	0	6	1

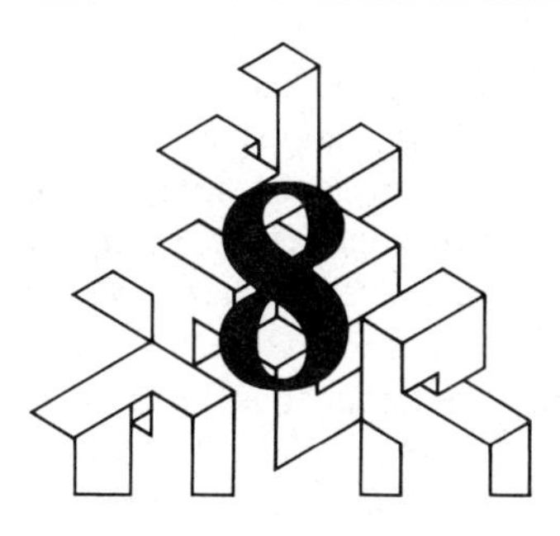

Resemblance Coefficients for Quantitative Attributes

OBJECTIVES

This chapter can help you to assess similarities successfully.

- First, it offers six resemblance coefficients for use in place of the Euclidean distance coefficient e_{jk} dealt with in Chapter 2. You can use any of these coefficients when the data matrix contains quantitative attributes—attributes measured on ordinal, interval, and ratio scales.
- Second, it explains how to view the columns of a data matrix as profiles of data. Deciding which objects are similar then boils down to deciding which data profiles are similar. Each resemblance coefficient is a different operational definition of similarity—each measures the similarity between the data profiles of two objects differently.
- Third, this chapter shows that standardizing the data matrix and choosing a resemblance coefficient are interrelated decisions that must be made in consideration of each other.
- Fourth, you can learn here how to cluster-analyze a data matrix when some of its values are missing.

8.1 DATA PROFILES

Before starting to discuss resemblance coefficients, let us select a data matrix that will help to illustrate their calculation. As shown in Figure 8.1,*a*, its four objects are the common stocks of four hypothetical companies. Its four attributes are the prices of the stocks at four points in time. Object 1 is the base case, denoted as *B*. Object 2's column values are larger than those of *B* by 15, denoted as $B + 15$. Object 3's values are twice those of *B*, denoted as $B \times 2$. Finally, object 4's values are a mirror image of those of *B*, denoted as *MIB*.

Figure 8.1,*b* shows the **data profiles** of the four objects. To get these we merely connect the values of an object's attributes with lines. We see that the data profiles of objects 1 and 2 are separated by a constant amount of 15. This is an example of an **additive translation** of two data profiles.

We see that the data profiles of objects 1 and 3 differ by the multiplication constant of 2. This is an example of a **proportional translation** of two data profiles.

Finally, we see that the data profiles of objects 1 and 4 differ by a 180-degree revolution. This is an example of a **mirror image translation**. Like a crankshaft, the data profile of object 1 is revolved about its value at attribute t_1, to produce object 4's profile. As a result, the value of 40 at t_2 revolves to the value of 0; the value of 25 at t_3 revolves to 15; and the value of 30 at t_4 revolves to 10.

Devised by Boyce (1969), these translations make a useful system for cataloging resemblance coefficients. Some of the resemblance coefficients that are discussed in this chapter ignore additive and/or proportional translations of data profiles when they measure similarity. For example, one of these coefficients, called the **coefficient of shape difference**, z_{jk}, measures objects 1 and 2 in Figure 8.1 as being maximally similar. It wholly ignores their separation and measures their similarity by comparing their shapes. Since their data profiles have the same shapes, they are accorded maximum similarity.

In real applications, data profiles are seldom exact additive or proportional translations of each other. Nevertheless, in many applications there may be a separation between profiles, which we call a **size displacement**, and which we may want to ignore. A size displacement occurs when the data profile of one object is, attribute for attribute, larger or smaller than that of another. When plotted, the two profiles do not cross. This is a useful concept because we may want to measure similarity based on the shapes of the profiles apart from size displacements between them. For example, if the objects are fossil bones and the attributes are measurements of the bones, we probably want to ignore size displacements between data profiles: after all, size displacements can incidentally arise because the animals were not

Attribute (time)	1 (B)	2 ($B+15$)	3 ($B \times 2$)	4 (MIB)	$\overline{X}_i$	S_i
	Object					
t_1	20	35	40	20	28.75	10.31
t_2	40	55	80	0	43.75	33.51
t_3	25	40	50	15	32.5	15.55
t_4	30	45	60	10	36.25	21.36
$\overline{X}_j$	28.75	43.75	57.50	11.25		
S_j	8.54	8.54	17.07	8.54		

(*a*)

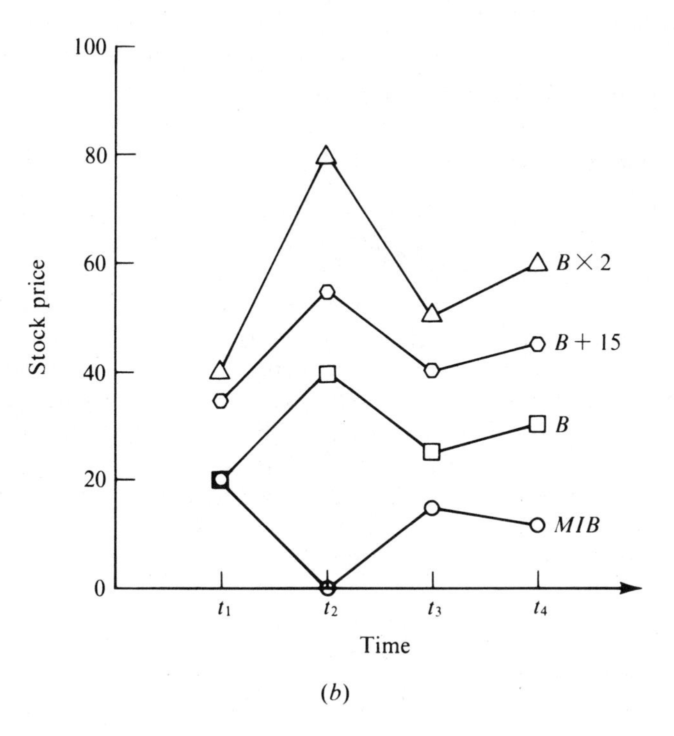

(*b*)

Figure 8.1. (*a*) Data matrix of common stocks of four hypothetical companies; (*b*) corresponding data profiles.

all the same age when they died. To reveal the genetic differences, we have to mask out incidental size displacements.

Unwanted size displacements can be masked out in several ways. We can standardize the data matrix in a certain way, or we can use a resemblance coefficient that we know is blind to additive and/or proportional translations and therefore should be partially blind to size displacements, or we can do both.

8.2 RESEMBLANCE COEFFICIENTS FOR QUANTITATIVE ATTRIBUTES

The six resemblance coefficients described in this section are designed to be used with data matrices whose attributes are measured on interval and/or ratio scales. However, practice has shown that they can also be used for data matrices whose attributes are measured on ordinal scales.

The Average Euclidean Distance Coefficient, d_{jk}

This coefficient is closely related to the Euclidean distance coefficient e_{jk}. As Section 8.6 shows, d_{jk} has the advantage that it can be used when the data matrix contains missing values, whereas e_{jk} cannot. To understand d_{jk}, let us first write e_{jk} as an equation, and then develop d_{jk} from this equation.

Recall from Chapter 7 that X_{ij} is the symbol for the value of the ith attribute measured on the jth object. It follows that X_{ik} is the value of the ith attribute measured on the kth object. Here, objects j and k are any pair of objects whose similarity we want to measure. Although in Chapter 2 we used the geometry of a two-dimensional attribute space, together with the Pythagorean theorem, to compute e_{jk}, we could have expressed it as an equation:

$$e_{jk} = \left[\sum_{i=1}^{2} (X_{ij} - X_{ik})^2 \right]^{1/2}.$$

In words, this means that to compute e_{jk} for any two objects j and k, we use the data in the jth and kth columns of the original data matrix (or standardized data matrix). Attribute by attribute we take the difference in their values. We square these differences, and after they are squared none are negative. We sum the squared differences, take the square root of the sum (signified by the exponent "1/2") and the result is e_{jk}.

Let us use this equation to compute e_{jk} for objects 1 and 5 of the data matrix on page 10 in Chapter 2. Recopied, the relevant section of the data matrix is

	Object	
Attribute	1	5
1	$X_{11} = 10$	$X_{15} = 5$
2	$X_{21} = 5$	$X_{25} = 10$

Thus,

$$e_{15} = \left[(X_{11} - X_{15})^2 + (X_{21} - X_{25})^2\right]^{1/2}$$

$$= \left[(10 - 5)^2 + (5 - 10)^2\right]^{1/2} = 7.07.$$

This agrees with the calculations done in Chapter 2.

For $n = 3$ attributes, the objects are in a three-dimensional attribute space typified in Figure 8.2. The Euclidean distance between objects j and k is

$$e_{jk} = \left[\sum_{i=1}^{3} (X_{ij} - X_{ik})^2\right]^{1/2}.$$

Adding an attribute adds another term to the sum.

The equation for e_{jk} generalizes to n attributes, although when n exceeds $n = 3$ we cannot visualize the objects in their attribute space. The generalized equation is

$$e_{jk} = \left[\sum_{i=1}^{n} (X_{ij} - X_{ik})^2\right]^{1/2}. \quad (8.1)$$

It is simply the square root of the sum of the squares of the differences of the values on the n attributes. The range of e_{jk} is $0 \leq e_{jk} < \infty$.

The **average Euclidean distance coefficient,** d_{jk}, is like e_{jk} except that we compute the average of the squares of the differences by dividing by n, as follows:

$$d_{jk} = \left[\sum_{i=1}^{n} (X_{ij} - X_{ik})^2/n\right]^{1/2}. \quad (8.2)$$

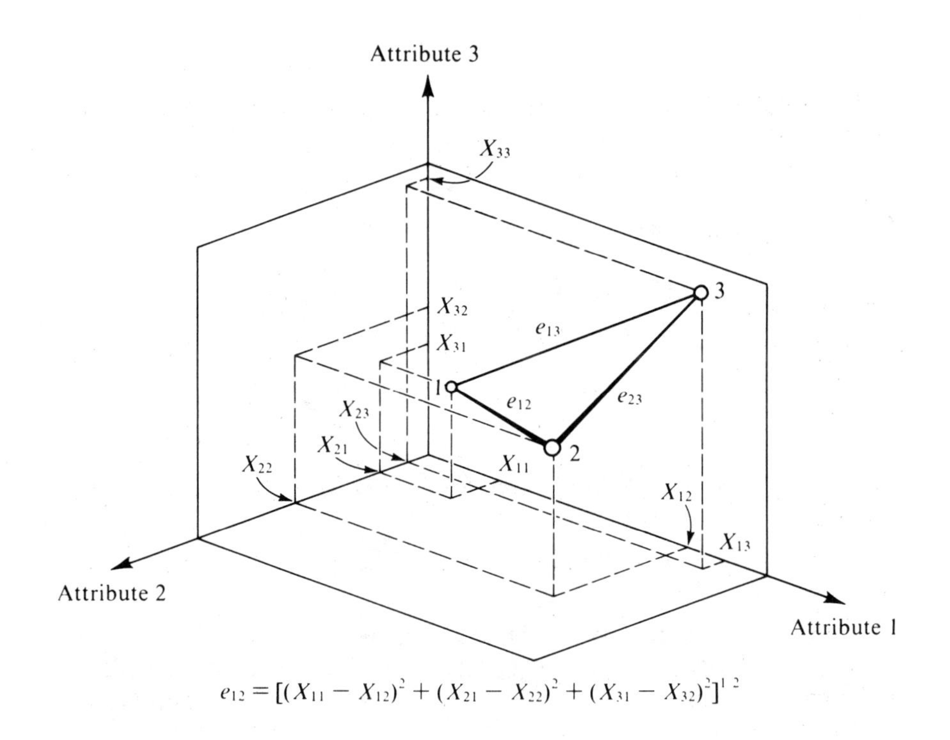

Figure 8.2. Geometry of the Euclidean distances between three objects in a three-dimensional attribute space.

It is defined on the range $0 \leq d_{jk} < \infty$. Comparing equations 8.1 and 8.2 shows that $d_{jk} = e_{jk}/n^{1/2}$. That is, d_{jk} is scaled down from e_{jk} by the constant $(1/n)^{1/2}$—by dividing e_{jk} by the square root of n. This means that two trees obtained from two cluster analyses of the same data matrix, one using e_{jk} and one using d_{jk}, and both using the same clustering method, will have the same relative topography. The objects will cluster in the same order and the values at which clustering occurs will differ by the constant $(1/n)^{1/2}$. Either tree would lead a researcher to the same conclusions.

Thus, when the data matrix contains no missing values, it makes no difference whether you use e_{jk} or d_{jk}. However, you can use d_{jk} when there are missing values, whereas you cannot use e_{jk}. Sneath and Sokal (1973, pp. 124–126) call d_{jk} the "average taxonomic coefficient." We have renamed it the "average Euclidean distance coefficient," which is more descriptive.

To illustrate its calculation, let us compute d_{jk} for objects 1 and 2 of the data matrix shown in Figure 8.1.

We get

$$d_{12} = \left[(20-35)^2 + (40-55)^2 + (25-40)^2 + (30-45)^2\right)/4\right]^{1/2}$$
$$= 15.$$

Rather than rewriting the coefficients as they are used with standardized data—that is, replacing X_{ij} and X_{ik} with Z_{ij} and Z_{ik}—we let this be implied.

The Coefficient of Shape Difference, z_{jk}

This dissimilarity coefficient is defined on the range $0.0 \leq z_{jk} \leq \infty$, and it is a function of d_{jk}. Created by Penrose (1953), it is written as

$$z_{jk} = \left\{[n/(n-1)]\left(d_{jk}^2 - q_{jk}^2\right)\right\}^{1/2}, \tag{8.3}$$

where

$$q_{jk}^2 = (1/n^2)\left(\sum_{i=1}^{n} X_{ij} - \sum_{i=1}^{n} X_{ik}\right)^2.$$

To calculate z_{jk} for objects 1 and 2 of the data matrix in Figure 8.1,a, we first calculate

$$\sum_{i=1}^{4} X_{i1} = 20 + 40 + 25 + 30 = 115,$$

and

$$\sum_{i=1}^{4} X_{i2} = 35 + 55 + 40 + 45 = 175.$$

Then we use $d_{12}^2 = (15)^2 = 225$ from the previous calculation. This gives

$$q_{12}^2 = \left[1/(4)^2\right](115 - 175)^2 = 225,$$

making

$$z_{12} = \{[4/(4-1)](225 - 225)\}^{1/2} = 0.0.$$

According to this, z_{jk} regards objects 1 and 2 as maximally similar. It ignores the additive translation of their profiles. That is, not only is z_{jk} equal to zero when the profiles of objects j and k are identical, but it is also equal to zero when the two profiles are displaced by the addition of any constant.

The Cosine Coefficient, c_{jk}

This is a similarity coefficient defined on the range $-1.0 \le c_{jk} \le 1.0$. The value $c_{jk} = 1.0$ indicates maximum similarity, but not necessarily identity of profiles. The value $c_{jk} = -1.0$ indicates minimum similarity—that is, maximum dissimilarity. Its equation is

$$c_{jk} = \frac{\sum_{i=1}^{n} X_{ij} X_{ik}}{\left(\sum_{i=1}^{n} X_{ij}^2\right)^{1/2} \left(\sum_{i=1}^{n} X_{ik}^2\right)^{1/2}}. \tag{8.4}$$

When objects j and k are viewed as two points in their attribute space and each point is connected by a line from the origin, then c_{jk} is the cosine of the angle between the two lines. Although we cannot visualize the angle in attribute spaces with more than $n = 3$ attributes, the concept of angle and its cosine does generalize and can be computed by using equation 8.4.

To illustrate the calculation of c_{jk} for objects 1 and 2 of Figure 8.1,a in the higher-ordered space of our example, which has $n = 4$ attributes, we first calculate the following sums:

$$\sum_{i=1}^{4} X_{i1} X_{i2} = (20)(35) + (40)(55) + (25)(40) + (30)(45) = 5{,}250;$$

$$\sum_{i=1}^{4} X_{i1}^2 = (20)^2 + (40)^2 + (25)^2 + (30)^2 = 3{,}525;$$

$$\sum_{i=1}^{4} X_{i2}^2 = (35)^2 + (55)^2 + (40)^2 + (45)^2 = 7{,}875.$$

Entering these into equation 8.4 gives

$$c_{12} = 5{,}250/\left[(3{,}525)^{1/2}(7{,}875)^{1/2}\right] = 0.996.$$

This shows that the similarity between objects 1 and 2 is just shy of the maximum possible—that is, just shy of $c_{jk} = 1.0$. Clearly, c_{jk} measures similarity by neglecting most of the size displacement between the data profiles. If two profiles have essentially the same shapes, then regardless of the size displacement, the coefficient c_{jk} will accord them a high degree of similarity.

The Correlation Coefficient, r_{jk}

This similarity coefficient is defined on the range $-1.0 \leq r_{jk} \leq 1.0$. The value $r_{jk} = 1.0$ indicates maximum similarity but not necessarily identity of profiles, and $r_{jk} = -1.0$ indicates maximum dissimilarity.

The coefficient r_{jk} is also known as the Pearson product-moment correlation coefficient; we first used it in Chapter 2. There we renamed it the cophenetic correlation coefficient to note its particular use. Now we use it as an index of similarity between objects in the data matrix. Of the various ways it can be written, one way that makes the calculations easier is

$$r_{jk} = \frac{\sum_{i=1}^{n} X_{ij}X_{ik} - (1/n)\left(\sum_{i=1}^{n} X_{ij}\right)\left(\sum_{i=1}^{n} X_{ik}\right)}{\left\{\left[\sum_{i=1}^{n} X_{ij}^2 - (1/n)\left(\sum_{i=1}^{n} X_{ij}\right)^2\right]\left[\sum_{i=1}^{n} X_{ik}^2 - (1/n)\left(\sum_{i=1}^{n} X_{ik}\right)^2\right]\right\}^{1/2}}. \tag{8.5}$$

Using the sums calculated previously in computing z_{jk} and c_{jk}, equation 8.5 gives

$$r_{12} = \frac{5{,}250 - (1/4)(115)(175)}{\left\{\left[3{,}525 - (1/4)(115)^2\right]\left[7{,}875 - (1/4)(175)^2\right]\right\}^{1/2}} = 1.0.$$

Coefficient r_{jk} regards objects 1 and 2 as maximally similar. It is wholly insensitive to their additive translation.

The Canberra Metric Coefficient, a_{jk}

This dissimilarity coefficient is defined on the range $0.0 \leq a_{jk} \leq 1.0$, where $a_{jk} = 0.0$ indicates maximum similarity, which occurs only when the profiles of objects j and k are identical. It is written as

$$a_{jk} = (1/n) \sum_{i=1}^{n} \frac{|X_{ij} - X_{ik}|}{(X_{ij} + X_{ik})}. \tag{8.6}$$

If we solve this equation with calculations for objects 1 and 2 in Figure 8.1,a, noting that $|X_{ij} - X_{ik}|$ means to take the absolute, or positive, value of the difference between the jth and kth objects for the ith attribute, we have

$$a_{12} = \left(\frac{1}{4}\right)\left[\frac{|20 - 35|}{(20 + 35)} + \frac{|40 - 55|}{(40 + 55)} + \frac{|25 - 40|}{(25 + 40)} + \frac{|30 - 45|}{(30 + 45)}\right]$$

$$= (1/4)(0.273 + 0.158 + 0.231 + 0.20) = 0.21.$$

Each term in the sum is scaled between 0.0 and 1.0, equalizing the contribution of each attribute to overall similarity. The leading multiplier, $1/n$, averages the n proportions.

The Bray-Curtis Coefficient, b_{jk}

This dissimilarity coefficient is defined on the range $0.0 \leq b_{jk} \leq 1.0$, where $b_{jk} = 0.0$ indicates maximum similarity, which occurs when the profiles are identical. It is written as

$$b_{jk} = \frac{\sum_{i=1}^{n} |X_{ij} - X_{ik}|}{\sum_{i=1}^{n} (X_{ij} + X_{ik})}. \tag{8.7}$$

With calculations for objects 1 and 2 in Figure 8.1,a, we have

$$b_{12} = \frac{|20 - 35| + |40 - 55| + |25 - 40| + |30 - 45|}{(20 + 35) + (40 + 55) + (25 + 40) + (30 + 45)}$$

$$= 60/290 = 0.21.$$

The agreement of the Canberra metric coefficient, a_{jk}, with the Bray-Curtis coefficient, b_{jk}, in these illustrative calculations is accidental.

There are several points about a_{jk} and b_{jk} that we should note.

To begin with, if a_{jk} and b_{jk} are to be meaningful, then all of the values of the original data matrix must be greater than or equal to zero. For this reason neither coefficient should be used when the data are standardized with equation 7.1, because that equation always makes some of the standardized data negative.

Second, for any attribute i in which $X_{ij} = X_{ik} = 0.0$, the denominator of the ith term of the sum in a_{jk} will be zero. In such cases, a_{jk} will be undefined. Coefficient b_{jk} gives more latitude here, but it does not work when a given pair of objects j and k have zero as the values of all n attributes. In such a case, the denominator of equation 8.7 will be zero and b_{jk} will be undefined.

Third, an equivalent formula for b_{jk} is

$$b_{jk} = 1.0 - \frac{2 \sum_{i=1}^{n} \min(X_{ij}, X_{ik})}{\sum_{i=1}^{n} (X_{ij} + X_{ik})}, \tag{8.8}$$

where $\min(X_{ij}, X_{ik})$ is the smaller of the two values X_{ij} and X_{ik}. With this equivalent formula, and using the same data for objects 1 and 2 as were used earlier, we have

$$b_{12} = 1.0 - 2(20 + 40 + 25 + 30)/(55 + 95 + 65 + 75)$$

$$= 0.21.$$

This is the same result we got by using equation 8.7.

Some researchers define b_{jk} as a similarity coefficient rather than as a dissimilarity coefficient as we have defined it here. They take b_{jk} to be the complement of b_{jk} as we have defined it. That is, they define it as

$$2 \sum_{i=1}^{n} \min(X_{ij}, X_{ik}) / \sum_{i=1}^{n} (X_{ij} + X_{ik}).$$

When defined this way, it is also scaled between 0.0 and 1.0, but 1.0 is now its point of maximum similarity.

Any similarity coefficient can be converted to a dissimilarity coefficient, or vice versa, either by subtracting it from a constant or by multiplying it by -1.0. This merely changes the direction of its scale. A cluster analysis using either form of the coefficient gives a tree that has the same topography and is scaled in the same proportions. The values of the scale on the tree depend of course on which form is used, but this does not affect any conclusions drawn from the tree.

Finally, a_{jk} uses a sum of normalized terms to equally weight the contribution of each attribute to overall resemblance, whereas b_{jk} weights them unequally. The following data shows the consequences of this for two objects, 1 and 2.

i	X_{i1}	X_{i2}
1	1	2
2	3	5
3	4	0
4	1,000	1

The Canberra metric coefficient gives

$$a_{12} = (1/4)\left(\frac{|1-2|}{(1+2)} + \frac{|3-5|}{(3+5)} + \frac{|4-0|}{(4+0)} + \frac{|1{,}000-1|}{(1{,}000+1)}\right) = 0.65.$$

Each term in the sum is a proportion, and so no term dominates. But with the Bray-Curtis coefficient, the fourth term dominates and by itself almost wholly determines the value of b_{jk}:

$$b_{12} = \frac{|1-2| + |3-5| + |4-0| + |1{,}000-1|}{(1+2) + (3+5) + (4+0) + (1{,}000+1)} = 0.99.$$

The fourth term makes the numerator and denominator almost equal, assuring that b_{jk} will be near 1.0, the limit of dissimilarity.

8.3 HOW RESEMBLANCE COEFFICIENTS ARE SENSITIVE TO SIZE DISPLACEMENTS BETWEEN DATA PROFILES

This section explains more about how the resemblance coefficients described so far are sensitive to size displacements between data profiles. Some coefficients largely ignore size displacements in computing similarity, and some do not. Knowing this is crucial for choosing the correct resemblance coefficient for a given research goal.

To begin, let us cluster-analyze the data matrix of common stocks shown in Figure 8.1,*a*, by using four of the resemblance coefficients presented in Section 8.2—namely, d_{jk}, z_{jk}, c_{jk}, and r_{jk}. Figure 8.3 shows the four resemblance matrices and trees.

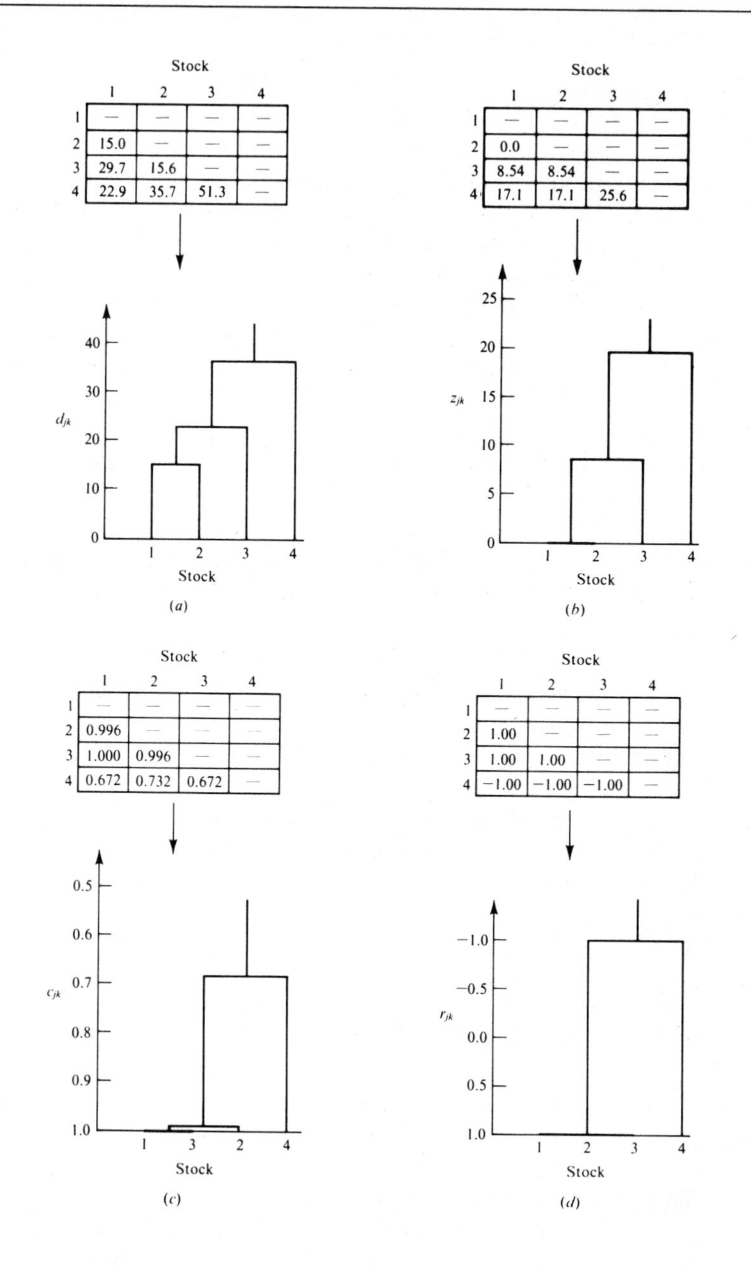

Figure 8.3. Resemblance matrices and trees obtained from the data matrix of common stocks shown in Figure 8.1*a* and the UPGMA clustering method; *a*, *b*, *c*, and *d* use the resemblance coefficients d_{jk}, z_{jk}, c_{jk}, and r_{jk}, respectively.

We see that when d_{jk} is used, no pair of objects is maximally similar. Thus, d_{jk} is sensitive to additive, proportional, and mirror image translations of these data profiles. In general, the coefficient is sensitive in this way.

When z_{jk} is used, objects 1 and 2, whose data profiles differ by an additive translation, are maximally similar. However, z_{jk} is sensitive to proportional translations (objects 1 and 3) and mirror image translations (objects 1 and 4). Thus, only when a size displacement of data profiles can be approximated by an additive translation does z_{jk} adequately ignore the size displacement.

When c_{jk} is used, it totally ignores the proportional translation between objects 1 and 3, as well as most of the additive translation between objects 1 and 2. Thus, c_{jk} does a good job of ignoring size displacements that can be approximated by proportional or additive translations of data profiles.

Finally, when r_{jk} is used, objects 1, 2 and 3 are pairwise maximally similar. Coefficient r_{jk} is wholly insensitive to both additive and proportional translations. Thus it ignores size displacements of data profiles that can be approximated by either kind of translation, or even translations that are combinations of the two.

The coefficient r_{jk} accords objects 1 and 4 maximum dissimilarity—that is, $r_{14} = -1.0$. This is why we included the mirror image translation in the data matrix; r_{jk} always judges any data profile and its mirror image as being maximally dissimilar.

Although we have not included a_{jk} and b_{jk} in this discussion, both are sensitive to additive, proportional, and mirror image translations. Table 8.1 summarizes the sensitivities of the resemblance coefficients presented in this chapter. Let us see how to use this information to help choose a resemblance coefficient for a cluster analysis.

After you have stated your research goal, you should be able to decide whether you want the size displacements between data profiles to affect their similarity or want instead to ignore size displacements as much as possible. In situations where size displacements should affect similarity, coefficients like d_{jk}, a_{jk}, and b_{jk} are logical alternatives. Coefficients like z_{jk}, c_{jk}, and r_{jk} are poorer alternatives because by degrees they are blind to size displacements. Conversely, in situations where size displacements should not be allowed to affect similarity, coefficients like c_{jk} and r_{jk} are good alternatives. And z_{jk} would be a good choice if you felt the size displacements could be approximated by additive translations; on the other hand, a possible problem with z_{jk}, as Figure 8.3,*b* shows, is that it is sensitive to proportional translations.

Figure 8.3 also shows that all of the resemblance coefficients d_{jk}, z_{jk}, c_{jk}, and r_{jk} accord a high degree of dissimilarity between data profiles that differ by a mirror image translation. In effect this means that when one profile tends to be the reverse of the other, these coefficients see them as being quite dissimilar.

Table 8.1 Sensitivities of Resemblance Coefficients to Additive and Proportional Translations of Data Profiles.

Coefficient	Range	Ignores: Additive translations	Ignores: Proportional translations
Dissimilarity			
e_{jk}	$0.0 \leq e_{jk} \leq \infty$	No	No
d_{jk}	$0.0 \leq d_{jk} \leq \infty$	No	No
a_{jk}	$0.0 \leq a_{jk} \leq 1.0$	No	No
b_{jk}	$0.0 \leq b_{jk} \leq 1.0$	No	No
z_{jk}	$0.0 \leq z_{jk} \leq \infty$	Yes	No
Similarity			
c_{jk}	$-1.0 \leq c_{jk} \leq 1.0$	No	Yes
r_{jk}	$-1.0 \leq r_{jk} \leq 1.0$	Yes	Yes

Researchers have used d_{jk} and r_{jk} more often than other coefficients. This is probably because once a research goal is fixed, a researcher can usually only discriminate between the two cases of (1) wanting size displacements between data profiles to contribute to their dissimilarity, and (2) not wanting size displacements to contribute.

For example, consider a data matrix of common stocks described by their prices over time. Suppose that our research goal is to test the hypothesis that the stocks of companies in similar industries—let us say electronics, banking, and computers, for examples—have price trends that parallel each other. Clearly we should select a resemblance coefficient that will indicate that two companies are similar when their price trends are parallel, and dissimilar when not parallel. This means the coefficient used ought to be insensitive to size differences among profiles, for they are arbitrarily set by the sizes of the companies and their histories of giving stock dividends. Thus, r_{jk} would be a good choice.

Knowing how each resemblance coefficient is sensitive to additive, proportional, and mirror image translations between data profiles, or, in a cruder sense, knowing how each is sensitive to size displacements, is a first step toward choosing the right resemblance coefficient for a given problem. You can refine the process of choosing as follows. First, make up several small data matrices, each an example of an alternative way that your research could turn out. Make the data matrices small enough that you don't need cluster analysis, but instead can base your hypothetical research conclusions on the direct examination of these matrices. After intuitively cluster-analyzing the small data matrices, try to find a resemblance coefficient that, when used in a formal cluster analysis, will give results that match your intuition. The goal is to find a mathematical resemblance coefficient that "sees" similarity in the same way you do. Once found, you can trust it to act for you in judging the similarities among objects in a real data matrix that is too large to solve by intuition.

The resemblance coefficients presented in Section 8.2 provide enough choice for almost any application. Still, there are many more mentioned in Sneath and Sokal (1973). Some, like the Mahalanobis generalized distance coefficient, are much more difficult to compute than those presented here. Partly because of this, partly because Sneath and Sokal cover these other coefficients, and partly because the literature review for this book showed that researchers do not widely use these other coefficients, we will not discuss them further. This does not mean, however, that they lack merit.

8.4 EXAMPLES OF APPLICATIONS USING QUANTITATIVE RESEMBLANCE COEFFICIENTS

The more you read the applications literature of cluster analysis, the better you will understand the situations for which each resemblance coefficient is most appropriate. To get you started, here are some illustrative applications.

Applications Using e_{jk} or d_{jk}. The example in Chapter 2 in which we used cluster analysis to identify clusters of similar experimental units is a good use of e_{jk} or d_{jk}. Recall that the experimental units were plots of land. The experiment was to measure the plant growth resulting from a proposed fertilizer treatment. The attributes were natural properties of the plots that could themselves affect plant growth. Clearly we would like the experimental units to be identical or nearly so. Thus, we should use a coefficient that is sensitive to size displacements among data profiles. It would be wrong to use a coefficient like z_{jk}, c_{jk}, or r_{jk}, because they would partly ignore size displacements that might exist.

Another appropriate use of d_{jk} is Fanizza and Bogyo's (1976) cluster analysis of 19 almond varieties grown in southern Italy. They described the varieties with 12 attributes of commercial importance—average kernal weight, first bloom date, nut thickness, and others. Their research goal was to cut the resulting tree into clusters of similar varieties, in order to be able to select one or more varieties from each cluster and then outcross these dissimilar varieties to produce new varieties. In this case, the closer to being identical two varieties are, the more they should be judged similar. Thus, a coefficient like d_{jk} is a good operational definition of similarity. A coefficient like r_{jk} would be a bad choice because it would define similarity in a way that the laws of genetics do not recognize.

Applications Using z_{jk}. Boyce (1969) used z_{jk} to measure the resemblance between the skulls of hominoids (the family that includes apes and man). The skulls were described by measurements, but since the size of a skull is a function of a hominoid's age, differences in size would not be helpful in assessing the resemblance. Boyce felt that z_{jk} would ignore most of the differences in the sizes of the skulls and at the same time would be sensitive to differences in their shapes. Outside of Boyce's application, z_{jk} has been little used.

Applications Using c_{jk}. The cosine coefficient has found its greatest use in geology. Imbrie and Purdy (1962) see the use of c_{jk} as being a natural result of geologists' customarily representing samples of rocks as vectors giving elemental compositions. In the mathematics of linear algebra the usual method for comparing two vectors is to compute the cosine of the angle between them, and to use it to index their agreement. Imbrie and Purdy (1962, pp. 253–272), Harbaugh and Merriam (1968, p. 165), Thompson (1972), and Bell (1976) give examples of use of this coefficient in geology.

Applications Using r_{jk}. Beal et al. (1971) used r_{jk} for the following cluster analysis. The objects were 221 New York State communities. The attributes were 13 community development measures such as population size, median family income, and number of flush toilets per capita, to name a few. For an R-analysis the researchers used r_{jk} to assess the similarities among community development measures. Because the measures had no standard basis—some, like population size were a function of a community's land area, while others were on a per capita basis—it was necessary to ignore size displacements among the data profiles of the attributes.

In another example, Johnson and Albani (1973) cluster-analyzed samples of foraminiferal populations (species of aquatic microorganisms) collected at 44 locations in Broken Bay, Australia. They described each sample by the abundances of 144 species encountered. Their research goal was to make a specific-purpose classification of samples and relate it to the locations of the samples in the bay. Because the data profiles of the samples were displaced from one another by size differences owing to the densities of the foraminiferal populations, which varied with sampling location, these researchers used r_{jk}. Incidently, the clusters of samples fell into a spatial correspondence with the inner, intermediate, and outer parts of Broken Bay. The discovery of this relation constituted the achievement of the research goal.

Applications Using a_{jk} and b_{jk}. These coefficients find their most frequent use in ecology. Notably, the Bray-Curtis coefficient, b_{jk}, has been used more widely than the Canberra metric coefficient, a_{jk}.

If you are planning an application of b_{jk} in ecological research, the following examples of its use may interest you: Schmitt and Norton's (1972) classification of nematode-infested sites in the Iowa prairies; Tietjen's (1977) classification of sediment samples from Long Island Sound, based on abundances of nematode species; and Doyle and Feldhausen's (1981) classification of sediments in the Gulf of Mexico. Bock et al. (1978) classified Audubon Society bird counts by regions of North America; Armstrong (1977) cluster-analyzed vegetation habitats of small mammals in the Colorado Rocky Mountains; and Thilenius (1972) made a classification of deer habitat in the ponderosa pine forest of Black Hills, South Dakota, finding the classification to be spatially correlated with deer densities.

Sneath and Sokal (1973, pp. 125–126), Clifford and Stevenson (1975, pp. 57–60), and Orlóci (1975) describe other applications of a_{jk} and b_{jk}.

You should note in your reading that the Bray-Curtis coefficient is sometimes called Czekanowski's coefficient and Sorenson's coefficient.

The following Research Application compares the clusters obtained by using four of the coefficients described earlier in this chapter on both standardized and unstandardized data.

RESEARCH APPLICATION 8.1

Using a discriminant function analysis, McHenry (1973) showed that a fossil hominoid humerus (the arm bone extending from shoulder to elbow) discovered by R.E.F. Leaky in East Rudolf, Kenya, was exceptional when compared with the humeri of chimpanzees, gorillas, orangutans, and man (*Homo*). To develop the discriminant function, he obtained many samples of the humeri of man, of *Pan troglodytes* chimpanzees, of *Pan paniscus* chimpanzees, of male gorillas, of female gorillas, of male orangutans, and of female orangutans. These species represented the seven classes of the dependent variable. On each humerus, he measured 18 dimensions, giving the values of the 18 independent variables. The first seven columns of Table 8.2 show the resulting data matrix.

Following the fitting of the discriminant function, McHenry entered the measurements made on the East Rudolf fossil, the eighth column of Table 8.2, into the discriminant function to predict its class membership. He found it to be exceptional, not fitting well into any of the seven classes. And as expected, the analysis also showed that the two chimpanzee classes were similar, that the male and female gorillas were similar, and that the male and female orangutans were similar. Man was somewhat dissimilar to all the other animals in this regard.

Let us see how cluster analysis fares on this problem, and in the process learn more about the properties of resemblance coefficients. Figure 8.4, *a*, *b*, *c*, and *d* show four trees respectively produced by using the resemblance

Table 8.2 Data Matrix of Humeri for Various Species of Hominoids. Numbers in Parentheses Indicate the Size of the Samples Used to Produce the Average Measurements Shown in the Respective Columns. (Data Taken from McHenry (1973), Table 1; Copyright 1973 by the American Association for the Advancement of Science.)

		Chimpanzee		Gorilla		Orangutan		
Measurement	*Homo* (63) 1	*Pan troglodytes* (42) 2	*Pan paniscus* (16) 3	Male (25) 4	Female (41) 5	Female (13) 6	Male (21) 7	East Rudolf 8
Trochlear width	21.9	21.6	19.5	34.1	28.6	24.9	21.0	24.7
Trochlear anterior-posterior diameter	15.0	16.1	14.4	25.9	19.9	17.4	13.9	14.6
Lateral trochlear ridge, anterior-posterior diameter	22.3	26.8	24.0	38.9	29.4	31.4	24.3	24.4
Capitular width	15.7	18.2	16.0	24.1	20.0	19.7	16.3	17.3
Capitular height	19.0	21.4	19.1	32.9	25.3	27.6	21.5	25.6
Articular surface width	38.8	44.6	39.5	69.6	53.8	51.5	41.8	43.6
Biepicondylar width	55.7	62.1	56.1	103.9	79.0	72.6	58.9	71.2
Trochlear to medial epicondyle	38.6	43.7	41.6	75.9	57.2	55.0	43.7	49.7
Trochlear to supracondylar ridge	36.6	40.4	39.8	70.8	54.2	53.7	43.1	47.1
Capitate to lateral epicondyle	26.5	30.8	29.4	50.2	39.7	39.3	31.7	34.3
Olecranon fossa width	25.0	25.2	23.5	42.1	31.9	29.2	23.0	29.9
Olecranon fossa depth	6.7	9.3	8.2	14.9	11.4	9.2	7.4	8.1
Olecranon fossa, width of medial wall	9.3	11.9	11.8	17.8	13.6	16.5	14.0	13.3
Olecranon fossa, width of lateral wall	14.8	19.3	17.1	30.6	23.7	21.9	18.0	18.6
Shaft anterior-posterior diameter	14.7	17.4	15.2	25.1	20.6	18.5	14.4	17.3
Medial epicondyle width	12.0	12.6	11.5	19.2	14.9	16.6	12.2	13.3
Total length	291.2	300.0	279.9	454.0	379.0	371.2	326.2	328.0
Shaft circumference	57.4	70.7	60.4	108.0	86.0	77.5	61.0	84.0

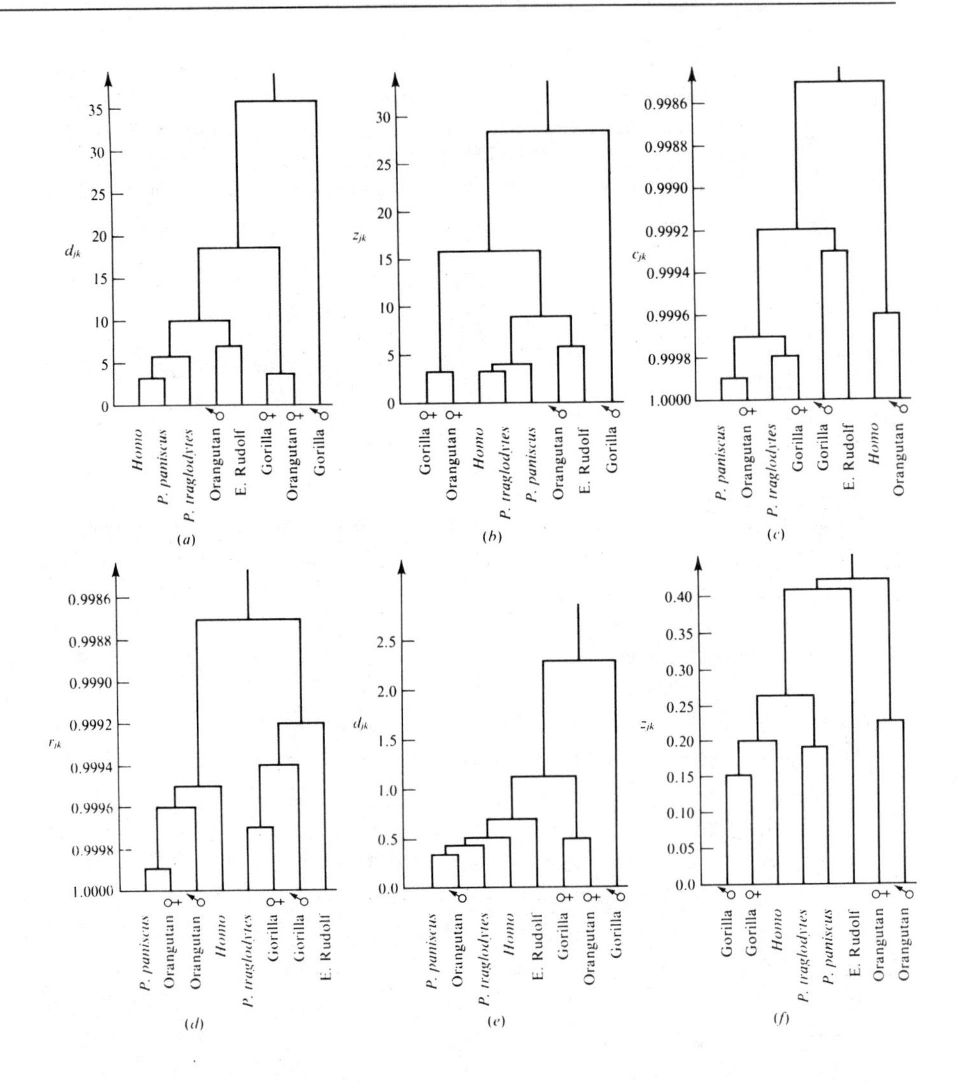

Figure 8.4. Cluster analyses of the data matrix shown in Table 8.2; the UPGMA clustering method was used. Trees *a*, *b*, *c*, *d* were produced by using the resemblance coefficients d_{jk}, z_{jk}, c_{jk}, and r_{jk}, respectively, with the unstandardized data matrix. Trees *e*, *f*, *g*, *h* were produced by applying these respective resemblance coefficients to the standardized data matrix (not shown), obtained with equation 7.1.

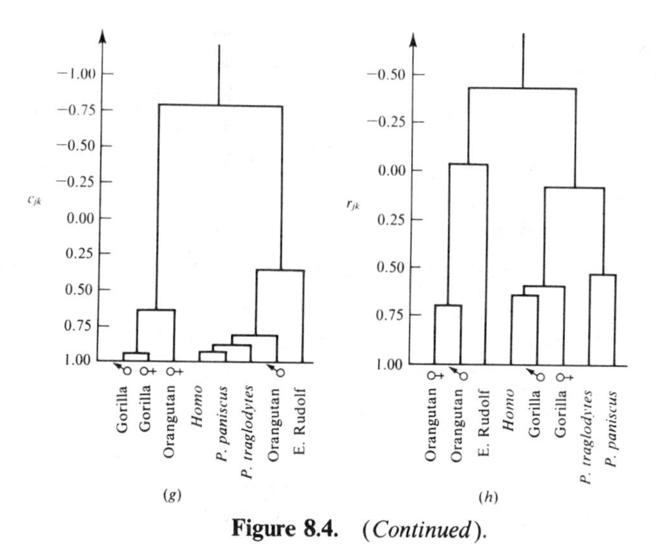

Figure 8.4. (*Continued*).

coefficients d_{jk}, z_{jk}, c_{jk}, and r_{jk} with the data matrix of Table 8.2. Figures 8.4,*e*, *f*, *g*, and *h* show four trees produced by using the same resemblance coefficients, but with the standardized data matrix (not shown) obtained by using equation 7.1.

We want to ignore size displacements among profiles because the bones are from hominoids of different sizes. On the basis of the shapes of the profiles we would expect the chimps to cluster together, the gorillas to cluster, and the orangutans to cluster. *Homo* should be more like our first cousins, the gorillas, than like our second cousins, the chimps and orangutans.

In the unstandardized cases, d_{jk} (in part *a* of the figure) fails to capture our intuitive understanding. It sees similarity as more dependent on size differences than on shape differences. Consequently it places the largest animal, the male gorilla, in the most unique position. The other coefficients, shown in parts *b*, *c*, and *d*, partly ignore the size displacements, and while the trees that result from their use fall more in line with expectations, they fall somewhat short.

Standardizing with equation 7.1 equalizes the contributions of the attributes to overall resemblance. Look, for example, at the attributes "olecranon fossa depth" and "total length." These attributes hardly contribute equally to resemblance when the data are unstandardized.

Figure 8.4,*e* shows the tree produced by applying d_{jk} to the standardized data. Overall size differences still exist; the male gorilla, consistently larger on all 18 measurements, remains consistently larger after standardizing. But "total length" no longer dominates the other attributes. The other trees, in parts f, g, and h, are more in line with the expected conception of hereditary similarity, because the coefficients z_{jk}, c_{jk}, and r_{jk} adjust, each in its own way, for size displacements among the standardized profiles. The tree produced by using r_{jk} seems most reasonable. The chimps and orangutans each cluster; the gorillas cluster closely and *Homo* falls in with them; the East Rudolf fossil is unique.

8.5 HOW STANDARDIZING CAN COMPENSATE FOR UNWANTED SIZE DISPLACEMENTS IN DATA PROFILES

Standardizing a data matrix partially removes unwanted size displacements between data profiles. To see how, let us return to an example presented earlier and standardize the data matrix of Figure 8.1,a in two ways. When we use the column-standardizing function given by equation 7.5, we get the

	Object			
Attribute (time)	1 (B)	2 ($B + 15$)	3 ($B \times 2$)	4 (MIB)
t_1	−1.02	−1.02	−1.02	1.02
t_2	1.32	1.32	1.32	−1.32
t_3	−0.44	−0.44	−0.44	0.44
t_4	0.15	0.15	0.15	−0.15

(a)

	Object			
Attribute (time)	1 (B)	2 ($B + 15$)	3 ($B \times 2$)	4 (MIB)
t_1	0.17	0.20	0.17	0.44
t_2	0.35	0.31	0.35	0.00
t_3	0.22	0.23	0.22	0.33
t_4	0.26	0.26	0.26	0.22

(b)

Figure 8.5. (a) Standardized data matrix obtained by standardizing the data matrix of Figure 8.1a, using equation 7.5; (b) standardized data matrix obtained by using equation 7.8.

standardized data matrix shown in Figure 8.5,*a*. We see that it completely compensates for the additive and proportional translations in the original data matrix, and now objects B, $B + 15$, and $B \times 2$ have identical profiles while *MIB* remains a mirror image of B.

The column-standardizing function given by equation 7.8 also does a good job of removing size displacements between profiles. Using it on the original data matrix of Figure 8.1,*a* gives the standardized data matrix shown in Figure 8.5,*b*. There we see that objects B and $B + 15$ have nearly the same data profiles when standardized, and objects B and $B \times 2$ have exactly the same profiles.

This suggests three strategies for ignoring size displacements among the data profiles of the original data matrix. The first is to apply a coefficient such as z_{jk}, c_{jk}, or r_{jk} to the unstandardized data matrix because by degrees these coefficients ignore size displacements. The second strategy is to column-standardize the data with an equation such as 7.5 or 7.8, which removes much of the size displacements, and then use a coefficient such as d_{jk} on the standardized data matrix to detect the residual differences between the profiles. The third strategy is to column-standardize the data as just mentioned and then use a resemblance coefficient such as z_{jk}, c_{jk}, or r_{jk}, and in this way doubly compensate for size displacements. That is, two back-to-back partial compensations can be expected to be more effective than either one used alone.

8.6 HOW TO HANDLE MISSING DATA MATRIX VALUES

Missing values of data can occur for two reasons: either a measurement is made and then lost, or a measurement cannot be made at all. We signify missing values by using the symbol "N" in the data matrix. For instance, consider the data matrix shown in Table 8.3, which contains five missing values.

When we standardize this data matrix we act as if the N's were not present—that is, we skip over them. Taking attribute 1, for example, and using equation 7.1 to standardize it, only the values 10, 5, and 4 enter into the calculation of $\overline{X}_1$ and S_1, giving $\overline{X}_1 = 6.3$ and $S_1 = 3.2$. The values in row 1 of the standardized data matrix become

$$Z_{11} = (10 - 6.3)/3.2 = 1.16; \qquad Z_{12} = (5 - 6.3)/3.2 = -0.41;$$

$$Z_{13} = \text{N}; \qquad Z_{14} = (4 - 6.3)/3.2 = -0.72.$$

The same rule of skipping the N's applies to the calculation of resemblance coefficients. Only those attributes are used for which there are values of data for both of the objects being compared. For example, let us use the data matrix shown in Table 8.3 and compute the coefficient d_{jk} for

Table 8.3 Example of a data matrix in which missing values are coded with "N."

	Object			
Attribute	1	2	3	4
1	10	5	N	4
2	N	N	2	19
3	15	4	N	0
4	5	N	2	30
5	3	5	5	4

d_{12} and d_{13}. For the two objects being compared, strike out all attributes containing an entry of "N." Compute the resemblance coefficient on the basis of the remaining attributes. Note that the number of attributes n must be redefined as the effective number of attributes remaining after the attributes with "N" in them have been struck out. Thus, in computing d_{12}, $n = 3$ is used; in computing d_{13}, $n = 2$ is used. Therefore we have

$$d_{12} = \left\{\left[(10 - 5)^2 + (15 - 4)^2 + (3 - 5)^2\right]/(3)\right\}^{1/2} = 7.1;$$

$$d_{13} = \left\{\left[(5 - 2)^2 + (3 - 5)^2\right]/(2)\right\}^{1/2} = 2.5.$$

An analogous scheme applies for R-analyses.

It is now clear why we favor d_{jk} over e_{jk} for general use. Equation 8.1 shows that e_{jk} tends to increase when the number of attributes n is increased simply because more terms are added in the summation. It follows that when two objects that share missing values are compared, fewer terms are added in the summation of e_{jk} than when two objects that have no missing values are compared. This is an undesirable feature of e_{jk}: missing values tend to produce lower values of e_{jk}, and hence tend to produce higher similarity artificially.

But equation 8.2 shows that d_{jk} is computed as an average on a "per effective attribute" basis. Thus, missing values do not artificially lower the value of d_{jk}. Indeed, with the exception of e_{jk}, all of the resemblance coefficients described in this chapter have this built-in averaging property and can be used as has just been described when there are missing values.

Naturally we prefer that a data matrix contain no missing values. Still, one or two values are sometimes lacking. In such cases we should go ahead and perform the cluster analysis, making sure that the conclusions we draw are insensitive to plausible values that the missing data might have assumed.

SUMMARY

This chapter has given you six more resemblance coefficients to supplement the Euclidean distance coefficient treated in Chapter 2. Any of these can be used with quantitative attributes.

The coefficients are organized according to how sensitive they are in detecting size displacements between data profiles of objects.

Depending on your research goal, you will want either to count size displacements between data profiles as affecting the similarity between objects, or to ignore size displacements and base the similarity of the objects on their shapes. Thus, knowledge of your research goal and of how the resemblance coefficients are sensitive to size displacements are crucial to selecting the best resemblance coefficient for your particular cluster analysis.

To some extent, size displacements in data profiles can be "standardized away." Thus, both standardizing and choosing a resemblance coefficient are options you have for ignoring size displacements between data profiles.

Finally, missing values in the data matrix are not fatal. The values in the resemblance matrix can still be computed even when some (preferably only a few) values are missing from the data matrix.

EXERCISES

(Answers to exercises are given in the back of the book.)

8.1. From the research of Reynolds (1979) we take the following data matrix in which the objects are years, and the attributes are the number of active grey heron nests at different locations in Britain.

	Year					
Location	1972 (1)	1973 (2)	1974 (3)	1975 (4)	1976 (5)	1977 (6)
Brownsea Island	124	114	127	114	101	107
High Halstow	167	174	178	182	185	195
Walthamstow	83	105	113	101	111	115
Islington	63	58	48	42	51	38
Tabley	91	95	88	102	103	103
Claughton	106	113	109	124	107	127

To check your ability to calculate the value of a resemblance coefficient, calculate for objects 1 and 6 the following: d_{16}, z_{16}, c_{16}, r_{16}, a_{16}, b_{16}.

8.2. Cain and Harrison (1958) proposed a resemblance coefficient, m_{jk}, which they called the "mean character difference":

$$m_{jk} = (1/n) \sum_{i=1}^{n} |X_{ij} - X_{ik}|/\text{RMAX}_i.$$

The function RMAX_i reads, "the maximum value in the ith row of the data matrix." Is this a similarity coefficient, or a dissimilarity coefficient? What is its range? Assume $X_{ij} \geq 0$ for all i, j.

8.3. Can d_{jk}, z_{jk}, c_{jk}, r_{jk}, a_{jk}, and b_{jk} be calculated when the number of attributes n is as few as $n = 1$?

8.4. Sometimes there is no natural direction in the scale of an attribute—its direction is arbitrary. This causes a problem which Cohen (1969, p. 281) summarizes in the context of psychological research: "...we can score high for extraversion *or* for introversion, high for good *or* bad, and high for agreement with 'It is easy to recognize someone who once had a serious mental illness' *or* for agreement with 'It is *difficult* to recognize someone who once had a serious mental illness.' Since the direction of measurement is arbitrary, none of these element reflections should change the substantive conclusions from the results of the data analysis." As an illustration of Cohen's comments, consider the following two data matrices. The attributes are measured on a 9-point scale (running from 1 through 9). The matrices differ only on attribute 3, where the values appear to trend in opposite directions.

	Object			
Attribute	1	2	3	4
1	7	3	7	6
2	1	4	9	3
3	7	5	1	2

	Object			
Attribute	1	2	3	4
1	7	3	7	6
2	1	4	9	3
3	3	5	9	8

Which of the resemblance coefficients d_{jk}, d_{jk}, c_{jk}, r_{jk}, a_{jk}, and b_{jk} are invariant under the change in the direction of coding?

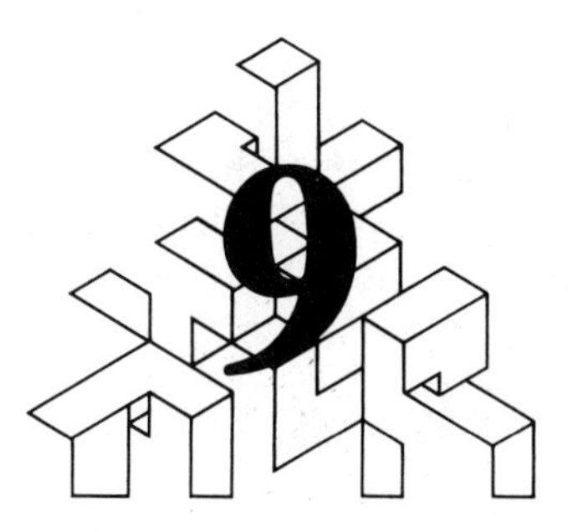

Clustering Methods

OBJECTIVES

While the UPGMA clustering method, treated in Chapter 2, is the one researchers favor in their applications, other clustering methods are also useful. The main alternatives to UPGMA are

- The SLINK method.
- The CLINK method.
- Ward's minimum variance method.

This chapter

- Explains these clustering methods in detail, giving examples of each.
- Mentions briefly other lesser-used clustering methods.
- Shows how the execution of a clustering method depends on whether a dissimilarity coefficient or a similarity coefficient is used.

9.1 SLINK CLUSTERING METHOD

SLINK is short for **single linkage clustering method**. Whereas the UPGMA clustering method defines the similarity between any two clusters as the arithmetic average of the similarities between the objects in the one cluster and the objects in the other, SLINK does this differently. With SLINK we evaluate the similarity between two clusters as follows.

First we list all pairs of "spanning objects" between the two clusters. By a pair of spanning objects we mean that one of the objects is in one cluster and the other is in the other cluster. Then from this list, we find the most similar pair of spanning objects and call their similarity the similarity of the two clusters. The difference in how SLINK defines the similarity between two clusters means that with SLINK we need to revise the resemblance matrix differently between clustering steps.

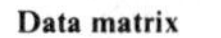

Data matrix

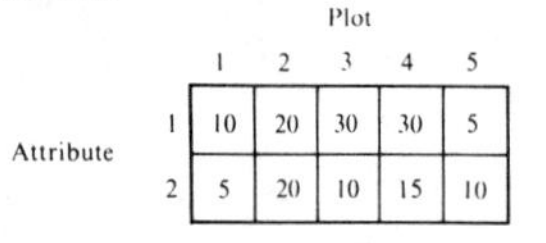

Attribute \ Plot	1	2	3	4	5
1	10	20	30	30	5
2	5	20	10	15	10

Initial resemblance matrix

	1	2	3	4	5
1					
2	18.0				
3	20.6	14.1			
4	22.4	11.2	5.00		
5	7.07	18.0	25.0	25.5	

Second resemblance matrix

	1	2	5	(34)
1				
2	18.0			
5	7.07	18.0		
(34)	20.6	11.2	25.0	

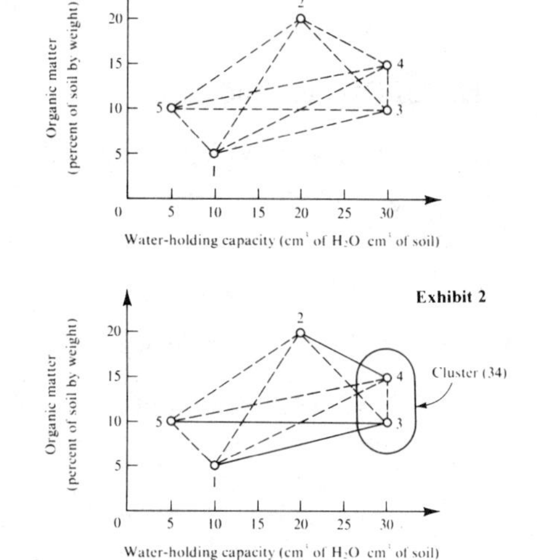

Figure 9.1. Cluster analysis using the SLINK clustering method; the data matrix is the one used in Chapter 2 to illustrate the UPGMA clustering method. The figure shows the data matrix, and the resemblance matrices and attribute space plots for each step of the clustering process. Dashed lines in the attribute space plots show the Euclidean distances between objects. Heavy lines show the minimum distances between clusters, which SLINK uses to measure their similarity.

Recall from Chapter 2 that at the start of the clustering method each object is a cluster that contains only itself. When this is the case, we will use the words "cluster" and "object" interchangeably.

Let us now demonstrate the SLINK clustering method by using the Euclidean distance coefficient and reworking the plot-fertilizer example of Chapter 2. As with UPGMA, the first step is to find the two most similar objects. These objects are those whose Euclidean distance coefficient e_{jk} is smallest. Figure 9.1 shows, in both the initial resemblance matrix and the attribute space plot, that objects 3 and 4 are the closest and thus are the most similar. We write the clustering step as

Step 1. Merge clusters 3 and 4, giving 1, 2, (34), and 5 at the value of $e_{34} = 5.0$.

This changes the attribute space plot in Figure 9.1 from that labeled Exhibit 1 to Exhibit 2.

Third resemblance matrix

	2	(34)	(15)
2			
(34)	11.2		
(15)	18.0	20.6	

Fourth resemblance matrix

	(15)	(234)
(15)	—	—
(234)	18.0	

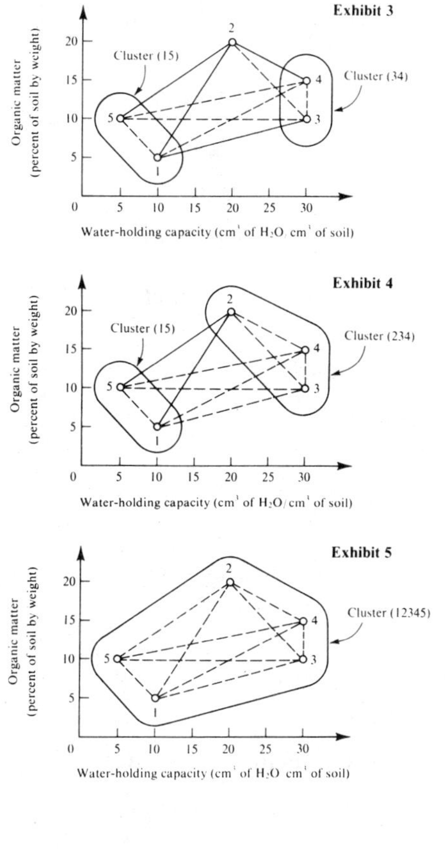

Figure 9.1. (*Continued*).

Next we revise the resemblance matrix by calculating $e_{1(34)}$, $e_{2(34)}$, and $e_{5(34)}$ and transcribing the remaining coefficients from the initial resemblance matrix to the second resemblance matrix. We calculate the following:

$$e_{1(34)} = \min(e_{13}, e_{14}) = \min(20.6, 22.4) = 20.6;$$

$$e_{2(34)} = \min(e_{23}, e_{24}) = \min(14.1, 11.2) = 11.2;$$

$$e_{5(34)} = \min(e_{35}, e_{45}) = \min(25.0, 25.5) = 25.0.$$

For example, $e_{1(34)} = \min(e_{13}, e_{14})$ means that we find the minimum of the two values e_{13} and e_{14} and assign it as the similarity between clusters 1 and (34). Exhibit 2 shows these three minimum spanning values as heavy lines.

Using the second resemblance matrix, we find the smallest value there, which stands for the most similar pair of objects; the second clustering step becomes

Step 2. Merge clusters 1 and 5, giving 2, (34), and (15) at the value of $e_{15} = 7.07$.

Now we revise our matrix once again, to get the third resemblance matrix:

$$e_{2(15)} = \min(e_{12}, e_{25}) = \min(18.0, 18.0) = 18.0;$$

$$\begin{aligned} e_{(15)(34)} &= \min(e_{13}, e_{14}, e_{35}, e_{45}) \\ &= \min(20.6, 22.4, 25.0, 25.5) = 20.6 \end{aligned}$$

Using the third resemblance matrix, the third clustering step is

Step 3. Merge clusters 2 and (34), giving (15) and (234) at the value of $e_{2(34)} = 11.2$.

Then we calculate the lone value, $e_{(15)(234)}$, of the fourth resemblance matrix:

$$\begin{aligned} e_{(15)(234)} &= \min(e_{12}, e_{13}, e_{14}, e_{25}, e_{35}, e_{45}) \\ &= \min(18.0, 20.6, 22.4, 18.0, 25.0, 25.5) \\ &= 18.0 \end{aligned}$$

This brings us to the final step.

Step 4. Merge clusters (15) and (234), giving (12345) at the value of $e_{(15)(234)} = 18.0$.

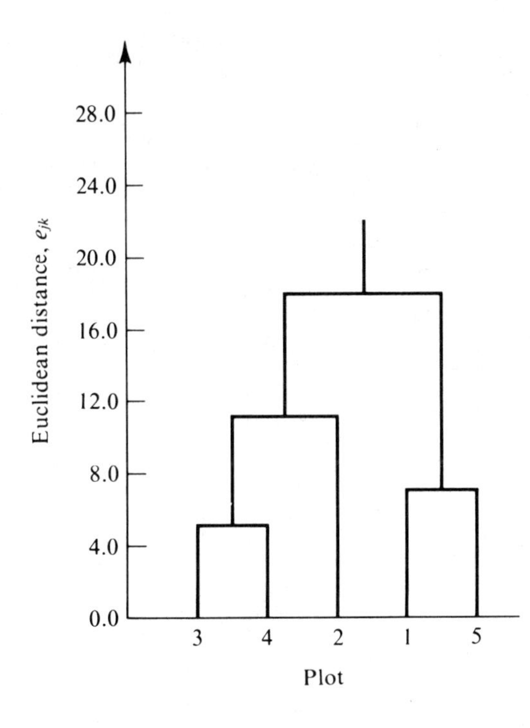

Figure 9.2. Tree produced by using the SLINK clustering method on the data shown in Figure 9.1.

Figure 9.2 shows the tree resulting from the four clustering steps. We will postpone discussing this tree, and discussing SLINK in general, until we have discussed the CLINK clustering method.

9.2 CLINK CLUSTERING METHOD

CLINK, short for **complete linkage clustering method**, is the antithesis of SLINK. With CLINK the similarity between any two clusters is determined by the similarity of their two most dissimilar spanning objects. Thus, in revising the resemblance matrix values between clustering steps, we must find the spanning objects having the maximum value of e_{jk}.

Let us make this clear by using CLINK to rework the previous example. We proceed exactly as we did with SLINK except that we replace all min(e_{jk}) by max(e_{jk}) when revising the resemblance matrix. The procedure can be followed in Figure 9.3. We begin as we began with UPGMA and SLINK by finding the two most similar objects (clusters).

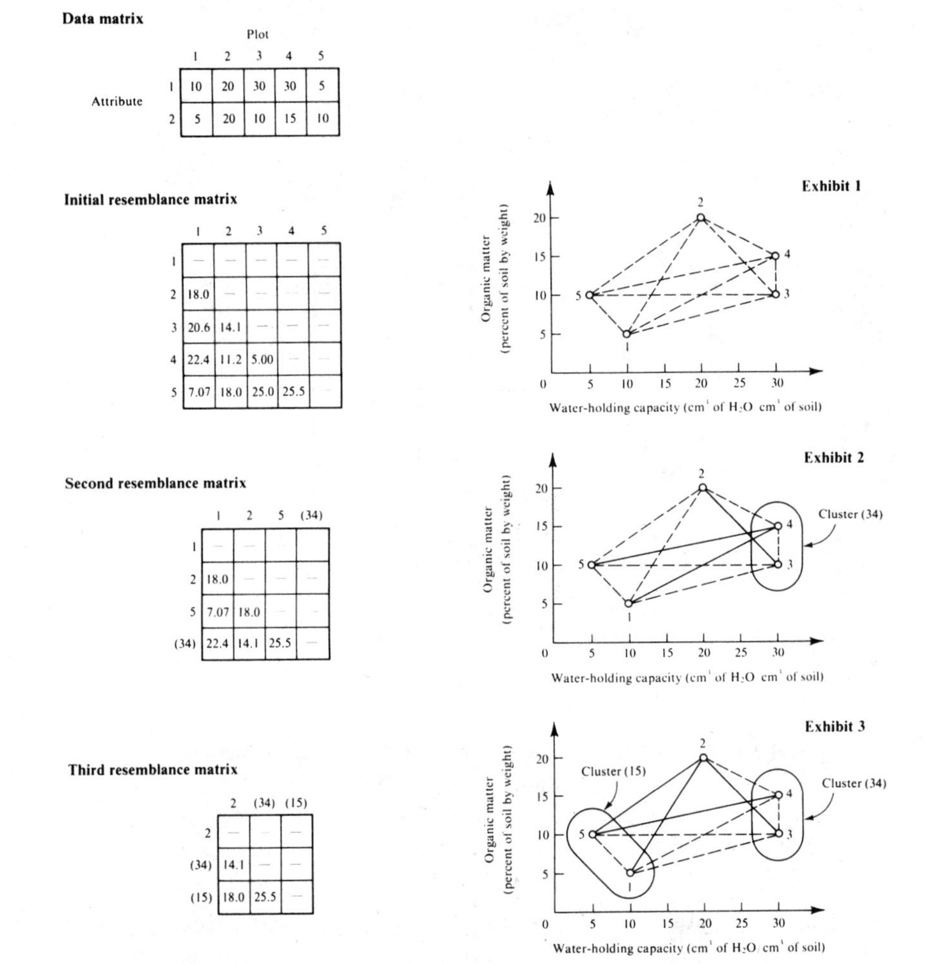

Data matrix

Attribute \ Plot	1	2	3	4	5
1	10	20	30	30	5
2	5	20	10	15	10

Initial resemblance matrix

	1	2	3	4	5
1	—	—	—	—	—
2	18.0	—	—	—	—
3	20.6	14.1	—	—	—
4	22.4	11.2	5.00	—	—
5	7.07	18.0	25.0	25.5	—

Second resemblance matrix

	1	2	5	(34)
1	—	—		
2	18.0	—	—	—
5	7.07	18.0	—	—
(34)	22.4	14.1	25.5	—

Third resemblance matrix

	2	(34)	(15)
2	—	—	—
(34)	14.1	—	—
(15)	18.0	25.5	—

Figure 9.3. Cluster analysis using the CLINK clustering method; the data matrix is the one used in Chapter 2 to illustrate the UPGMA clustering method. The figure shows the data matrix, and the resemblance matrices and attribute space plots for each step of the clustering process. Dashed lines in the attribute space plots show the Euclidean distances between objects. Heavy lines show the maximum distances between clusters, which CLINK uses to measure their similarity.

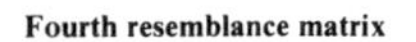
Fourth resemblance matrix

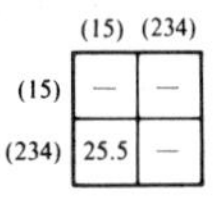

	(15)	(234)
(15)	—	—
(234)	25.5	—

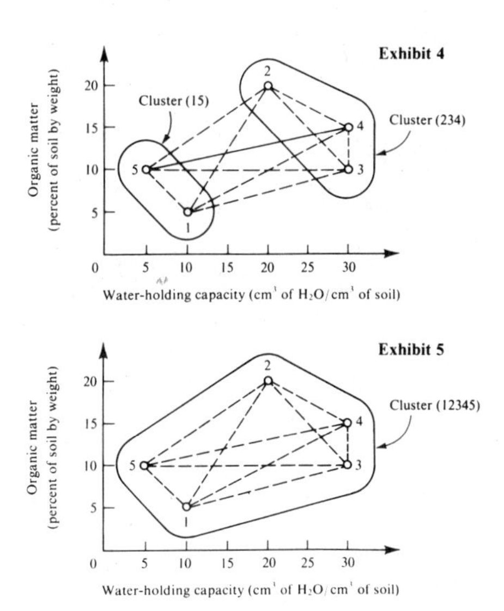

Figure 9.3. (*Continued*).

Step 1. Merge clusters 3 and 4, giving 1, 2, (34), and 5 at the value of $e_{35} = 5.0$.

The revised values for the second resemblance matrix are calculated as

$$e_{1(34)} = \max(e_{13}, e_{14}) = \max(20.6, 22.4) = 22.4;$$

$$e_{2(34)} = \max(e_{23}, e_{24}) = \max(14.1, 11.2) = 14.1;$$

$$e_{5(34)} = \max(e_{35}, e_{45}) = \max(25.0, 25.5) = 25.5.$$

Exhibit 2 shows these three maximum spanning values as heavy lines. Next we form another cluster by finding the most similar pair of clusters—the pair whose value of e_{jk} is smallest.

Step 2. Merge clusters 1 and 5, giving 2, (34), and (15) at the value of $e_{15} = 7.07$.

The revised values of the second resemblance matrix that will be used in the third are

$$e_{2(15)} = \max(e_{12}, e_{25}) = \max(18.0, 18.0) = 18.0;$$

$$e_{(15)(34)} = \max(e_{13}, e_{14}, e_{35}, e_{45})$$

$$= \max(20.6, 22.4, 25.0, 25.5) = 25.5.$$

Step 3. Merge clusters 2 and (34), giving (15) and (234) at the value of $e_{2(34)} = 14.1$.

Now we revise the third resemblance matrix to form the fourth:

$$e_{(15)(234)} = \max(e_{12}, e_{13}, e_{14}, e_{25}, e_{35}, e_{45})$$

$$= \max(18.0, 20.6, 22.4, 18.0, 25.0, 25.5)$$

$$= 25.5.$$

This brings us to the final step.

Step 4. Merge clusters (15) and (234), giving (12345) at the value of $e_{(15)(234)} = 25.5$.

Figure 9.4 shows the tree that results from the four clustering steps.

Now let's compare the three cluster analyses of the plot-fertilizer data matrix. The tree produced by using UPGMA, shown in Chapter 2 (page 21), and those produced by using SLINK and CLINK, shown in Figures 9.2 and 9.4, are much alike. Apart from the values at which objects form clusters, the order in which objects merge is the same in the three trees. This is coincidental—a consequence of the data in our original matrix. The three clustering methods can, and often do, differ in the order in which they merge objects.

In general, trees produced by these clustering methods will merge objects at different values of the resemblance coefficient, even in cases where they merge the objects in the same order. The examples just given illustrate this. Moreover, SLINK tends to produce compacted trees; CLINK, extended trees; and UPGMA, trees intermediate between these extremes.

When choosing among these three clustering methods, most researchers select UPGMA. They consider that UPGMA is based on a middle-of-the-road philosophy cast between the two extremes. To them this is safer and thus more appealing because extremes, rightly or wrongly, connote risk.

It turns out that a theoretical reason supports this intuition. The theoretical reason for favoring UPGMA is that it tends to give higher values

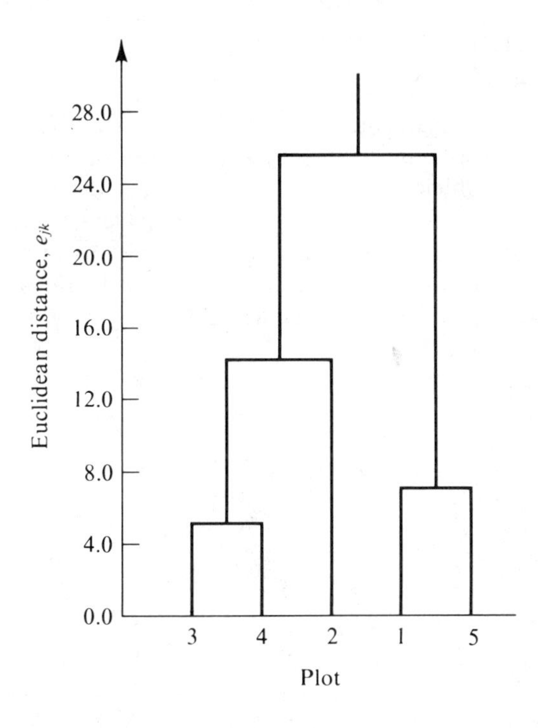

Figure 9.4. Tree produced by using the CLINK clustering method on the data shown in Figure 9.3.

of the cophenetic correlation coefficient (Farris, 1969). This means that on average UPGMA produces less distortion in transforming the similarities between objects into a tree.

The three methods are often referred to by aliases. SLINK is called "the minimum method" and "the nearest neighbor method." UPGMA is called "the group average method" and "the average linkage method." CLINK is called "the maximum method" and "the furthest neighbor method."

In a cluster analysis for research into play in young children, Hughes (1979) used both the SLINK and CLINK clustering methods to investigate the stability of the clusters. She argued that since SLINK and CLINK are based on opposite philosophies, if they give essentially the same trees for a given data matrix, then the clusters are well-defined in their attribute space and in a sense are "real." Conversely, if the two clustering methods give entirely different trees, the clusters are only weakly defined and perhaps are artifacts of the clustering method.

9.3 CLUSTERING WITH SIMILARITY AND DISSIMILARITY COEFFICIENTS

Let us see how the mathematical steps of the SLINK, UPGMA, and CLINK clustering methods are influenced by whether a similarity coefficient or a dissimilarity coefficient is used.

With all three methods, the two most similar clusters are merged at each clustering step. This means that when a dissimilarity coefficient is being used, the two clusters with the *smallest* value of the dissimilarity coefficient are merged. But when a similarity coefficient is being used, the two clusters with the *largest* value are merged.

When a dissimilarity coefficient is used, the vertical scale of the tree has the smallest value at its bottom ("branches") and the largest at its top

Which clusters to join at each clustering step?

	Resemblance coefficient	
	Dissimilarity coefficient	Similarity coefficient
	Join the two clusters identified with the smallest dissimilarity coefficient value.	Join the two clusters identified with the largest similarity coefficient value.

How to revise the resemblance coefficients after a clustering step?

		Resemblance coefficient	
		Dissimilarity coefficient	Similarity coefficient
Clustering method	SLINK	$e_{(jk)(mn)} = \min(e_{jm}, e_{jn}, e_{km}, e_{kn})$	$r_{(jk)(mn)} = \max(r_{jm}, r_{jn}, r_{km}, r_{kn})$
	UPGMA	$e_{(jk)(mn)} = 1/4(e_{jm} + e_{jn} + e_{km} + e_{kn})$	$r_{(jk)(mn)} = 1/4(r_{jm} + r_{jn} + r_{km} + r_{kn})$
	CLINK	$e_{(jk)(mn)} = \max(e_{jm}, e_{jn}, e_{km}, e_{kn})$	$e_{(jk)(mn)} = \min(r_{jm}, r_{jn}, r_{km}, r_{kn})$

Figure 9.5. Differences in the SLINK, UPGMA, and CLINK clustering methods that depend on whether a dissimilarity coefficient or a similarity coefficient is being used. To show how the resemblance coefficient is revised, cluster (jk) containing objects j and k, and cluster (mn) containing objects m and n, are used; e_{jk} is used as an example of a dissimilarity coefficient, and r_{jk} is used as an example of a similarity coefficient.

("trunk," referenced to the way the trees are oriented in Figures 9.2 and 9.4). But when a similarity coefficient is used, the direction is reversed and the scale has the largest value at its bottom and the smallest at its top.

Another difference is in how the resemblance matrix is revised between clustering steps. For the SLINK method, we take the *minimum* of the values of the *dissimilarity* coefficients of the objects that span the two clusters. But we take the *maximum* of the values when a *similarity* coefficient is used. The reason for the about-face is this. SLINK's philosophy measures the similarity between two clusters in terms of the most similar objects between those separate clusters. When a *dissimilarity* coefficient is used, the most similar spanning members have the smallest value; hence we must search for the *minimum* value when revising the resemblance matrix. But when a *similarity* coefficient is used, the most similar spanning members have the largest value; hence we must search for the *maximum* value.

With CLINK, all of this is reversed. Its philosophy takes the similarity between two clusters to be the similarity between their most dissimilar spanning objects. When a *dissimilarity* coefficient is used, the most dissimilar spanning objects have the largest value; hence we must search for the *maximum* value when revising the resemblance matrix. But when a *similarity* coefficient is used, the most dissimilar spanning objects have the smallest value; hence we must search for the *minimum* value.

For UPGMA, the resemblance matrix is revised by averaging the resemblance coefficients regardless of whether a dissimilarity coefficient or similarity coefficient is used.

Figure 9.5 summarizes these differences. Exercise 9.1 uses the SLINK and CLINK clustering methods with a similarity coefficient to illustrate the details.

9.4 WARD'S MINIMUM VARIANCE CLUSTERING METHOD

After UPGMA, Ward's minimum variance clustering method is the most often used. To explain it, we will depart from the format of computing the resemblance matrix and revising it at each clustering step. Although we could do it this way, and indeed computer programs for cluster analysis *do*, it can be understood more easily if all the calculations are done directly with the data matrix.

Like the other clustering methods, Ward's method follows a series of clustering steps that begin with t clusters, each containing one object, and it ends with one cluster containing all objects. At each step it makes whichever merger of two clusters that will result in the smallest increase in the value of an index E, called the sum-of-squares index, or variance. This means that at each clustering step we must try all possible mergers of two clusters, compute the value of E for each, and select that one whose value of E is the smallest. Then we go on to the next clustering step and repeat the process.

For each tentative set of clusters, E is computed as follows. First we calculate the mean of each cluster. The cluster mean is a fictitious object whose attribute values are the average of the attribute values for the objects in the given cluster.

Second, we compute the differences between each object in a given cluster and its cluster mean. For example, suppose a cluster contains three objects, each described by five attributes. For the first object we would compute five differences, namely the differences in the values from it to its cluster mean, for each of the five attributes. These differences could be positive or negative. For the second and third objects we would do the same, thus ending up with (3)(5) = 15 differences for the cluster.

Third, for each cluster we square each of the differences we have computed above. We add these for each cluster, giving a sum-of-squares value for each cluster.

Finally, we compute the value of E by adding the sum-of-squares values for all the clusters.

Let us demonstrate Ward's method by using the data matrix shown in Figures 9.1 and 9.3. At the start, or "zeroth" clustering step, there are $t = 5$ clusters each containing one object. The cluster mean for any of the five clusters has the same attribute values as the one object contained in the cluster. Hence, the value of E at this step is always zero. But we will compute it for completeness:

$$E = \underbrace{(10-10)^2 + (5-5)^2}_{\text{Cluster 1}} + \underbrace{(20-20)^2 + (20-20)^2}_{\text{Cluster 2}}$$

$$+ \underbrace{(30-30)^2 + (10-10)^2}_{\text{Cluster 3}} + \underbrace{(30-30)^2 + (15-15)^2}_{\text{Cluster 4}}$$

$$+ \underbrace{(5-5)^2 + (10-10)^2}_{\text{Cluster 5}} = 0.0.$$

To begin Step 1, we must first compute E for each of the ten possible mergers at "Step 1" in Table 9.1. For example, take the first one: (12), 3, 4, 5. We must first calculate the cluster mean for (12). It is $(10 + 20)/2 = 15$ on attribute 1, and $(5 + 20)/2 = 12.5$ on attribute 2. The other cluster means stay as they were when we computed E at the zeroth clustering step.

For the first possible merger the value of E is

$$E = \underbrace{(10-15)^2 + (5-12.5)^2 + (20-15)^2 + (20-12.5)^2}_{\text{Cluster (12)}}$$

$$+ \underbrace{(30-30)^2 + (10-10)^2}_{\text{Cluster 3}} + \underbrace{(30-30)^2 + (15-15)^2}_{\text{Cluster 4}}$$

$$+ \underbrace{(5-5)^2 + (10-10)^2}_{\text{Cluster 5}} = 162.5.$$

Table 9.1 An example illustrating the application of Ward's minimum variance clustering method to the data matrix shown in Figures 9.1 and 9.3. At each clustering step the possible partitions of objects into clusters and the corresponding values of the index E are shown. The optimal partition at each clustering step is signified by the symbol "*."

Step	Possible partitions				E
1	(12)	3	4	5	162.5
	(13)	2	4	5	212.5
	(14)	2	3	5	250.
	(15)	2	3	4	25.
	(23)	1	4	5	100.
	(24)	1	3	5	62.5
	(25)	1	3	4	162.5
	(34)	1	2	5	12.5*
	(35)	1	2	4	312.5
	(45)	1	2	3	325.0
2	(34)	(12)	5		175.0
	(34)	(15)	2		37.5*
	(34)	(25)	1		175.
	(134)	2	5		316.7
	(234)	1	5		116.7
	(345)	1	2		433.3
3	(234)	(15)			141.7*
	(125)	(34)			245.9
	(1345)	2			568.8
4	(12345)				650.

However, when we calculate E for each of the other nine possible mergers we find that the set of clusters 1, 2, (34), 5 gives the smallest value possible; for this merger, E is

$$E = \underbrace{(10-10)^2 + (5-5)^2}_{\text{Cluster 1}} + \underbrace{(20-20)^2 + (20-20)^2}_{\text{Cluster 2}}$$

$$+ \underbrace{(30-30)^2 + (10-12.5)^2 + (30-30)^2 + (15-12.5)^2}_{\text{Cluster (34)}}$$

$$+ \underbrace{(5-5)^2 + (10-10)^2}_{\text{Cluster (5)}} = 12.5.$$

Note that in this calculation the cluster mean of (34) is $(30 + 30)/2 = 30$ for attribute 1, and $(10 + 15)/2 = 12.5$ for attribute 2. Table 9.1 shows the value of E for each possible set of clusters, and $E = 12.5$ is the smallest. This decides the outcome at step 1:

Step 1. Merge clusters 3 and 4, giving 1, 2, (34), and 5 at the value of $E = 12.5$.

Now we can begin Step 2. With Ward's method, objects merged at previous clustering steps are never unmerged. Thus, at the beginning of step 2 there are six possible mergers of two clusters. Table 9.1 shows the six possibilities at "Step 2," and the set of clusters 2, (34), (15) is chosen because it gives the smallest value of E, namely:

$$E = \underbrace{(20-20)^2 + (20-20)^2}_{\text{Cluster 2}}$$

$$+ \underbrace{(30-30)^2 + (10-12.5)^2 + (30-30)^2 + (15-12.5)^2}_{\text{Cluster (34)}}$$

$$+ \underbrace{(10-7.5)^2 + (5-7.5)^2 + (5-7.5)^2 + (10-7.5)^2}_{\text{Cluster (15)}}$$

$$= 37.5.$$

Note that in addition to the cluster means of clusters 2 and (34), which we carried down from the previous step, we computed the mean of cluster (15).

It is $(10 + 5)/2 = 7.5$ for attribute 1, and $(5 + 10)/2 = 7.5$ for attribute 2. Thus,

Step 2. Merge clusters 1 and 5, giving 2, (34), and (15) at the value of $E = 37.5$.

Now on to step 3. The value of $E = 141.7$ is the minimum of the three possible partitions shown in Table 9.1, and it is calculated as

$$E = \underbrace{(10-7.5)^2 + (5-7.5)^2 + (5-7.5)^2 + (10-7.5)^2}_{\text{Cluster (15)}}$$

$$\underbrace{+(20-26.7)^2 + (20-15)^2 + (30-26.7)^2 + (10-15)^2 + (30-26.7)^2 + (15-15)^2}_{\text{Cluster (234)}}$$

$$= 141.7.$$

Note that we had to compute the cluster mean for cluster (234). It is $(20 + 30 + 30)/3 = 26.7$ for attribute 1, and $(20 + 10 + 15)/3 = 15$ for attribute 2. Thus,

Step 3. Merge clusters 2 and (34), giving (15) and (234) at the value of $E = 141.7$.

This brings us to the final step. Only one other merger is possible. The cluster mean for attribute 1 is $(10 + 20 + 30 + 30 + 5)/5 = 19$, and it is $(5 + 20 + 10 + 15 + 10)/5 = 12$ for attribute 2. The value of E is calculated as

$$\underbrace{E = (10-19)^2 + (5-12)^2 + (20-19)^2 + (20-12)^2 + (30-19)^2 + (10-12)^2 + (30-19)^2 + (15-12)^2 + (5-19)^2 + (10-12)^2}_{\text{Cluster (12345)}} = 650.$$

Thus,

Step 4. Merge (15) and (234), giving (12345) at the value of $E = 650$.

Figure 9.6 shows the tree that Ward's method gives. Since the clustering index E is a function of the *squares* of the attribute differences, and since the attribute differences increase as more objects are forced into clusters as the method progresses, it follows that E increases nonlinearly as we work up through the clustering steps. Our example illustrated as much. E increased through the progression: $E = 0.0$, $E = 12.5$, $E = 37.5$, $E = 141.7$, and $E = 650$. This gives trees computed with Ward's clustering method a well-defined look in which clusters usually jump out at the eye. The popularity of Ward's method partly stems from this characteristic: researchers like methods that give what look like well-defined clusters.

This is not of real significance, however. Because E is a function of the square of differences, it tends to increase nonlinearly as the number of clusters becomes smaller. The effect of the tree being well defined would be even more pronounced if we took the fourth power of the differences when computing E, instead of taking the second power. For this reason it is better to plot the tree obtained with Ward's method not as a function of E, but as the square root of E, as this will undo the nonlinearity. Such a tree would not be so well defined, but neither would its look be mathematically contrived. However, this suggestion has not been taken by practitioners, and

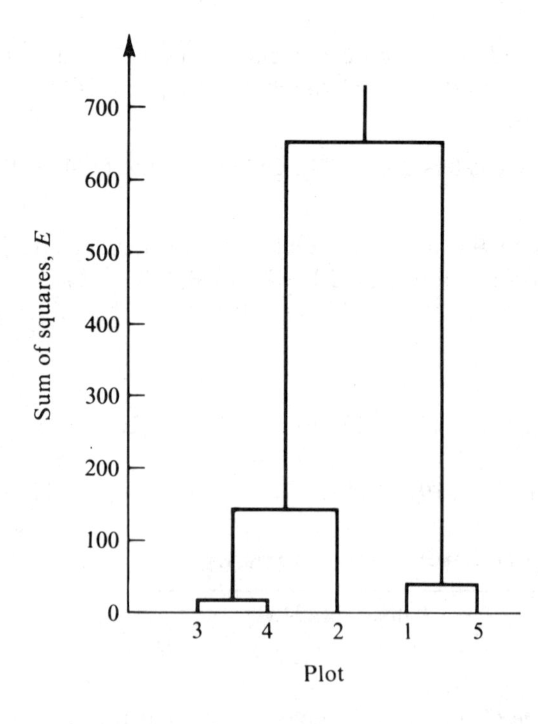

Figure 9.6. Tree produced by applying Ward's minimum variance clustering method to the data matrix shown in Figures 9.1 and 9.3.

they continue to use E (or the increase in E from one clustering step to the next) for the scale of their trees. Bear in mind, however, that this is a kind of mathematical plastic surgery.

While we have explained Ward's clustering method in words, Anderberg (1973, p. 142) explains it mathematically.

Ward's method can, of course, be used with standardized data.

There are several other points to note about Ward's method.

First, the index E is a built-in dissimilarity coefficient sensitive to additive and proportional translations of data profiles. In selecting Ward's method, you lose your option of picking a resemblance coefficient: the index E must be used since it is an inalterable part of the method.

The second point is that Ward's method does not guarantee an optimal partitioning of objects into clusters. That is, there may be other partitions that give a value of E that is less than the one obtained by using this method. Because the objects merged at any step are never unmerged at subsequent steps, the finding of the minimum value of E at each step is conditioned on the set of clusters already formed at prior clustering steps. This value of E may be larger than the true minimum—larger than the one that could be found if the unmerging of clusters were permitted. But using the less-than-optimal solution offered by Ward's method greatly reduces the computations required by an optimal method, and it usually gives a near-optimal solution that is good enough for most purposes. And depending on the data, it sometimes gives the optimal solution.

One stimulus to the current wide use of Ward's clustering method was the early availability of Veldman's (1967) computer program for Ward's method.

9.5 OTHER CLUSTERING METHODS

Referring to a 1973 survey of users of cluster analysis, Blashfield (1976) states that UPGMA, Ward's, and SLINK accounted for three-fourths of the published uses of cluster analysis. While the survey is dated and slanted towards numerical taxonomists, it probably remains typical of current usage. Several of the lesser-used clustering methods, however, appear occasionally in applications and thus deserve brief mention. These are the **WPGMA method**, the **centroid method**, and the **flexible method**. They are fully described by Sneath and Sokal (1973).

WPGMA

WPGMA, standing for "**weighted pair-group method using arithmetic averages**," requires taking a weighted average of the resemblance coefficients when revising the resemblance matrix. For example, suppose we are using a resemblance coefficient whose value is denoted by R_{jk}. Under the UPGMA clustering method, the value of this coefficient between, say, clusters (124) and (35), is

$$R_{(124)(35)} = (1/6)R_{13} + (1/6)R_{15} + (1/6)R_{23} + (1/6)R_{25} + (1/6)R_{34} + (1/6)R_{45}.$$

In computing the average, each term is given equal weight, namely 1/6. But under the WPGMA clustering method, the coefficient is

$$R_{(124)(35)} = w_{13}R_{13} + w_{15}R_{15} + w_{23}R_{23} + w_{25}R_{25} + w_{34}R_{34} + w_{45}R_{45},$$

where the weights, w_{jk}, are generally unequal.

The WPGMA method prescribes the values of the weights, w_{jk}, in a way that gives a larger weight to those objects admitted to their clusters at a more recent clustering step. This means that the earlier an object joins a cluster in the sequence of clustering steps, the less weight its resemblance with objects outside its cluster is given when the resemblance between clusters is evaluated.

Biological numerical taxonomists created WPGMA. Its logic was meant to allow the resulting tree of specimens to be interpreted according to their evolutionary lines of development. But WPGMA has drifted into disuse. Furthermore, it never was a logical method for making nonbiological classifications that lack an evolutionary basis.

Centroid Method

In this method, the resemblance between two clusters is equal to the resemblance between their centroids, where a cluster's centroid is its center of mass. By analogy, each object is like a star fixed in its attribute space. Each star carries the same unit mass, and the centroid of a cluster of stars is its center of mass. The resemblance between two clusters is measured by the intergalactic distance between their centroids. Most often e_{jk} is used for a resemblance coefficient with the centroid method, but certain other distance-type measures are acceptable.

While intuitively appealing, the centroid clustering method is not used much in practice, partly owing to its tendency to produce trees with reversals. Reversals occur when the values at which clusters merge do not increase from one clustering step to the next, but decrease instead. Thus, the tree can collapse onto itself and be difficult to interpret.

Flexible Method

Created by Lance and Williams (1967), the flexible clustering method subsumes all of the above-mentioned clustering methods—UPGMA, SLINK, CLINK, Ward's, WPGMA, and centroid—into a common mathematical formula. By choosing certain values of the formula's three parameters, the user can "dial in" to the method he wants. And since the parameters are continuous, they can be chosen to produce an infinite variety of clustering methods beyond those mentioned above.

The flexible clustering method is the basis for many computer programs for cluster analysis because it can be compactly programmed yet made to simulate the well-known methods. For example, Mather's (1976) computer program for hierarchical cluster analysis uses the flexible clustering method to simulate the other clustering methods. But as a clustering method in its own right, where the user specifies parameter values to create individual, tailor-made clustering methods, it is seldom used. Apparently potential users are stymied by not knowing how to choose the right values for the parameters from the infinite set of possibilities.

9.6 CHAINING

Chaining is a term that describes the following situation: at the first clustering step, two objects merge to form one cluster; and throughout the remaining clustering steps this cluster grows progressively larger through the annexation of lone objects that have not yet clustered. An analogy is a black hole in astronomy that grows by sucking in matter. Figure 9.7,*b* shows what a tree looks like when formed by chaining.

To see why chaining can occur, look at the data matrix and its corresponding attribute space plot shown in Figure 9.7,*a*. Note the progression of increasing distance from objects 1 to 2, from 2 to 3, from 3 to 4, and from 4 to 5. When we use the Euclidean distance coefficient e_{jk} and the SLINK clustering method on these data, the tree shown in Figure 9.7,*b* chains. First the nearest objects 1 and 2 merge. Then at the next clustering step object 3 is closest to (12), by virtue of SLINK's nearest neighbor criterion for computing the similarity between two clusters. So 3 merges with (12) to form (123). At the next clustering step, object 4 is closest to (123), so it merges to form (1234). Finally at the last step, object 5 merges with (1234).

On the other hand, Figure 9.7,*c* shows that when the same data matrix is cluster-analyzed by using e_{jk} and CLINK, chaining does not occur. Here is why. At the first clustering step objects 1 and 2 merge as before. But by virtue of CLINK's furthest neighbor criterion, at the next step object 3 is closest to object 4. So they merge to give (34). And so on. As a result,

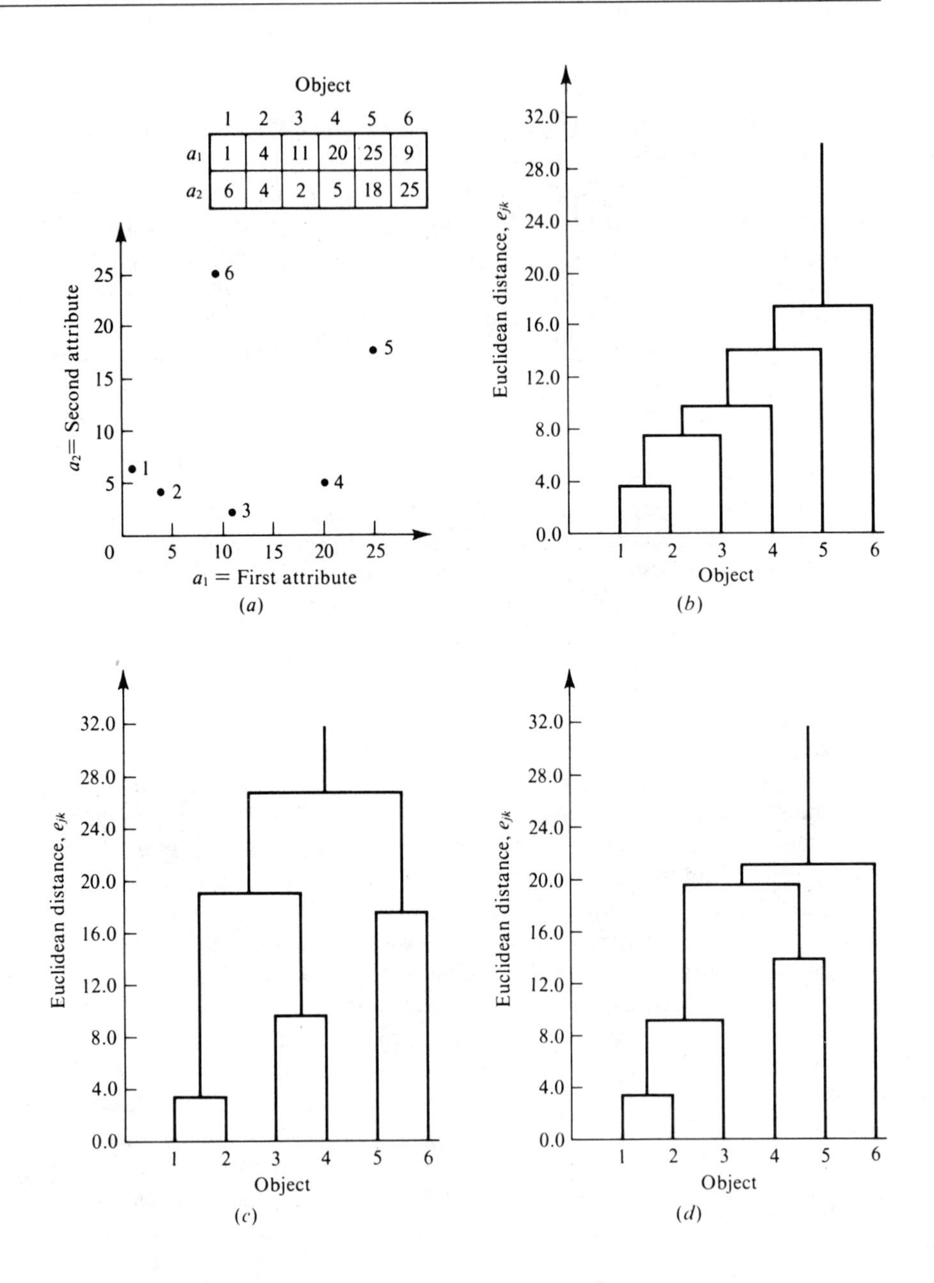

Figure 9.7. (a) Data matrix and attribute space plot. (b) Tree produced from this data matrix by using the SLINK clustering method; note that the tree illustrates chaining. (c) Tree produced by using the CLINK clustering method. (d) Tree produced by using the UPGMA clustering method.

chaining does not occur. Furthermore, Figure 9.7,d shows that UPGMA does not produce chaining either.

Whether or not chaining occurs depends on the data, the resemblance coefficient, and the clustering method. Because SLINK embodies the nearest neighbor philosophy, it is the most likely to cause chaining. And because CLINK is the antithesis of SLINK, CLINK is the least likely to cause chaining. Both UPGMA and Ward's method should produce results between these extremes. However, it is possible to contrive a data matrix that will produce chaining with each of these clustering methods.

When UPGMA is used as the main method, it is still often informative to perform a second cluster analysis using e_{jk} or d_{jk} for the resemblance coefficient and SLINK for the clustering method. If the objects are strung out in the attribute space in the manner shown in Figure 9.7,a, then the second cluster analysis should detect this quality of the data and signify it by chaining. And of course, chaining does not have to be unbroken to be informative about data. Partial chaining—chains that occupy separate branches of the tree—will also tell you something about how the objects are distributed in the attributed space.

SUMMARY

There are four main clustering methods. In applications, researchers use UPGMA the most often, and Ward's method the second most often, followed by SLINK and CLINK. This book recommends that the UPGMA clustering method be used because (1) it can be used with any resemblance coefficient whereas Ward's method cannot, and because (2) it judges the similarity between pairs of clusters in a manner less extreme than either SLINK or CLINK. However, this recommendation is not categorical, and problems are bound to arise for which one of the other methods is better suited.

The calculations required for SLINK, CLINK, and UPGMA depend upon whether a dissimilarity coefficient or a similarity coefficient is used. Moreover, if the data are such that they promote chaining, then use of e_{jk} or d_{jk} as a resemblance coefficient, and SLINK as a clustering method, is most likely to verify the chaining in a tree.

EXERCISE

(Answers to exercises are given in the back of the book.)

9.1. Imagine that Sue, Saint, Marie, and Mich—a family of quadruplets—take their college entrance tests, and their scores on the verbal, mathematical, and logic tests are as follows:

	Quadruplet			
Test	Sue	Saint	Marie	Mich
Verbal	490	320	480	440
Mathematics	570	500	520	560
Logic	320	410	450	440

Using the correlation coefficient r_{jk} (equation 8.5) for a resemblance coefficient, and using the unstandardized data matrix, cluster-analyze the quadruplets with the SLINK and CLINK clustering methods. Draw and compare the trees.

Because r_{jk} is a similarity coefficient, follow the procedures given in Section 9.3.

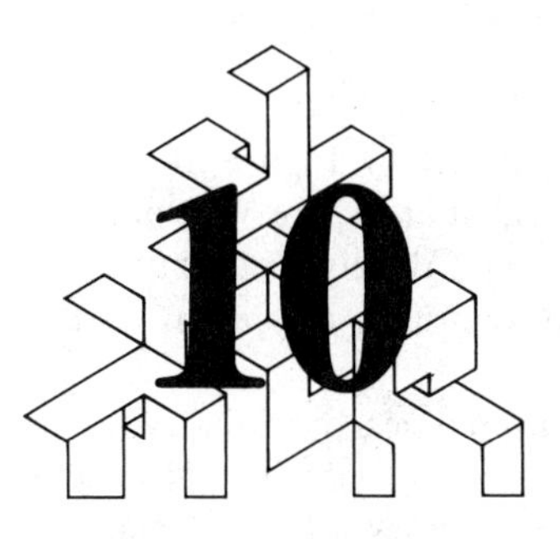

Resemblance Coefficients for Qualitative Attributes

OBJECTIVES

Qualitative attributes are those measured on nominal scales.

- The simplest are **binary attributes**. They have two states—for examples, presence/absence, yes/no, male/female.
- The more complex are **multistate attributes**. They have more than two states—for example, red/white/blue.

This chapter

- Presents resemblance coefficients for qualitative attributes.
- Discusses their properties.
- Illustrates their application.

10.1 QUALITATIVE RESEMBLANCE COEFFICIENTS

We will use the section of the data matrix shown in Figure 10.1,*a* to illustrate how each resemblance coefficient is calculated. We will pretend that the two objects j and k are medical patients, and the eight attributes are symptoms of diseases. The code "1" means a symptom is present in the patient and "0" means it is absent.

The two-by-two table in Figure 10.1,*b* summarizes the four types of matches among these data. There are $a = 2$ cases of 1–1 matches (attributes for which both objects are coded "1"); $b = 1$ cases of 1–0 matches (attributes for which object j is coded "1" and k is coded "0"); $c = 2$ cases of 0–1 matches (for which j is coded "0" and k is coded "1"); and $d = 3$ cases of 0–0 matches (for which both objects are coded "0").

We will use the symbol C_{jk} to stand for any qualitative resemblance coefficient that measures the resemblance between objects j and k. To compute C_{jk} we compare the two columns of data for the given objects, j and k, computing the values of a, b, c, and d. Each coefficient will be a different function of these values.

One simplification that comes with using qualitative attributes is that the data matrix never needs to be standardized. The resemblance matrix is computed directly from the data matrix.

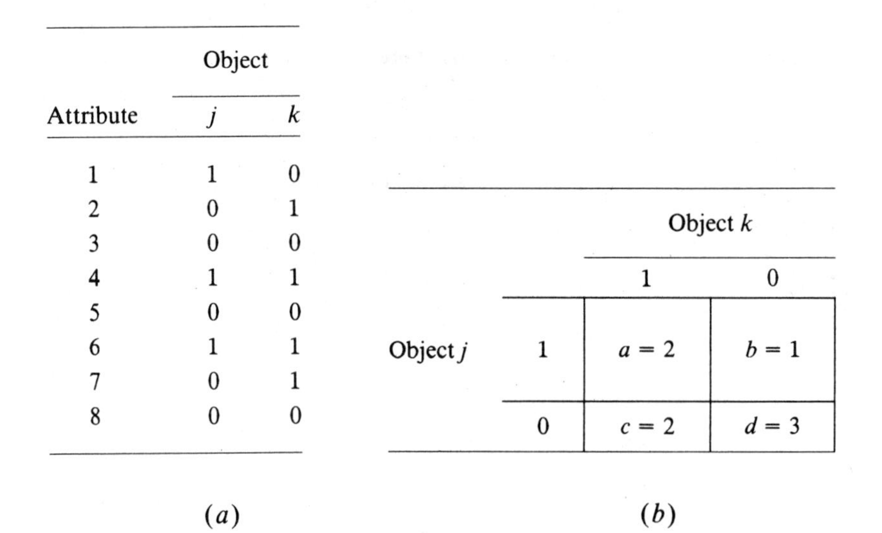

Attribute	Object j	Object k
1	1	0
2	0	1
3	0	0
4	1	1
5	0	0
6	1	1
7	0	1
8	0	0

(*a*)

		Object k: 1	Object k: 0
Object j	1	$a = 2$	$b = 1$
	0	$c = 2$	$d = 3$

(*b*)

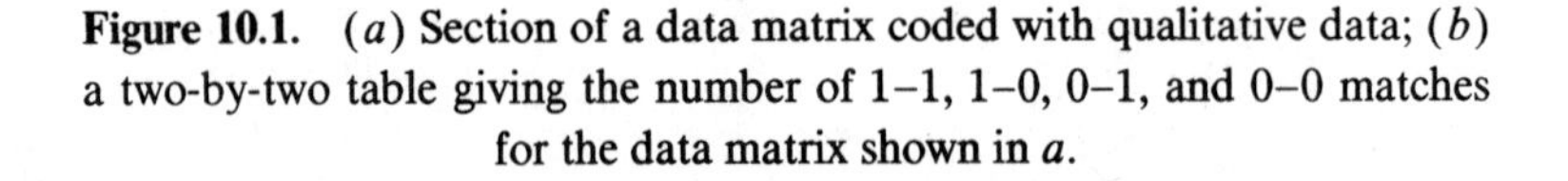

Figure 10.1. (*a*) Section of a data matrix coded with qualitative data; (*b*) a two-by-two table giving the number of 1–1, 1–0, 0–1, and 0–0 matches for the data matrix shown in *a*.

As we discuss the various resemblance coefficients in this chapter, we will use Table 10.1 (page 150) to summarize their equations and properties.

Jaccard Coefficient

This similarity coefficient is written

$$C_{jk} = \frac{a}{a + b + c} \qquad 0.0 \leq C_{jk} \leq 1.0. \tag{10.1}$$

Using the example data of Figure 10.1,*b*,

$$C_{jk} = \frac{2}{2 + 1 + 2} = 0.4.$$

The Jaccard coefficient indicates maximum similarity when the two objects have identical values, in which case $b = c = 0$ and $C_{jk} = 1.0$. It indicates maximum dissimilarity when there are no 1–1 matches, in which case $a = 0$ and $C_{jk} = 0.0$.

The Jaccard coefficient is sensitive to the direction of coding. That is, interchanging what "1" and "0" stand for usually changes the resemblance between objects.

The Jaccard coefficient has been widely used in numerical taxonomy where the objects are specimens, the attributes are their features, and "1" indicates the presence of a feature and "0" indicates its absence. It is not a function of *d*—that is, it ignores 0–0 matches. Leaving *d* out of the formulation has been supported by numerical taxonomists who argue that the joint lack of features between specimens should not be allowed to contribute to their similarity. For example, humans lack tails, fins, fur, and scales. Jointly they share the lack of a countless number of features such as these. Counting what humans lack as a basis for their similarity makes all people nearly perfectly similar to each other. Later in this chapter the merits of this argument are discussed.

The Jaccard coefficient has also been used in ecological cluster analysis where the objects are samples of land or water, the attributes are different species of organisms, and "1" and "0" code the presence and absence of species.

The following ecological cluster analyses are notable for their use of the Jaccard coefficient: Kaesler et al. (1977), samples of ocean water described by the presence and absence of benthic myodocopid Ostracoda; Cairns et al. (1974), samples of stream water based on the presence and absence of species of protozoa; Scott et al. (1977), samples of estuary water based on species of Foraminifera; Crossman et al. (1974), samples of river water near a power plant, described by the presence and absence of insect fauna, to assess the environmental impact of the power plant; and Williams and Schreiber (1976), species of weeds in the green foxtail complex, measured for the presence and absence of assorted morphological features. (Scientifically, the last of these studies is interesting for other reasons, too. It (1) hypothetico-deductively tested a theory of green foxtail evolution (Pohl's), and (2) showed that weed species clustered so that their similarities coincided with their tolerances to certain herbicides.)

Simple Matching Coefficient

This similarity coefficient is written

$$C_{jk} = \frac{a + d}{a + b + c + d} \qquad 0.0 \leq C_{jk} \leq 1.0. \tag{10.2}$$

Using the example data of Figure 10.1, b,

$$C_{jk} = \frac{2 + 3}{2 + 1 + 2 + 3} = 0.625.$$

The equation includes d in the numerator and denominator, and it is insensitive to the direction of coding. It has an intuitive meaning—namely, it is the proportion of 1–1 and 0–0 agreements in the n comparisons of attributes. Its maximum value of $C_{jk} = 1.0$ indicates perfect similarity and occurs when $b = c = 0$.

An example in which it is meaningful to count 0–0 agreements as contributing to similarity is the following. Suppose that pharmaceutical drugs being evaluated in a clinical trial are the objects, and the responses of n patients treated with the drugs are the attributes, coded "1" if a drug has a positive effect and "0" if negative. Since both the positive and negative effects of drugs are important for most medical research goals, both 1–1 and 0–0 matches should be made to contribute to similarity.

Yule Coefficient

This similarity coefficient is written

$$C_{jk} = \frac{ad - bc}{ad + bc} \qquad -1.0 \leq C_{jk} \leq 1.0. \tag{10.3}$$

Using the data of the example,

$$C_{jk} = \frac{(2)(3) - (1)(2)}{(2)(3) + (1)(2)} = 0.5.$$

The maximum value of $C_{jk} = 1.0$ indicates perfect similarity and occurs when either $b = 0$ or $c = 0$. The minimum value of $C_{jk} = -1.0$ indicates perfect dissimilarity and occurs when either $a = 0$ or $d = 0$. Furthermore, when $ad = bc$, then $C_{jk} = 0.0$, a value midway between the extremes. The Yule coefficient has been used mostly in psychological research. Bishop et al. (1975) have related it to probability theory.

Hamann Coefficient

This similarity coefficient is written as

$$C_{jk} = \frac{(a+d)-(b+c)}{(a+d)+(b+c)} \qquad -1.0 \le C_{jk} \le 1.0. \qquad (10.4)$$

Using the data of the example,

$$C_{jk} = \frac{(2+3)-(1+2)}{(2+3)+(1+2)} = 0.25.$$

The maximum value of $C_{jk} = 1.0$ indicates perfect similarity, and occurs when $b = c = 0$. The minimum value of $C_{jk} = -1.0$ indicates perfect dissimilarity and occurs when $a = d = 0$. And when $a + d = b + c$, then $C_{jk} = 0.0$, a value midway between the extremes.

The Hamann and Yule coefficients are mathematically related. Both have the same variables in their numerators and denominators, but the Hamann coefficient relates the variables by addition whereas the Yule coefficient relates them by multiplication. Holley and Guilford (1964) independently invented the Hamann coefficient and called it the G index of agreement.

Sorenson Coefficient

This similarity coefficient is written as

$$C_{jk} = \frac{2a}{2a+b+c} \qquad 0.0 \le C_{jk} \le 1.0. \qquad (10.5)$$

Using the data of the example,

$$C_{jk} = \frac{(2)(2)}{(2)(2)+1+2} = 0.57.$$

It ignores the 0–0 matches, d, but gives twice the weight to the 1–1 matches, a. In essence it makes the weight of a equal to the combined weights of b and c. The maximum value of $C_{jk} = 1.0$ indicates perfect similarity and occurs when $b = c = 0$. The minimum value of $C_{jk} = 0.0$ occurs when $a = 0$.

The Sorenson coefficient is equivalent to the Bray-Curtis coefficient, equation 8.8, when the data X_{jk} and X_{ik} are restricted to values of 0 and 1. To show this, recall that the Bray-Curtis coefficient is written

$$b_{jk} = \frac{\sum_{i=1}^{n} |X_{ij} - X_{ik}|}{\sum_{i=1}^{n} (X_{ij} + X_{ik})}.$$

When X_{ij} and X_{ik} are 0 or 1, the numerator is

$$\sum_{i=1}^{n} |X_{ij} - X_{ik}| = b + c,$$

and the denominator is

$$\sum_{i=1}^{n} X_{ij} + \sum_{i=1}^{n} X_{ik} = (a + b) + (a + c).$$

Thus,

$$b_{jk} = \frac{b + c}{(a + b) + (a + c)} = \frac{b + c}{2a + b + c} = 1 - \frac{2a}{2a + b + c}.$$

Thus, b_{jk}, a dissimilarity coefficient, is the complement of the Sorenson coefficient, a similarity coefficient. Knowing this, where the "2" in the Sorenson coefficient comes from is less mysterious. The Sorenson coefficient is also called the Czekanowski coefficient and the Dice coefficient, a consequence of its independent reinvention.

Rogers and Tanimoto Coefficient

This similarity coefficient is written as

$$C_{jk} = \frac{a + d}{a + 2(b + c) + d} \qquad 0.0 \le C_{jk} \le 1.0. \qquad (10.6)$$

Using the data of the example,

$$C_{jk} = \frac{2 + 3}{2 + 2(1 + 2) + 3} = 0.45.$$

Perfect similarity occurs when $b = c = 0$, making $C_{jk} = 1.0$; perfect dissimilarity occurs when $a = d = 0$, making $C_{jk} = 0.0$. This coefficient is quite like the simple matching coefficient except that its denominator doubly weights the 0–1 and 1–0 mismatches.

Sokal and Sneath Coefficient

This similarity coefficient is written as

$$C_{jk} = \frac{2(a + d)}{2(a + d) + b + c} \qquad 0.0 \leq C_{jk} \leq 1.0. \qquad (10.7)$$

Using the data of the example,

$$C_{jk} = \frac{2(2 + 3)}{2(2 + 3) + 1 + 2} = 0.77.$$

Perfect similarity occurs when $b = c = 0$, making $C_{jk} = 1.0$; perfect dissimilarity occurs when $a = d = 0$, making $C_{jk} = 0.0$.

This coefficient too is quite like the simple matching coefficient. This can be seen by dividing its numerator and denominator by 2, giving $(a + d)/[a + \frac{1}{2}(b + c) + d]$. Except for its denominator, in which mismatches receive half-weight, its form is like the simple matching coefficient.

Russell and Rao Coefficient

This similarity coefficient is written as

$$C_{jk} = \frac{a}{a + b + c + d} \qquad 0.0 \leq C_{jk} \leq 1.0. \qquad (10.8)$$

Using the data of the example,

$$C_{jk} = \frac{2}{2 + 1 + 2 + 3} = 0.25.$$

It expresses the proportion of 1–1 matches in the n comparisons. Perfect similarity, making $C_{jk} = 1.0$, requires that $b = c = d = 0$, in which case the attributes for both objects are all 1's. Perfect dissimilarity, making $C_{jk} = 0.0$, requires that $a = 0$—that is, there are no 1–1 matches. Except for the d variable in its denominator, it resembles the Jaccard coefficient.

Baroni-Urbani and Buser Coefficient

This similarity coefficient is written

$$C_{jk} = \frac{a + (ad)^{1/2}}{a + b + c + (ad)^{1/2}} \qquad 0.0 \le C_{jk} \le 1.0. \tag{10.9}$$

Using the data of the example,

$$C_{jk} = \frac{2 + [(2)(3)]^{1/2}}{2 + 1 + 2 + [(2)(3)]^{1/2}} = 0.60.$$

The inventors of this coefficient, Baroni-Urbani and Buser (1976), felt that the properties of its distribution were superior. That is, they showed that when the data in the attribute-vectors of two objects were randomly permutated (when the 0's and 1's in each column were randomly shuffled and the coefficient's value was recomputed after each shuffling), the distributions of the coefficient's values were more normal (bell-shaped) and continuous than the distributions of other coefficients. It is yet to be shown, however, that this mathematical superiority categorically translates into real-world superiority, as measured by how well research goals are achieved.

Sokal Binary Distance Coefficient

This dissimilarity coefficient is written,

$$C_{jk} = \left(\frac{b + c}{a + b + c + d}\right)^{1/2} \qquad 0.0 \le C_{jk} \le 1.0. \tag{10.10}$$

Using the data of the example,

$$C_{jk} = \left(\frac{1 + 2}{2 + 1 + 2 + 3}\right)^{1/2} = 0.61.$$

The minimum value of $C_{jk} = 0.0$ indicates perfect similarity and occurs when $b = c = 0$. The maximum value of $C_{jk} = 1.0$ indicates perfect dissimilarity, and occurs when $a = d = 0$. When X_{ij} and X_{ik} are binary (have only 0 and 1 values), this coefficient is equivalent to the average Euclidean distance coefficient, d_{jk}, equation 8.2. To see this, note that the numerator of $d_{jk}, \Sigma_{i=1}^{n}(X_{ij} - X_{ik})^2$, equals $b + c$; and its denominator, n, equals $a + b + c + d$.

Ochiai Coefficient

This similarity coefficient is written

$$C_{jk} = \frac{a}{[(a + b)(a + c)]^{1/2}}, \qquad 0.0 \leq C_{jk} \leq 1.0. \qquad (10.11)$$

Using the data of the example,

$$C_{jk} = \frac{2}{[(2 + 1)(2 + 2)]^{1/2}} = 0.58.$$

The maximum value of $C_{jk} = 1.0$ indicates perfect similarity and occurs when $b = c = 0$. The minimum value of $C_{jk} = 0.0$ indicates perfect dissimilarity and occurs when $a = 0$. The Ochiai coefficient is identical to the cosine coefficient, c_{jk}, equation 8.4, when the data are values of 0 and 1.

Phi Coefficient

This similarity coefficient, often denoted by its Greek symbol, ϕ, is written

$$C_{jk} = \frac{ad - bc}{[(a + b)(a + c)(b + d)(c + d)]^{1/2}}, \qquad -1.0 \leq C_{jk} \leq 1.0. \qquad (10.12)$$

Using the data of the example,

$$C_{jk} = \frac{(2)(3) - (1)(2)}{[(2 + 1)(2 + 2)(1 + 3)(2 + 3)]^{1/2}} = 0.26.$$

The maximum value of $C_{jk} = 1.0$ indicates perfect similarity and occurs whenever b and c are both 0. The minimum value of $C_{jk} = -1.0$ indicates perfect dissimilarity and occurs whenever a and d are both 0.

The phi coefficient is the equivalent of the correlation coefficient, r_{jk}, equation 8.5, when the data are values of 0 and 1. It has been much used in psychological research. Under certain assumptions its distribution is related to the chi-square distribution. Lindeman et al. (1980, pp. 83–86) discuss the phi coefficient and its related family of coefficients. Anderberg (1973, pp. 84–85) derives the Ochiai coefficient and the phi coefficient from c_{jk} and r_{jk} respectively.

Table 10.1 Equations and properties of some commonly used qualitative resemblance coefficients. The symbol "*" denotes the end of the range that corresponds to maximum similarity.

Coefficient	Range	Sensitive to direction of coding?
1. Jaccard $C_{jk} = \dfrac{a}{a+b+c}$	0–1*	Yes
2. Simple matching $C_{jk} = \dfrac{a+d}{a+b+c+d}$	0–1*	No
3. Yule $C_{jk} = \dfrac{ad-bc}{ad+bc}$	−1–1*	No
4. Hamann $C_{jk} = \dfrac{(a+d)-(b+c)}{(a+d)+(b+c)}$	−1–1*	No
5. Sorenson $C_{jk} = \dfrac{2a}{2a+b+c}$	0–1*	Yes
6. Rogers and Tanimoto $C_{jk} = \dfrac{a+d}{a+2(b+c)+d}$	0–1*	No
7. Sokal and Sneath $C_{jk} = \dfrac{2(a+d)}{2(a+d)+b+c}$	0–1*	No
8. Russell and Rao $C_{jk} = \dfrac{a}{a+b+c+d}$	0–1*	Yes
9. Baroni-Urbani and Buser $C_{jk} = \dfrac{a+(ad)^{1/2}}{a+b+c+(ad)^{1/2}}$	0–1*	No
10. Sokal binary distance $C_{jk} = \left(\dfrac{b+c}{a+b+c+d}\right)^{1/2}$	0*–1	No
11. Ochiai $C_{jk} = \dfrac{a}{[(a+b)(a+c)]^{1/2}}$	0–1*	Yes
12. Phi $C_{jk} = \dfrac{ad-bc}{[(a+b)(a+c)(b+d)(c+d)]^{1/2}}$	−1–1*	No

This ends our description of qualitative resemblance coefficients; these coefficients are compared in Table 10.1.

A few points need to be clarified and expanded, however.

Point 1. The Jaccard, simple matching, and Sorenson coefficients are much used. Perhaps this follows from their clear conceptual bases: all are meaningful proportions. The Jaccard coefficient is the proportion of 1–1 matches in a set of comparisons that ignores 0–0 matches. The simple matching coefficient is the proportion of agreements, either 1–1 or 0–0, in a set of n comparisons. The Sorenson coefficient, although less straightforward, is the analogue of the Jaccard coefficient when the number of 1–1 matches, a, is given twice its normal weight.

Point 2. As we would expect, the sharing of common variables, a, b, c, d, tends to correlate the coefficients. As an illustration, suppose the data matrix is such that there are no 0–0 matches ($d = 0$ for any pair of objects). Then clearly the Jaccard and simple matching coefficients have the same values, and a cluster analysis based on either one will give the same tree. And while $d = 0$ is improbable in a data matrix with a large n, often d is small relative to a, b, and c, making the two coefficients nearly alike.

Point 3. Because of their mathematical properties, in some cases the coefficients are perfectly correlated irrespective of the data used with them. Let us look at an example that shows the intrinsic correlation between the Hamann coefficient and the simple matching coefficient. The Hamann coefficient can be factored as

$$C_{jk} = \underbrace{\frac{a + d}{a + b + c + d}}_{\text{S.M.}} - \frac{b + c}{a + b + c + d}.$$

Note that the left-hand term is the simple matching coefficient (S.M.), and note also the identity,

$$\frac{a + d}{a + b + c + d} + \frac{b + c}{a + b + c + d} = 1.0.$$

Solving this latter equation for $(b + c)/(a + b + c + d)$ and substituting it into the former gives

$$C_{jk} = 2\frac{a + d}{a + b + c + d} - 1 = 2(\text{S.M.}) - 1.$$

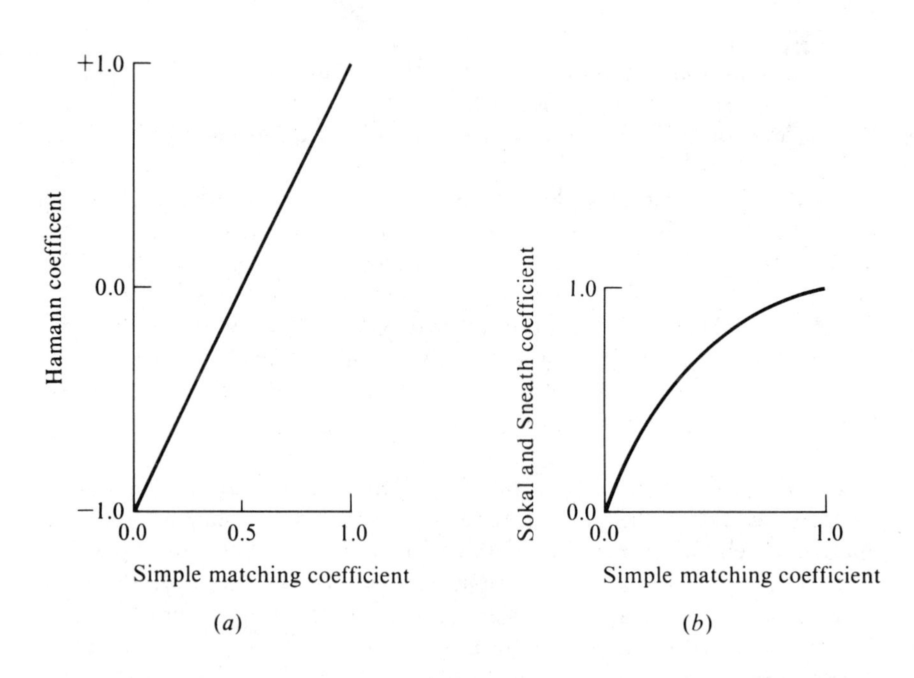

Figure 10.2. (*a*) The linear relation between the Hamann coefficient and the simple matching coefficient. (*b*) Idealized monotonic relation between the Sneath and Sokal coefficient and the simple matching coefficient.

Thus, the Hamann coefficient is twice the simple matching coefficient, minus 1.0. One is a linear combination of the other. The graph in Figure 10.2,*a* shows this clearly.

The Hamann coefficient merely stretches the simple matching coefficient out over the range -1.0 to $+1.0$ and centers it on 0.0. A given data matrix, when used with either coefficient, will produce trees with identical shapes except that they will be proportionally scaled.

Another case in which two or more resemblance coefficients are highly correlated is that in which they are monotonically related. A monotonic relation between two equations is one in which the two equations are related by an inequality. For instance, it is known that the Sokal and Sneath coefficient (S.S.) is always greater than or equal to the simple matching coefficient. That is, S.S. $\geq$ S.M. Idealized, their monotonic relation could look like that shown in Figure 10.2,*b*.

The SLINK and CLINK clustering methods (Chapter 9) carry monotonic relations through to the tree. For example, suppose we have a data matrix and use it to calculate two resemblance matrices by applying two resemblance coefficients that are monotonically related. We then use SLINK (or CLINK) to calculate the tree for each resemblance matrix. The two trees will be identical in the order in which objects cluster, although the relative levels of clustering will usually be different.

With many coefficients in use and with new ones being invented, it is difficult to check them pairwise for possible monotonic relatedness. But it is generally known that the Jaccard, Sorenson, and Baroni-Urbani and Buser coefficients are pairwise monotonically related, as are the simple matching, Sokal and Sneath, Roger and Tanimoto, and Hamann coefficients.

It is important for you to know which coefficients are monotonically related to each other because you might try several coefficients in alternative analyses of the same data matrix, get essentially the same trees, and think that this has happened because of the data rather than because of the intrinsic relations between the resemblance coefficients.

Point 4. Resemblance coefficients are most commonly expressed as functions of the variables a, b, c, d. There is, however, a second way of expressing them that uses different symbols to achieve the same results. For example, the Jaccard coefficient in this second system is written as

$$C_{jk} = \frac{C}{N_1 + N_2 - C},$$

where C is the number of 1–1 matches, N_1 is the total number of 1's in the vector of attributes for object j, and N_2 is the same for the vector of object k. To illustrate the calculation using the data given in Figure 10.1,a, we see that $C = 2$, $N_1 = 1 + 0 + 0 + 1 + 0 + 1 + 0 + 0 = 3$, and $N_2 = 0 + 1 + 0 + 1 + 0 + 1 + 1 + 0 = 4$. This makes $C_{jk} = 2/(3 + 4 - 2) = 0.4$, the same as we calculated earlier. Cheetham and Hazel (1969) and Simpson (1960) restate many of the resemblance coefficients discussed in this chapter in terms of this second system. The first system has the advantage of conceptual understanding, but the second is computationally easier.

Point 5. These and many more resemblance coefficients for qualitative attributes are described in the following references: Sneath and Sokal (1973, pp. 129–134), Anderberg (1973, chaps. 4, 5), Clifford and Stephenson (1975, chap. 6), Doran and Hodson (1975, chap. 6), Boesch (1977, section 5), Lorr (1976, Table 1), and Legendre and Legendre (1983, chap. 6). In a comprehensive review, Cheetham and Hazel (1969) list more than twenty resemblance coefficients for qualitative data and describe their properties, and Baroni-Urbani and Buser (1976) describe their coefficient as well as several others.

Point 6. While the choice of a resemblance coefficient should be as far as possible guided by logic, there is a tendency among researchers within fields to make habitual use of certain coefficients that others in their field are using, even when logical reasons are lacking. This may partly explain why geologists favor using the cosine coefficient and ecologists favor using the Bray-Curtis coefficient, and may explain why certain qualitative resemblance coefficients find their uses too. Some researchers choose to use a certain resemblance coefficient because others are using it, or because it is readily available as an option in a computer program that they have access to.

10.2 THE ROLE OF 0–0 MATCHES

Some of the resemblance coefficients include *d*, the number of 0–0 matches, in their numerators whereas others do not. Let us discuss the significance of this.

First, more often than not researchers use coefficients that include *d*. This preference is reflected in the twelve most-used resemblance coefficients shown in Table 10.1. Three of the coefficients—Jaccard, Sorenson, and Ochiai—do not include *d* but the other nine do.

Second, the resemblance coefficients that leave *d* out tend to be correlated with those that include it. Thus, cluster analyses using either type often give similar trees. For this reason the decision of whether to include *d* is often not critical.

Third, logic often makes it clear that the number of 0–0 matches should contribute to similarity. A simple example is two people who express their likes and dislikes as yes/no answers to a series of items. Most of us would say that what they dislike, as well as what they like, should count toward their similarity.

Earlier, when we discussed the Jaccard coefficient, we said that it did not include *d* because of the argument that there could be countless items that two objects jointly lacked, and it would be silly to say that this joint lack should contribute to their similarity. A way to circumvent this difficulty is as follows.

To begin, attributes that do not vary at all across a set of objects—those that are all 1's or 0's—cannot help to discriminate between classes of objects. Thus, an ecologist who bases a cluster analysis of streams on the presence/absence of aquatic insects living in them should not use as attributes a list of all known aquatic insects. Most of the insects in this list would not occur in any of the streams and their inclusion could not help to discriminate between the streams. A better way would be to use only insect species that appear in at least one of the streams being clustered. Then it would be reasonable to use a coefficient that allowed *d* to contribute to similarity. For instance, two of the streams that jointly lacked a certain

Congressman

Bill	Hunt (R) 1	Sandman (R) 2	Howard (D) 3	Thompson (D) 4	Frelinghuysen (R) 5	Forsythe (R) 6	Widnall (R) 7	Roe (D) 8	Helstoski (D) 9	Rodino (D) 10	Minish (D) 11	Rinaldo (R) 12	Maraziti (R) 13	Daniels (D) 14	Patten (D) 15
1 (Y)	0	0	0	1	1	1	1	0	1	1	1	1	0	1	N
2 (N)	1	0	N	0	0	0	1	0	0	0	0	0	1	0	0
3 (N)	1	N	0	N	1	1	N	1	1	1	1	1	1	1	1
4 (Y)	0	0	1	N	0	0	0	1	1	1	1	1	0	0	0
5 (Y)	0	N	1	N	0	0	0	1	1	1	1	1	1	1	0
6 (Y)	1	N	1	N	0	1	1	0	0	0	0	0	1	0	0
7 (Y)	0	0	1	1	1	1	1	1	N	1	1	1	1	1	1
8 (N)	1	1	0	0	1	0	0	N	0	0	0	0	1	0	0
9 (Y)	0	0	1	1	0	0	0	N	1	1	1	1	1	N	1
10 (Y)	0	0	1	1	0	0	0	1	1	1	1	1	0	0	1
11 (Y)	0	0	1	1	1	0	0	1	1	1	1	1	0	0	1
12 (Y)	0	N	1	N	0	0	0	1	1	1	1	1	0	1	1
13 (N)	1	N	0	1	N	0	1	0	0	0	0	0	0	0	1
14 (N)	0	N	0	0	0	0	1	0	0	0	0	0	0	0	0
15 (N)	1	N	0	0	1	0	1	0	0	0	0	0	1	0	0
16 (N)	N	N	0	N	1	1	1	0	0	N	0	0	1	N	0
17 (N)	N	N	0	0	1	0	1	0	0	0	0	0	0	0	0
18 (Y)	1	1	1	1	N	0	1	1	1	1	1	1	0	1	1
19 (Y)	0	0	1	1	N	0	0	1	1	0	0	1	0	0	1

(*a*)

Figure 10.3. (*a*) Qualitative data matrix of U.S. Congressmen described by their voting on 19 environmental bills. (*b*) Tree obtained by cluster-analyzing the data matrix; the simple matching coefficient and the UPGMA clustering method were used. (*c*) Tree obtained by cluster-analyzing the data matrix; the Jaccard coefficient and the UPGMA clustering method were used.

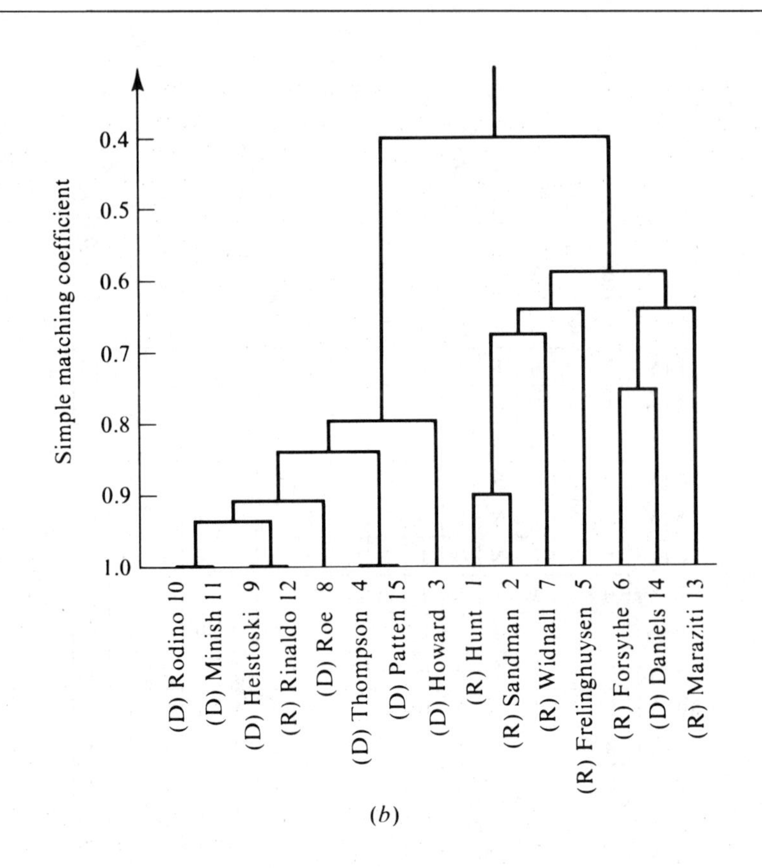

(*b*)

Figure 10.3. (*Continued*).

species might lack it for the same reason—for example, that their temperatures are too high.

The following Research Application illustrates some of the concepts we've discussed so far. It is an example of Research Goal 3 ("Test a hypothesis").

RESEARCH APPLICATION 10.1

Generally it is believed that Republicans and Democrats in the U.S. House of Representatives tend to vote along party lines. To test this hypothesis, we will use the data matrix shown in Figure 10.3,*a*. Its objects are 15 congressmen from New Jersey; its attributes, 19 environmental bills they voted on in 1973. A code of "1" signifies a "yes" vote; "0" is a "no" vote; and "N" indicates a "no comparison" owing to an abstention or absence. The bills are also labeled "Yes" (Y) or "No" (N) according to how the League of Conservation Voters recommended that the congressmen cast their

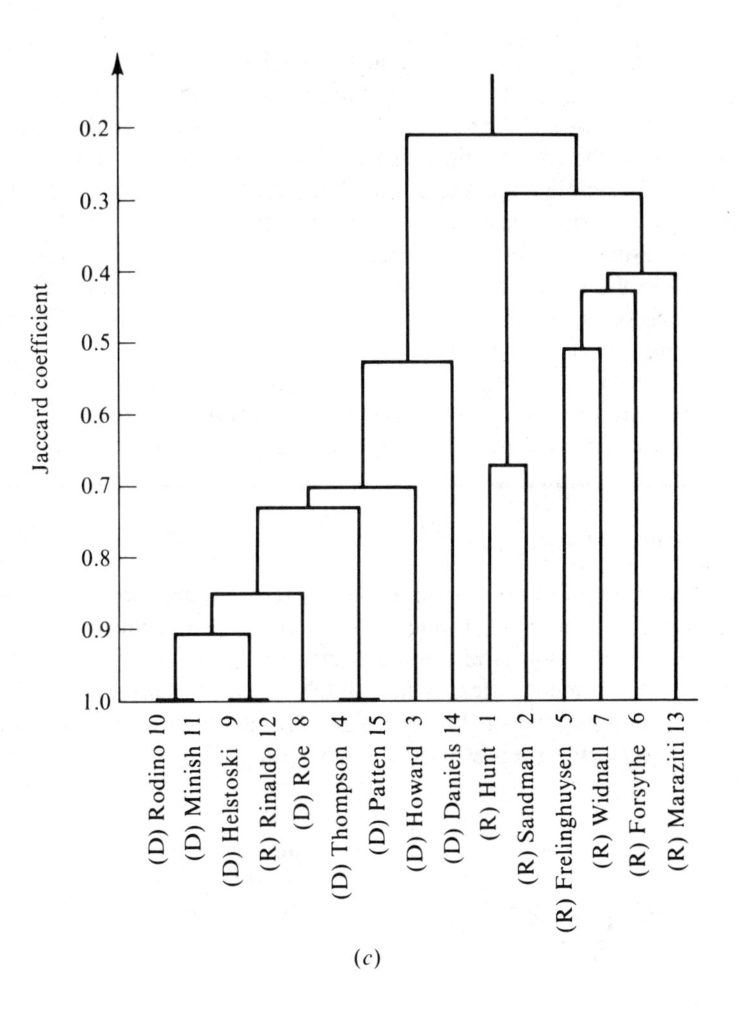

Figure 10.3. (*Continued*).

votes—that is, how a group with special knowledgeability of the natural environment felt the issues should be decided.

Of importance is whether we should use a resemblance coefficient that includes d, the number of 0–0 matches, in its formulation. On reflection it seems reasonable to include d: if two congressmen are against a bill for essentially the same reasons, this should contribute to the similarity of their voting patterns. Therefore, let us try the simple matching coefficient in a cluster analysis. This gives the tree shown in Figure 10.3,*b*. The congressmen separate according to their parties. The exceptions are Daniels (D), who clusters with the Republicans, and Rinaldo (R), who clusters with the Democrats.

Figure 10.3,*c* shows that the Jaccard coefficient gives basically the same results. Here, Rinaldo (R) is the only congressman who goes contrary to the hypothesis. Even though the Jaccard coefficient gives results that are more consistent with the hypothetical expectation that votes will be cast along party lines, the simple matching coefficient is the better choice on the logical grounds that d should be included. That the two analyses mostly agree is due to the natural correlation between the two coefficients, *using these data.*

The assumption that congressmen vote "no" for the same reasons is unproved, however. A serious researcher would want to look into this before deciding one way or the other.

An analysis like this might be useful to lobbyists. Daniels and Howard, for example, are the democrats nearest the borderline and perhaps the most easily influenced to vote with the Republicans.

10.3 MULTISTATE ATTRIBUTES

We can use dummy binary attributes to represent more than two states of an attribute. For example, Blong (1973) used cluster analysis to make a classification of hillside landslides. He measured one of the attributes by using three unordered states: a landslide's mode of failure could be (1) simple, (2) compound, or (3) complex. He used three dummy binary attributes a_1, a_2, a_3 to represent the three states, using this coding scheme:

Dummy attribute	Mode of failure		
	Simple	Compound	Complex
a_1	1	0	0
a_2	0	1	0
a_3	0	0	1

As an illustration, suppose we have a data matrix with four landslides. The mode of failure might be coded this way with three dummy attributes:

	Landslide			
Attribute	1	2	3	4
a_1	0	1	0	0
a_2	1	0	0	1
a_3	0	0	1	0

The first and fourth landslides are of the compound type, the second is simple, and the third is complex. For the mode-of-failure attribute, the value of the Jaccard coefficient between landslides 1 and 2 is $C_{12} = 0/(0 + 1 + 1) = 0.0$; that is, 1 and 2 are maximally dissimilar. Between 1 and 4 it is $C_{14} = 1/(1 + 0 + 0) = 1.0$; 1 and 4 are maximally similar. This accords with intuition.

But in general, coefficients that utilize d, the number of 0–0 matches, are unsatisfactory for multistate attributes. As a demonstration of this, observe that the simple matching coefficient between 1 and 2 is $C_{12} = (0 + 1)/(0 + 1 + 1 + 1) = 0.33$. That landslides 1 and 2 should have one-third of their features in common is illogical.

This method of coding will accommodate any number of states. For example, with five states there must be five dummy binary attributes. Moreover, Sneath and Sokal (1973, pp. 150–152) have adapted the coding scheme to represent multistate ordinal data.

10.4 THE INFORMATION IN QUALITATIVE ATTRIBUTES

Any quantitative attribute can be dichotomized and converted into a binary or multistate attribute. We can, for instance, measure the height of people in feet and then collapse the set of measurements into binary classes: those less than six feet tall, and those greater than six feet. Anderberg (1973, chap. 3) gives various strategies for collapsing quantitative attributes into qualitative ones (the strategies use cluster analysis itself to do the collapsing).

There is the general feeling that collapsing loses information in the data because the process is irreversible—that we can collapse data measured on a higher scale to a lower one but cannot reconstruct a higher scale from a lower one. According to this view, ratio scales are more informative than interval scales, interval scales are more informative than ordinal scales, and ordinal scales are more informative than nominal scales. For several reasons this view is not always right.

First, it neglects the constraints of research budgets. In general, it is cheaper to measure a qualitative attribute than a quantitative one. It may, therefore, be better to measure many qualitative attributes than a few quantitative ones. As an illustration, suppose a researcher's budget has a fixed number of dollars. Each value in the data matrix costs, say, $4.00 to measure on a ratio scale and $.50 to measure on a binary scale. In this case the researcher could collect eight times as much qualitative data, and would have extra money for measuring more things; this may more than compensate for the fact that each measurement is not as informative as it might be.

Researchers face a choice between an intensively detailed picture of the world painted with ratio-scaled data and a less detailed picture painted with binary data. In some cases the tradeoff between scope and vividness favors the use of qualitative attributes.

The second reason why quantitative attributes are not categorically better is that information is not solely a property of data. Information is very much a property of the researcher's interpretation of data. As we will argue in Chapter 20, data do not automatically inform the researcher. To have meaning, data must be interpreted in context. For example, a visitor from another world who should happen to look at a red traffic light would see data, not information. It would not inform the visitor to "stop" because the visitor would not know how to react.

Hazel (1970) has demonstrated the value of qualitative attributes in biostratigraphical research. Looking at cluster analyses of data matrices based on quantitative attributes, and also at their counterparts based on qualitative attributes, he concluded that for the common biostratigraphical research goals, qualitative data can serve well. While this is not a universal proof applicable to all research goals, it does make us mindful that qualitative data are not categorically inferior. And as a matter of course many attributes can only be measured on qualitative scales—for example, a patient's health.

Let us conclude this chapter with a Research Application. An example of Research Goal 6 ("Facilitate planning and management"), it illustrates a problem in industrial engineering that requires qualitative attributes and that 0–0 matches be included in the measurement of similarity.

RESEARCH APPLICATION 10.2

Suppose a factory produces t component parts each requiring manufacture by one or another of n machines. A problem is to avoid unnecessary routing of the components from machine to machine. How can the machines be arranged efficiently in spatial clusters so that a component can circulate through the machines in one cluster before being shipped on to the next?

To begin, we make a data matrix having components as objects and machines as attributes. A code of "1" means a given component must pass through a given machine; "0" means it need not. Figure 10.4,*a* shows a sample data matrix taken from King (1980) having $t = 6$ components and $n = 5$ machines. We perform two cluster analyses using the simple matching coefficient—a Q-analysis of components and an R-analysis of machines. Figure 10.4,*b* and *c* show the respective trees. Then we simultaneously rearrange the columns of the data matrix according to the Q-analysis and the rows according to the R-analysis. This gives the efficient groupings of components and machines shown in the doubly-rearranged data matrix, Figure 10.4,*d*. Even for small data matrices like this one the correct groupings are virtually impossible to see in the original data matrix.

Clearly, *d* should be included in both the Q-analysis and the R-analysis because the similarity between components should be based on both the

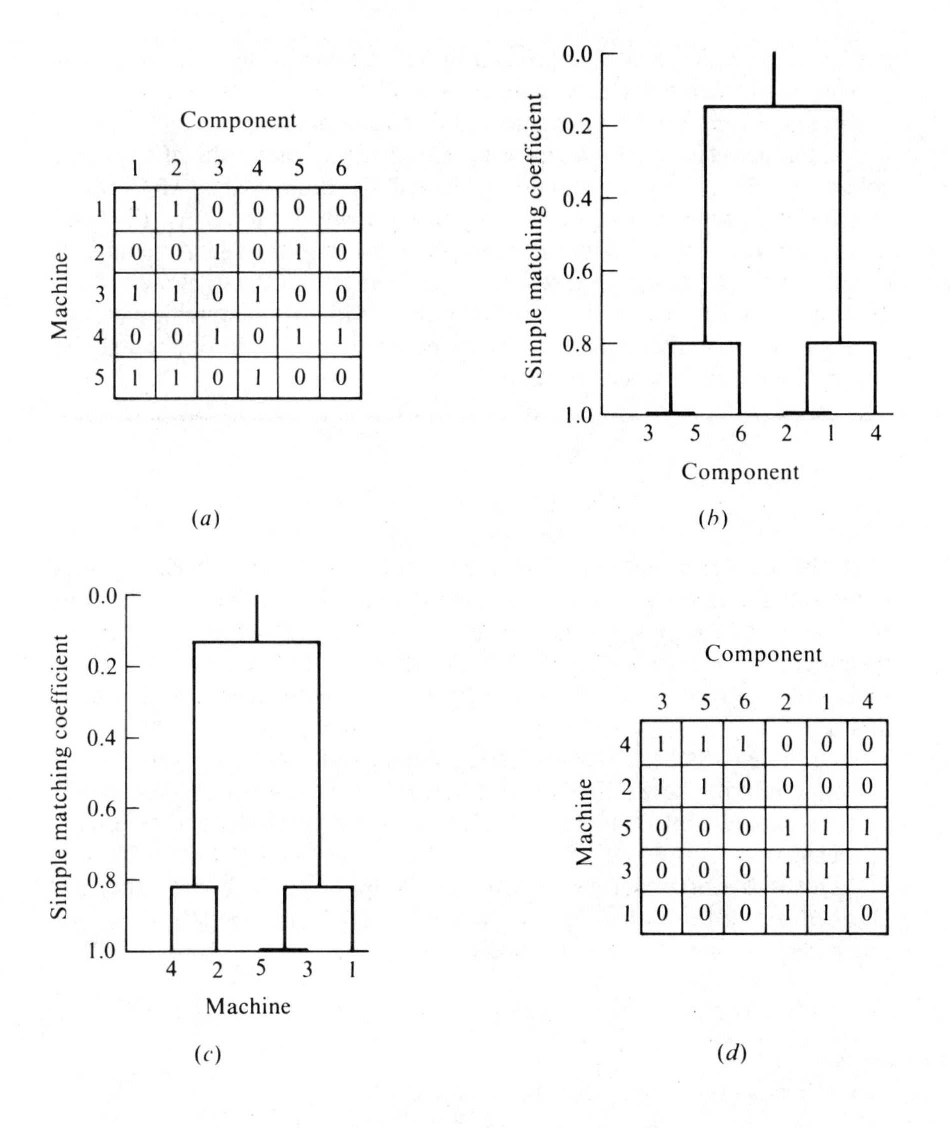

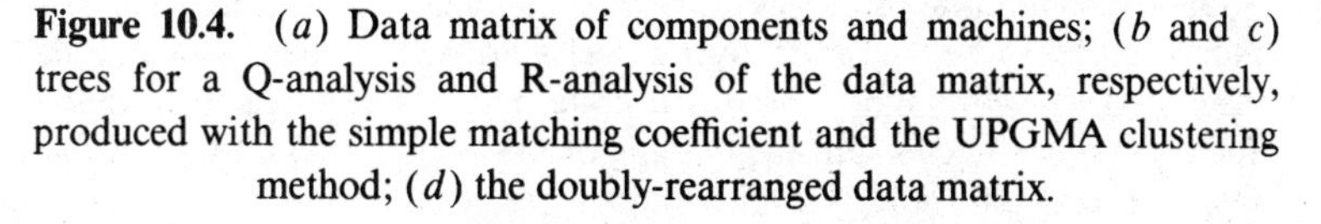
Figure 10.4. (*a*) Data matrix of components and machines; (*b* and *c*) trees for a Q-analysis and R-analysis of the data matrix, respectively, produced with the simple matching coefficient and the UPGMA clustering method; (*d*) the doubly-rearranged data matrix.

machines the components require and those they don't. The same argument applies for the similarity between machines.

King (1980) proposed a new method of cluster analysis, which he called the **rank order cluster algorithm**, for the simultaneous clustering of rows and columns of a data matrix containing 0's and 1's. Prior to this, McCormick et al. (1972) proposed a closely related method, which they called the **bond energy algorithm**. One of the BMDP programs described in Appendix 2, called the **block clustering method**, simultaneously clusters rows and columns and rearranges them into blocks of data. Although hierarchical cluster analysis can solve this type of problem, better solutions can doubtlessly be obtained with these special methods.

SUMMARY

Resemblance coefficients for qualitative attributes are used when all attributes in the data matrix are measured on nominal scales. The attributes can be binary (with two states), such as presence/absence, yes/no, or male/female. Or they can be multistate (with more than two states), such as blue/yellow/green. Both kinds of attributes are coded with values of 1 and 0.

Of these coefficients, some make the number of 0–0 matches contribute to the similarity between objects, whereas others do not. Whether 0–0 matches should contribute depends on the research goal and context of the problem.

Qualitative attributes are widely used in applications. Usually they are cheaper to measure than quantitative attributes. And certain kinds of attributes can only be measured qualitatively.

EXERCISES

(Answers to exercises are given in the back of the book.)

10.1. Suppose the admiral of the Swiss navy is planning to issue a Swiss navy knife. Not wanting to be surpassed by the Swiss army, he orders a unique design. The designer shows him a proposed knife, named the Yoeman Yodeler, and for comparison shows him the designs of five Swiss army knives. In the data matrix shown in Table 10.2, the first five designs are the Swiss army knives and the sixth is the Yoeman Yodeler. The code "1" means the knife possesses the given attribute; "0" means it does not.

Use the simple matching coefficient and the UPGMA clustering method to obtain the tree of the six knives. Does the Yoeman Yodeler stand apart as unique? Why is the simple matching coefficient a better choice than the Jaccard coefficient?

Table 10.2 Data matrix of Swiss Navy knives.

	Knife					
Attribute	Swiss Miss (1)	Wm. Tell (2)	Sherpa Helper (3)	Sunday Morning Special (4)	G.I. Josef (5)	Yoeman Yodeler (6)
Can opener	1	1	0	0	0	1
Reamer	0	1	1	1	1	0
Corkscrew	0	0	1	1	1	1
Sextant	0	0	0	0	0	1
Scissors	1	1	0	1	0	0
Small screwdriver	0	0	0	1	1	1
Large screwdriver	0	0	1	1	1	0
Magnifying glass	0	0	0	0	0	1
Tweezers	0	1	1	1	1	0
Toothpick	1	1	1	1	1	0
Wood saw	0	0	0	1	1	0
Compass	0	0	0	0	1	1

10.2. Show mathematically that the Jaccard coefficient and Sorenson coefficient are monotonically related.

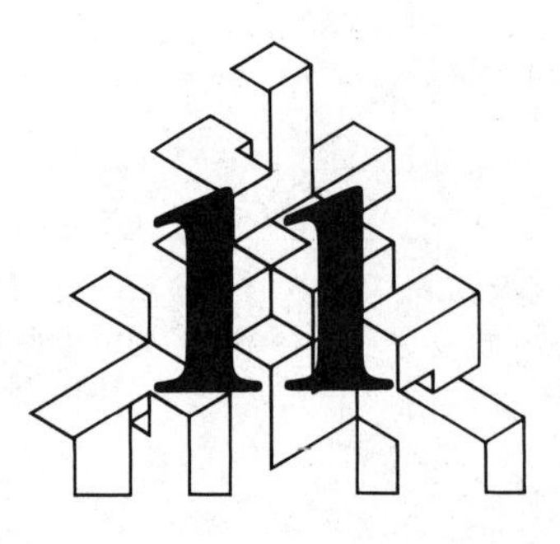

Special Resemblance Coefficients for Ordinal-Scaled Attributes

OBJECTIVES

- Any of the resemblance coefficients covered in Chapter 8 can be used with ordinal-scaled attributes.
- There are also resemblance coefficients specially suited for this purpose.
- They are rank correlation coefficients, which cluster analysis has borrowed from statistics.
- The one we discuss in this chapter is called **Kendall's tau coefficient**.

11.1 KENDALL'S TAU COEFFICIENT, τ_{jk}

Kendall's tau coefficient, τ_{jk}, is a similarity coefficient defined on the range $-1.0 \leq \tau_{jk} \leq 1.0$. It measures the correlation between two sets of rankings, and its equation is written as

$$\tau_{jk} = \frac{S_{jk}}{(0.5)(P)(P-1)} \tag{11.1}$$

We shall use the data matrix shown in Table 11.1 to help explain its calculation. The eight objects of the data matrix are regions of Canada, and the 13 attributes are cultural values that people living in the regions consider important. In all, 2,000 people gave their rankings for their regions, and Table 11.1 is the consensus (data are from Financial Times of Canada, 1977).

Using these data, we can compute a resemblance matrix in which τ_{jk} measures the similarity of each pair of regions. To illustrate, we will do this for British Columbia and Alberta. As shown below, we first write the names of the two objects at the left. It doesn't matter which one we write first. Then we write the ranks for the first object *in increasing order* from left to right (notice that this differs from Table 11.1). Thus, the ranks of British Columbia run from 1 through 13. The corresponding ranks for Alberta are, however, copied from Table 11.1. For example, where British Columbia was given a rank of 1 ("The importance of a good education"), so was Alberta; where British Columbia was given a rank of 2 ("Freedom of the press"), Alberta was given a 3; where British Columbia was given a rank of 3 ("Respect for the courts or judicial system"), Alberta was given a 4—and so on.

1. British Columbia	1	2	3	4	5	6	7	8	9	10	11	12	13
2. Alberta	1	3	4	2	7	6	8	5	11	10	13	12	9

Since British Columbia is object 1 and Alberta is object 2, we are calculating τ_{12}. To do this, we first calculate S_{12} in equation 11.1.

The value of S_{12} is accumulated by working across the second row shown above, from right to left. Starting with the number 9 at the far right, we count how many numbers to its right have larger ranks, and how many have smaller ranks. Then we compute the difference in these counts. For the number at the far right (9 in this case), this will always be zero because there are no numbers to its right. So, at the start of the accumulation process, S_{12} is 0.

Table 11.1 Data matrix giving ranks of 13 values for eight regions of Canada.

Item	Region							
						Quebec		
	British Columbia	Alberta	Saskatchewan	Manitoba	Ontario	Total Quebec	Extreme Separatist	Maritimes
The importance of a good education	1	1	1	1	1	1	1	1
Respect for the courts or judicial system	3	4	2	4	2	3	7	4
Freedom of the press	2	3	3	2	3	4	2	2
Home ownership, being able to own a piece of property	4	2	4	3	4	5	5	3
Finding job security	5	7	6	5	5	2	3	5
Respect for authority	8	5	5	7	6	6	9	6
Commitment to the free enterprise system	6	6	7	6	7	8	11	7
Commitment to the work ethic	7	8	8	8	8	7	8	9
Religion	13	9	9	9	10	13	13	8
Politics	10	10	10	11	9	10	4	11
Achieving a position of power	12	12	12	10	11	9	12	12
The importance of unions in our society	11	13	13	12	12	11	6	10
Women's rights movement	9	11	11	13	13	12	10	13

Then we move to the next number to the left, 12. We count how many numbers to its right have larger values (there are none) and how many have smaller values (there is 1, the number 9). We take the difference $0 - 1 = -1$, and add this to the running total of S_{12}. This makes $S_{12} = 0 + (-1) = -1$.

Then we move to the next number to the left, 13. It has no larger values to its right, but has two smaller values (12 and 9). The difference is $0 - 2 = -2$. The new running total is now $S_{12} = 0 + (-1) + (-2)$.

We continue on like this until we reach the value of 1 at the far left of the second row. You can verify that the result of this is

$$S_{12} = 0 + (-1) + (-2) + 1 + 0 + 5 + 4 + 5 + 4 + 9 + 8 + 9 + 12$$
$$= 54.$$

From this, the value of τ_{12} follows directly. The variable P in equation 11.1 is simply the number of rows in the data matrix, 13 in this case. Using these values in equation 11.1 gives

$$\tau_{jk} = \frac{54}{(0.5)(13)(13 - 1)} = 0.69.$$

This indicates a fairly high degree of similarity between the rankings of peoples' values for British Columbia and Alberta because it falls nearer to the maximum value of $\tau_{jk} = 1.0$ than it does to the minimum value of $\tau_{jk} = -1.0$.

When the ranks of objects j and k are the same, attribute for attribute, then $\tau_{jk} = 1.0$. When the ranks are completely opposite in the way they run —the smallest of one is paired with the largest of the other, and so on—then $\tau_{jk} = -1.0$. And when there is no tendency for the ranks to run with or against each other, then $\tau_{jk} = 0.0$, a halfway point between the possible extremes of similarity.

Siegel (1956), Hays (1963), Kendall (1970), and Lindeman et al. (1980) illustrate the calculation of Kendall's tau coefficient, and they cover the special treatment required when there are ties in the rankings. Ghent (1963) discusses the use of Kendall's tau in cluster analysis. Spearman's rank correlation coefficient is an alternative to Kendall's tau and has also been used as a resemblance coefficient in applications of cluster analysis. Although we will not discuss it, the books just mentioned do.

11.2 AN EXAMPLE OF KENDALL'S TAU COEFFICIENT

Let us continue on with the example begun in the preceding section and compute the resemblance matrix of τ_{jk} values for the data matrix in Table 11.1. Then, using the UPGMA clustering method, we will perform a cluster analysis that helps to answer the following questions: Do British Columbians, who have the most socialistic of the provincial governments, hold values similar to other Canadians? Are the values held by the Maritime provinces more similar to neighboring Quebec, or more similar to the rest of Canada? How much different are Quebecers in general compared to the segment of the population termed "extreme separatist?" Do similarities among provinces reflect their spatial proximities?

Figure 11.1 shows the resulting tree. We see that the values the people hold are related to where they live. The neighboring provinces of Alberta and Saskatchewan are similar, as are Manitoba and Ontario. British Columbia and Quebec each stand somewhat alone. The extreme separatists of Quebec are dissimilar to all other provinces. And despite their geographi-

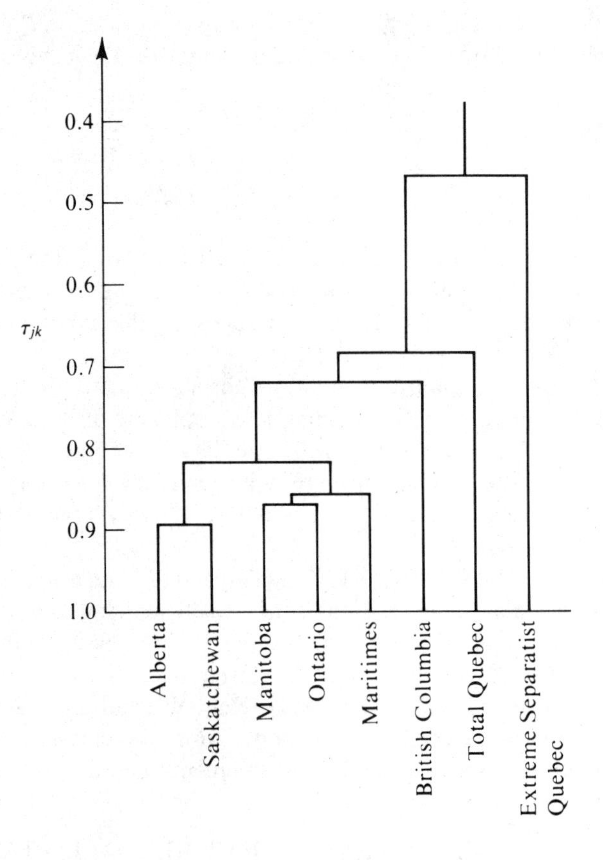

Figure 11.1. Tree resulting from a cluster analysis of the data matrix given in Table 11.1. Kendall's tau coefficient τ_{jk} and the UPGMA clustering method were used.

cal isolation, the Maritime provinces are most like the mainstream provinces typified by Ontario. Examination of the data matrix confirms these relationships.

Before closing, let us note that to use a rank correlation coefficient we must be careful of which way the ranks run. The data matrix of Table 11.1 can be used for a Q-analysis of provinces, but it is unsuitable for an R-analysis of the peoples' values. An R-analysis would require the ranks to run from 1 to 8 on each attribute across the objects. Such data could be obtained by having a group of experts rank the provinces on each attribute —for example, rank the provinces according to how much they thought the people living there valued a good education, valued job security, and so forth. With this data matrix we could use equation 11.1 to compute τ_{jk} for all pairs of values, in which case P in the equation would be the number of objects in the data matrix.

Statisticians have developed rank correlation coefficients to allow them to make statistical inferences about correlations in populations of ordinal-scaled data. However, cluster analysis is concerned with the description of similarities in samples, not with making statistical inferences. So the advantage that rank correlation coefficients have for making inferences is of little use for us. For this reason we may regard rank correlation coefficients as alternatives to the resemblance coefficients presented in Chapter 8, to be considered when all of the attributes are ordinal-scaled. As a means of measuring similarities they are superior only to the extent that actual applications show them to lead to superior insights, which of course must be decided case by case.

SUMMARY

Kendall's tau is one of the rank correlation coefficients developed in statistics that can be used as a resemblance coefficient when the attributes are ordinal-scaled. There are restrictions, however, on the direction the ranks can run. For a Q-analysis of objects, the attributes must be ranked on each object from 1 to n. For an R-analysis of attributes, the objects must be ranked on each attribute from 1 to t.

When used with data matrices meeting these restrictions, Kendall's tau coefficient holds no categorical advantages over the resemblance coefficients covered in Chapter 8, which also can be used with ranked data. All are alternatives when the data consist of ranks.

EXERCISE

11.1. Use the data matrix in Table 11.1 to compute the value of Kendall's tau coefficient for British Columbia (object 1) and Ontario (object 5). You should get $\tau_{15} = 0.74$.

Resemblance Coefficients for Mixtures of Quantitative and Qualitative Attributes

OBJECTIVES

In this chapter we discuss two important questions:

- What do you do when your data matrix contains both quantitative and qualitative attributes?
- How do you obtain a resemblance matrix for data that are mixed in this way?

12.1 STRATEGIES

There are at least a half-dozen strategies for dealing with mixed quantitative and qualitative data, and they are as follows:

1. Ignore the Problem. The simplest way to handle mixtures of quantitative and qualitative attributes is to ignore their difference. That is, pretend that the qualitative attributes are really quantitative attributes, and then use a resemblance coefficient suitable for quantitative attributes.

This simplistic solution may appear absurd, but in fact it often works. Take, for example, a cluster analysis of hardware items—nails, screws, tacks, staples. Suppose we describe their sizes and shapes with a mixture of quantitative and qualitative attributes. When the data matrix is standardized by using equation 7.1, and the average Euclidean distance coefficient d_{jk} is used for the resemblance coefficient, a cluster analysis will usually give clusters of hardware items in groups that match our intuitive distinctions between these objects. The small nails will cluster together, as will the large nails, the small screws, the large screws, and so forth. The results will not be very sensitive to our having treated qualitative attributes as if they were quantitative attributes.

2. Perform Separate Analyses. Another strategy is to partition the data matrix into two sets of attributes, one quantitative and one qualitative. Then compute a resemblance matrix for each set with appropriate resemblance coefficients, independently cluster-analyze both, and hope that the resulting trees are essentially the same. If the two kinds of attributes "say the same thing," then either one can be used.

This long-shot strategy seems to work more often than can be attributed to chance. Perhaps this is because the quantitative and qualitative data for any given problem result from the same natural phenomena. Both quantitative and qualitative attributes reflect the same underlying nature, and this appears in the agreement of the two trees.

3. Make the Quantitative Attributes Qualitative. Reduce all quantitative attributes to qualitative attributes by dichotomizing the quantitative attributes. This gives a single set of qualitative attributes—those obtained by reduction, and those measured naturally. They can then be analyzed as a set by using a resemblance coefficient for qualitative attributes.

This strategy reduces the quantitative attributes to the lowest common denominator. It runs the risk of losing information that the quantitative attributes may contain, but while this may be information in the sense of mathematics, it is not always information in the sense that matters to your research goal. How well this strategy achieves you research goal will determine if the mathematical loss is a consequential loss.

4. Use Gower's General Resemblance Coefficient. There is a resemblance coefficient called **Gower's general coefficient of similarity** that gives a measure of resemblance between pairs of objects when the attributes are a mixture of quantitative and qualitative (both binary and multistate) attributes. Gower's coefficient is easy to understand through its limiting cases: when all of the data matrix's attributes are binary, it reduces to the Jaccard coefficient; when all are multistate, it reduces to the simple matching coefficient; and when all are quantitative, it becomes a distance-type coefficient similar to the Euclidean distance coefficient, called the **city block metric**. In effect, Gower's coefficient welds together, in weighted fashion, these three coefficients. To use Gower's coefficient, the data matrix is first partitioned into the three kinds of attributes (some of the partitions may be empty). The data in each partition are then fed into an appropriate part of the coefficient's formula, and all are mathematically blended into one value that is an overall measure of similarity.

In practice Gower's coefficient has seldom been used, partly because it is not widely understood, and partly because its three built-in resemblance coefficients are only suitable for a subset of all research goals. In point of fact, the literature review for this book found only two applications that used Gower's coefficient. And they were not independent: Betters and Rubingh (1978), and Radloff and Betters (1978), used it in making a classification of wildlands. In addition to Gower's (1971) original paper, Doran and Hodson (1975) and Everitt (1980) describe this coefficient and illustrate its calculation.

5. Form a Combined Resemblance Matrix. Closely related to Gower's coefficient is the more general **combined resemblance matrix approach**, a method invented for this book. As its name suggests, we partition the data matrix's n attributes into p subsets. Using an appropriate resemblance coefficient, we compute a resemblance matrix for each subset. Then we convert all that are similarity matrices to dissimilarity matrices by multiplying their values by -1.0. Then we standardize the values in each resemblance matrix, and weight the standardized resemblance matrices together to get a combined resemblance matrix. Symbolically, we write this as

$$R = w_1 R_1 + w_2 R_2 + \cdots + w_p R_p = \sum_{q=1}^{p} w_q R_q, \qquad (12.1)$$

where R is the combined resemblance matrix; R_q is the resemblance matrix for the qth partition of attributes, converted to a dissimilarity coefficient if necessary, and standardized; w_q is the weight for the qth resemblance matrix; and $0.0 < w_q \leq 1.0$ and $\Sigma_{q=1}^{p} w_q = 1.0$—that is, the weights are nonnegative and sum to 1.0.

You must supply the weights w_q, and this may prove difficult. One method of estimating the weights is to base them on the proportion of each kind of attribute present. For example, say there are $n = 100$ attributes, and the first partition contains 12 of these. Then the weight for the first partition is $w_1 = 12/100 = 0.12$. Or if you want to give more weight to the quantitative attributes than their frequency of occurrence suggests, you are free to concoct a greater weight.

One advantage this procedure holds over Gower's coefficient is that one kind of attribute can be partitioned and the subsets handled separately with different resemblance coefficients. For instance, binary attributes can be partitioned into two sets: a set in which 0–0 matches do not count toward resemblance (in which case, the Jaccard coefficient may be appropriate) and a set in which 0–0 matches do count (in which case the simple matching coefficient may be a good choice).

When you use Gower's coefficient, you accept Gower's views of how three resemblance coefficients should be weighted together into one formula. Essentially, Gower's coefficient gives a combined resemblance matrix according to his prescription. The combined resemblance matrix approach just described is more flexible and allows any prescription to be used. By tailoring a combined resemblance matrix to your research goal, you should be able to do as well or better than you could by using Gower's coefficient —for the same reason that a custom-tailored suit fits better than one off the rack. Let us now illustrate the combined resemblance matrix approach.

12.2 AN EXAMPLE OF THE COMBINED RESEMBLANCE MATRIX APPROACH

The following example clarifies our explanation of the combined resemblance matrix approach. Although it partitions the data matrix into only two sets of attributes, it illustrates the principles that apply to problems with any number of partitions.

Figure 12.1,*a* shows a data matrix that contains 5 objects and 10 attributes. Attributes 1 through 4 are quantitative and make up the first partition. Using the Euclidean distance coefficient e_{jk} with them, we calculate the resemblance matrix, *b*. Attributes 5 through 10 are qualitative and make up the second partition. Using the simple matching coefficient with them, we calculate the resemblance matrix, *c*.

Next, we unstring these two resemblance matrices into two lists. Table 12.1, columns 1 and 2, shows the unstrung resemblance matrices of Figure 12.1,*b* and *c*. Note that values in column 2 have been multiplied by -1.0, converting them from values of a similarity coefficient to values of a dissimilarity coefficient.

Next we standardize columns 1 and 2. To do this, we use their means ($\bar{x}$) and standard deviations (s.d.); these are shown below their respective

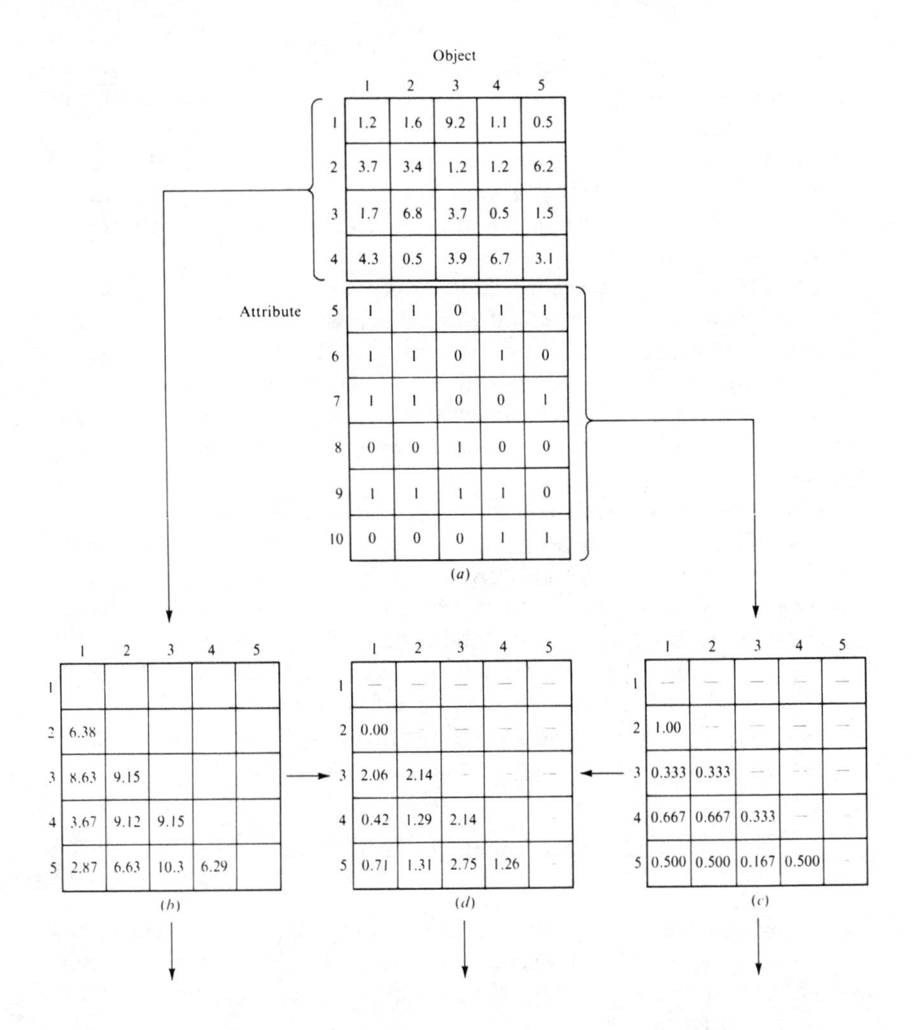

Object

Attribute	1	2	3	4	5
1	1.2	1.6	9.2	1.1	0.5
2	3.7	3.4	1.2	1.2	6.2
3	1.7	6.8	3.7	0.5	1.5
4	4.3	0.5	3.9	6.7	3.1
5	1	1	0	1	1
6	1	1	0	1	0
7	1	1	0	0	1
8	0	0	1	0	0
9	1	1	1	1	0
10	0	0	0	1	1

(a)

	1	2	3	4	5
1					
2	6.38				
3	8.63	9.15			
4	3.67	9.12	9.15		
5	2.87	6.63	10.3	6.29	

(b)

	1	2	3	4	5
1	—	—	—	—	—
2	0.00	—	—	—	—
3	2.06	2.14	—	—	—
4	0.42	1.29	2.14	—	—
5	0.71	1.31	2.75	1.26	—

(d)

	1	2	3	4	5
1	—	—	—	—	—
2	1.00	—	—	—	—
3	0.333	0.333	—	—	—
4	0.667	0.667	0.333	—	—
5	0.500	0.500	0.167	0.500	—

(c)

Figure 12.1. An example of the combined resemblance matrix approach. (a) Data matrix; (b) a resemblance matrix using e_{jk} and quantitative attributes 1 to 4; (c) a resemblance matrix using the simple matching coefficient and qualitative attributes 5 to 10; (d) the combined resemblance matrix, from Table 12.1; (e, f, g) trees produced by using the UPGMA clustering method and resemblance matrices b, d, and c, respectively.

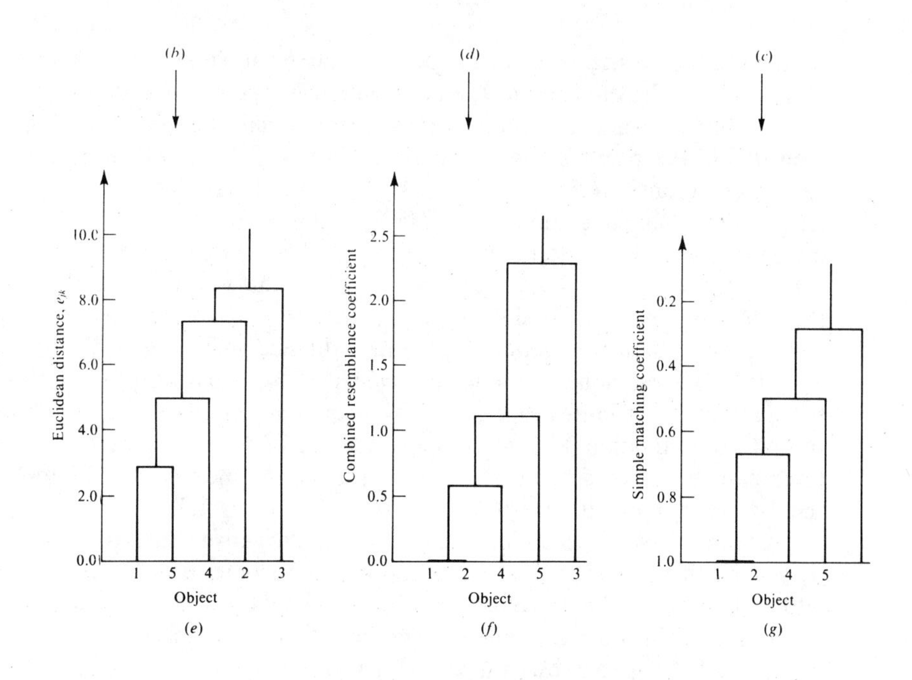

Figure 12.1. (*Continued*).

Table 12.1 Calculations of the combined resemblance matrix shown in Figure 12.1, *d*.

Column					
1	2	3	4	5	6
6.38	−1.000	−0.335	−2.12	−1.41	0.00
8.63	−0.333	0.566	0.71	0.65	2.06
3.67	−0.667	−1.42	−0.71	−0.99	0.42
2.87	−0.500	−1.74	0.00	−0.70	0.71
9.15	−0.333	0.77	0.71	0.73	2.14
9.12	−0.667	0.76	−0.71	−0.12	1.29
6.63	−0.500	−0.24	0.00	−0.10	1.31
9.15	−0.333	0.77	0.71	0.73	2.14
10.30	−0.167	1.24	1.41	1.34	2.75
6.29	−0.500	−0.37	0.00	−0.15	1.26
$\bar{x}$: 7.217	−0.500	0.0	0.0		
s.d.: 2.496	0.2357	1.0	1.0		

columns. Then, using an analogue of equation 7.1 to standardize them, we obtain column 3 from column 1, and column 4 from column 2. To illustrate, we take the first value in column 1 (6.38), from it subtract its mean (7.217), then divide the result by its standard deviation (2.496), and get the first value in column 3: $(6.38 - 7.217)/2.496 = -0.335$. To illustrate again, we take the first value in column 2 (-1.000), from it subtract its mean (.5), then divide the result by its standard deviation (.2357), and get the first value in column 4: $[-1.000 - (0.5)]/0.2357 = -2.12$. In this way we generate all of the values in columns 3 and 4.

Suppose that we want to give resemblance matrix *b* a weight of $w_1 = 0.4$ and resemblance matrix *c* a weight of $w_2 = 0.6$, in proportion to the number of attributes each partition of the data matrix contains. To do this, we use equation 12.1 to compute the list of values in the combined resemblance matrix, column 5. For example, taking the value at the top of the list, we compute $(0.4)(-0.335) + (0.6)(-2.12) = -1.41$.

Column 6, which is optional, eliminates the negative values in column 5 by scaling the smallest value, -1.41, so that it becomes 0.0. To do this, we add 1.41 to each value in column 5. Finally, we take the values in column 6 and put them into the combined resemblance matrix, Figure 12.1,*d*.

Figure 12.1 also shows the trees for several cluster analyses done with the UPGMA clustering method. Resemblance matrix *b*, based on the quantitative attributes, gives the tree *e*. Resemblance matrix *c*, based on the qualitative attributes, gives the tree *g*. And the combined resemblance matrix *d*, based on the mixture of quantitative and qualitative attributes, gives the tree *f*. Resemblance matrix *c* contributes more weight to the combined resemblance matrix *d* than resemblance matrix *b* does; possibly this is why its tree, *g*, agrees well with tree *f*, obtained from the combined resemblance matrix.

The computer programs CLUSTAR and CLUSTID, described in Appendix 2, can perform the combined resemblance matrix approach.

SUMMARY

The strategies for cluster-analyzing data matrices that contain both quantitative and qualitative attributes include the following:

First, ignore the difference in the qualitative attributes. Treat them as if they were quantitative, and use one of the resemblance coefficients for quantitative attributes presented in Chapter 8.

Second, partition the two kinds of attributes into separate data matrices, cluster-analyze each, and try to reach a consensus by studying the two trees.

Third, dichotomize the quantitative attributes, making them qualitative. Then use one of the resemblance coefficients for qualitative attributes presented in Chapter 10.

Fourth, use Gower's general resemblance coefficient. It can handle mixtures of quantitative and qualitative attributes.

Fifth, use a combined resemblance matrix approach; this computes a separate resemblance matrix for each type of attribute, combines them into a single resemblance matrix, and cluster-analyzes it.

Bypassing the Data Matrix

OBJECTIVES

- Our aim in this chapter is to show that for some applications it is best to begin the cluster analysis with the resemblance matrix, thus bypassing the data matrix.
- We discuss the situations that call for this and give examples.

13.1 REASONS FOR BYPASSING THE DATA MATRIX

Suppose you have read a journal article or a book that published a resemblance matrix but didn't give the original data matrix or the tree, and you want to obtain the tree to further your understanding. For example, the book by Laumann (1973) gives various resemblance matrices that describe social structures. The objects of one of them are 22 ethnic groups, and the pairwise measure of dissimilarity among groups indexes the friendship choices between ethnic groups. The original data were obtained from questionnaires completed by over 1,000 respondents. From these data, the number of friends each respondent had in each ethnic group was tallied in the original data matrix. Since the data matrix was too large to be included in the book, the author instead published the resemblance matrix, which contained only $(22)(21)/2 = 231$ values.

The author did not apply cluster analysis to these values, but instead used another multivariate method that did not produce a tree. If you want to reexamine the similarity structure of these friendship choices by using cluster analysis, you must bypass the original data matrix and apply a clustering method directly to the published resemblance matrix.

In point of fact, Vanneman (1977) used cluster analysis in this way to reexamine a resemblance matrix of occupational groups given in Laumann's book. The objects of the resemblance matrix were 16 occupational groups, such as managers, clerical workers, construction foremen, and so on. The matrix contained values of a dissimilarity coefficient that indexed the friendship choices between occupational groups. Vanneman used the tree resulting from a cluster analysis, together with the results from other analyses, to help choose between two competing hypotheses: (1) the embourgeoisement hypothesis that the white collar working classes are blending with the blue collar working classes, and (2) the proletarianization hypothesis that the two classes are segregating. The cluster analysis supported the proletarianization hypothesis.

A second reason for bypassing the original data matrix is that, whereas it may be possible to construct that matrix, it is inconvenient to do so. That is, it may be easier to summarize the raw data directly in the form of a resemblance matrix than to arrange it first as a data matrix and then compute the resemblance matrix.

A third reason is that sometimes it is impossible to construct an initial data matrix, yet we can obtain a resemblance matrix directly from other measurements. For example, suppose that for a marketing study we want to know which brands of automobiles (or any other product) consumers perceive as being similar. And suppose further that we cannot decide the right type of attribute to use, and so we cannot construct a data matrix. We could, nonetheless, formulate a resemblance matrix directly by asking a sample of consumers to rank the cells of the resemblance matrix—1, 2, 3, and so on—from the most similar pair of automobiles to the least similar pair, as they perceive them. This lets each consumer use his own intrinsic definition of overall similarity. When each consumer's resemblance matrix is obtained and the sample is averaged into one, the resulting resemblance matrix can be cluster-analyzed to show the similarities among automobiles.

The first of the above-mentioned reasons—reanalyzing a published resemblance matrix—needs no more discussion, but the other two do.

13.2 WHEN USING THE DATA MATRIX IS INCONVENIENT

Let us now examine two studies of the behavior of groups of animals, for which it was convenient to bypass the original data matrix and to compute the resemblance matrix directly from the raw field observations. In the first study, Morgan et al. (1976) studied the associations of 11 adult male chimpanzees. The researchers used the time each pair among the 11 were observed feeding together as a measure of their behavioral similarity. An initial data matrix could have been constructed. Objects would be chimpanzees. Attributes would be the n time periods of observation. The ij-cell values of the data matrix would be the amount of time the jth chimp was observed feeding in the ith observational period. It was easier, however, to make a direct tally of the times that each pair of chimps was seen feeding together. After normalizing these raw data to adjust for the time each chimp was in the feeding area, the researchers cluster-analyzed the resemblance matrix. The resulting tree showed the hierarchy of friendship choices made while feeding.

This example makes it convenient to digress momentarily, to discuss asymmetric resemblance matrices. In the appendix of their paper, Morgan et al. discuss the cluster analysis of asymmetric resemblance matrices, which are also formed by bypassing the original data matrix. An asymmetric resemblance matrix has elements on, above, and below its diagonal. They arise, for example, in the analysis of **Markov chains of events**. For instance, suppose a blackbird has a repertoire of three song types: A, B, C. As it sings, a blackbird may chain together song types in a sequence that could look like this:

$$BBCACAACBBBBCCABCACBCCBBCA\ldots$$

If we record the chains for a number of birds, we can make an asymmetric matrix of the percentage of the time a given song type follows another. It might look like this:

Preceding Call	Following Call		
	A	*B*	*C*
A	36	48	16
B	42	30	28
C	22	35	43

Note that each row sums to 100 percent. For example, in recording the singing habits of a group of blackbirds, 48 percent of the time when an *A*-type call was sung, a *B*-type immediately followed.

The matrix is asymmetric because the value in cell *ij* is not the same as the value in cell *ji*. To cluster-analyze asymmetric matrices like this one, the matrices must first be converted to a symmetric form having all the values below the diagonal. There are a number of methods for doing the conversion. Morgan et al. (1976) give one method; Shepard (1963) and Levine and Brown (1979) give others.

Another example of asymmetric resemblance matrices is the input/output matrices that economists use to show the amount of trade between pairs of trading entities—cities, counties, states, or countries. For instance, the exports of country *A* to country *B* do not in general equal the exports of country *B* to country *A*, and thus the matrix is asymmetric. But when reduced to symmetric form by calculating net exports, the countries can be cluster-analyzed to see which have similar patterns of trade.

Converting an asymmetric matrix to a symmetric matrix is only valid when what is given up by the conversion does not matter for the achievement of the research goal. To simplify—necessary so that the cluster analysis can go on—removes some of the detail. Whether this matters depends on the research goal, a decision that must be made case by case.

13.3 WHEN USING THE DATA MATRIX IS IMPOSSIBLE

Confusion data record the frequency with which different stimuli are confused with each other. As Levine (1977) observes, "Confusion between stimuli represents a measure of their similarity. It is plausible to assume that the more two stimuli are confused, the greater the degree of perceived similarity between them."

To illustrate this, Levine showed experimental subjects ten different stimuli—each of the numerical digits ranging from 0 to 9, for a variable time of 4 to 20 msec. Figure 13.1,*a* shows the observed confusion matrix (the number of each of the correct and incorrect responses elicited). For example, when 2 was presented, 330 times it was judged correctly, and 13 times it was judged incorrectly to be a 3. To cluster-analyze these data, Levine first converted the asymmetric confusion matrix to a symmetric resemblance matrix. He then cluster-analyzed it by using the CLINK clustering method and found that the tree showed the 0 and the 8 were the most similar pair; the 2 and the 7 were the next most similar; 5 and 9 were next, and 1 and 4 followed. As the tree in Figure 13.1,*b* shows, there are two main clusters. One contains mostly the numbers constructed from curved

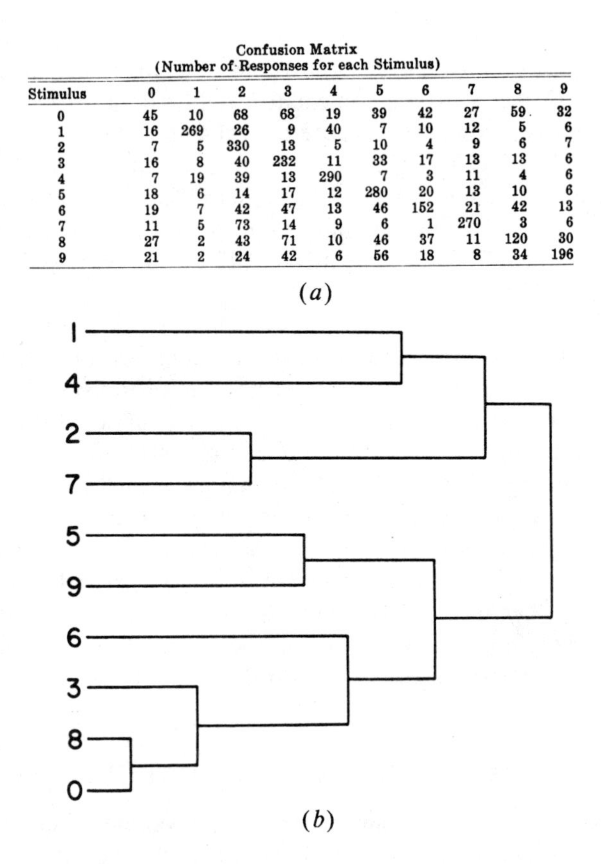

Confusion Matrix
(Number of Responses for each Stimulus)

Stimulus	0	1	2	3	4	5	6	7	8	9
0	45	10	68	68	19	39	42	27	59	32
1	16	269	26	9	40	7	10	12	5	6
2	7	5	330	13	5	10	4	9	6	7
3	16	8	40	232	11	33	17	13	13	6
4	7	19	39	13	290	7	3	11	4	6
5	18	6	14	17	12	280	20	13	10	6
6	19	7	42	47	13	46	152	21	42	13
7	11	5	73	14	9	6	1	270	3	6
8	27	2	43	71	10	46	37	11	120	30
9	21	2	24	42	6	56	18	8	34	196

(*a*)

(*b*)

Figure 13.1. (*a*) A confusion matrix of digits obtained from experimental subjects. (*b*) Tree produced from a cluster analysis of the confusion matrix showing the perceived similarities among digits. (The confusion matrix is Table 1 and the tree is Figure 5 in Levine (1977); reproduced by permission.)

lines; the other contains the numbers constructed from straight lines. Levine goes on to retroductively derive a hypothesis that explains these results.

The engineering of remote sensing satellites designed to collect earth resources data provides another example. Satellites must be able to identify classes of resources as they are identified by ground mapping or by aerial photographs. For example, Fox et al. (1983) generated a confusion matrix in a study of Landsat earth resources data. First a forested area in California was mapped by using data from aerial photographs according to an existing classification of 16 land types. Then it was remapped by using Landsat data and the same 16 land types. The confusion matrix was generated by taking 2,496 plots of land and identifying each of them according to (1) the aerial photograph data and (2) the Landsat data. The rows of the confusion matrix were the 16 land classes based on aerial photograph data, and the 16 columns were the 16 land classes based on Landsat data.

Had the two methods agreed perfectly, all of the plots would have fallen along the diagonal of the confusion matrix. For example, if a plot was identified into a given class on the basis of aerial photograph data, it would have been identified into the same class on the basis of Landsat data. Although this tended to happen, the two methods were not identical. Some confusion was observed in which a plot placed in one class according to the aerial photograph data was placed into another class according to the Landsat data. The value of a cluster analysis of the confusion matrix is that it will show those land classes that are similarly confused. This is the first step in tracking down the reasons for the confusion to make the satellite-based identification more nearly agree with the aerial photographic identification.

Other studies based on confusion matrices have examined the similarities between spoken letters of the alphabet (Morgan, 1973), and the similarities between letters and numbers in the signals of the International Morse Code (Rothkopf, 1957; Shepard, 1963). Studies like these have application in the planning of technological devices that interface with human operators. Nuclear reactors, airplanes, even elevators speak to their operators with numerical stimuli, and their operators cannot afford to mistake what is said. If the possible sources of confusion can be identified through human factors engineering research, then the chance that confusion will occur in their use can be minimized by their designs.

Besides the confusion matrix approach, another method of obtaining a resemblance matrix directly from human testing is the **sorting method**. It works this way. The names of objects to be compared are written on cards. Each subject is asked to sort the cards according to how similar he perceives the objects to be, and he places cards for similar objects in the same pile. This is repeated with a set of subjects, and the frequencies with which two objects are sorted into the same pile becomes the index of their similarity. A cluster analysis of the resemblance matrix, so constructed, shows the similarity structure for the objects. This sorting method, and a variation of it called the **triads method**, are discussed by Miller and Johnson-Laird (1976).

For a psychological analysis of the similarities among verbal concepts, Miller (1969) used the sorting method to derive a resemblance matrix for the similarities among 48 common nouns. The subjects sorted the names of nouns they felt were similar into piles, and from this the resemblance matrix was obtained. The cluster analysis revealed the features of these nouns that the subjects had intuitively based their judgments on. And in a more extensive study along the same lines, Fillenbaum and Rapoport (1971) investigated the similarities among the names of colors, of kinship terms, of pronouns, of words of emotion like "love," "fear," "pride," or "envy"; of verbs, of prepositions, and of conjunctions (where they found that "but," "still," "however," and "yet" formed one cluster, and that "and," "too," "also," and "as well as" formed another).

Moskowitz and Klarman (1977) used the sorting method in a planning study for the design of meal menus. Subjects sorted foods on the basis of their compatibility for being eaten together. When they cluster-analyzed the resemblance matrix of food compatibilities, the tree's branches showed compatible and incompatible foods. Thus, foods that complement each other on the same menu were identified.

The idea obviously has more general applications that extend to the compatibility of most human activities. Planners of recreation in wilderness areas could assess the compatibility of different forms of recreation on the same land; architects could assess the compatibility of public behaviors and accordingly design buildings to minimize conflicts; and so on.

SUMMARY

Some cluster analyses can, or must, be started with the resemblance matrix. The data matrix is bypassed because either it is inconvenient to use it, or it is impossible to use it.

The analysis of confusion data is an example of the latter case. Here, pairs of stimuli are presented to subjects and the extent to which they confuse them is taken as a measure of the similarity of the stimuli. In this way a resemblance matrix is obtained directly from experiment.

Asymmetric resemblance matrices in which the similarity between objects j and k is not usually the same as between objects k and j can be handled by various methods that reduce them to symmetric form.

Matrix Correlation

OBJECTIVES

This chapter explains

- How to compare resemblance matrices and cophenetic matrices.
- How research goals can be addressed by comparing resemblance and cophenetic matrices.
- Alternatives to the cophenetic correlation coefficient.

In Chapter 2 we learned about the cophenetic correlation coefficient, $r_{X,Y}$. We learned that when the resemblance matrix is unstrung into a list of values, X, and when the cophenetic matrix (essentially, the tree in matrix form) is unstrung into a corresponding list, Y, then the value of the correlation coefficient between these lists, $r_{X,Y}$, indexes the agreement between the resemblance matrix and the tree.

Now we will learn that unstringing matrices into lists and comparing them in pairs by computing $r_{X,Y}$, where X and Y can be the lists for any two matrices describing pairwise resemblance of the same objects, is a general idea. For the special case where X and Y are lists for a resemblance matrix and a cophenetic matrix, respectively, $r_{X,Y}$ is called the cophenetic correlation coefficient. The usual custom in cluster analysis is to denote it by r_{coph}. For all other cases, $r_{X,Y}$ is called the matrix correlation coefficient. For example, the following cases arise in practice:

1. Suppose that we collect two data matrices having a common set of objects. As an illustration, take the case where the common objects are a set of forested plots. One data matrix has attributes of vegetation, and the other has attributes of soil properties. For each data matrix we compute a resemblance matrix and unstring the two resemblance matrices into two lists, X and Y. It follows that the matrix correlation coefficient $r_{X,Y}$ is an index of the agreement between the two kinds of attributes: vegetation and soil properties.

2. Suppose that we collect two data matrices having a common set of objects, and as above each is based on a different type of attribute. Then, by using the same resemblance coefficient and the same clustering method, we cluster-analyze both data matrices to obtain their trees. We unstring the cophenetic matrices for the two trees into two lists, X and Y. It follows that $r_{X,Y}$ is an index of the agreement between the two different kinds of attributes, *at the level of the tree*.

Note that whereas the first example compares the effects of the different kinds of attributes at the level of the resemblance matrices, the second example compares the effects at the level of the trees. Since the tree is further along in the sequence of steps of the cluster analysis, and is produced by the clustering method, it is usually preferable to make the comparison at the level of the resemblance matrices. This excludes from the comparison any distortion caused by the clustering method.

3. Suppose that for a given data matrix we compute two resemblance matrices using two different resemblance coefficients—for example one using d_{jk} and another using r_{jk}. Then we unstring the resemblance matrices into lists, X and Y. Since the resemblance matrices are based on the same data matrix and differ only because they use different resemblance coefficients, it follows that $r_{X,Y}$ indexes the agreement between the two resemblance coefficients, d_{jk} and r_{jk} in this case.

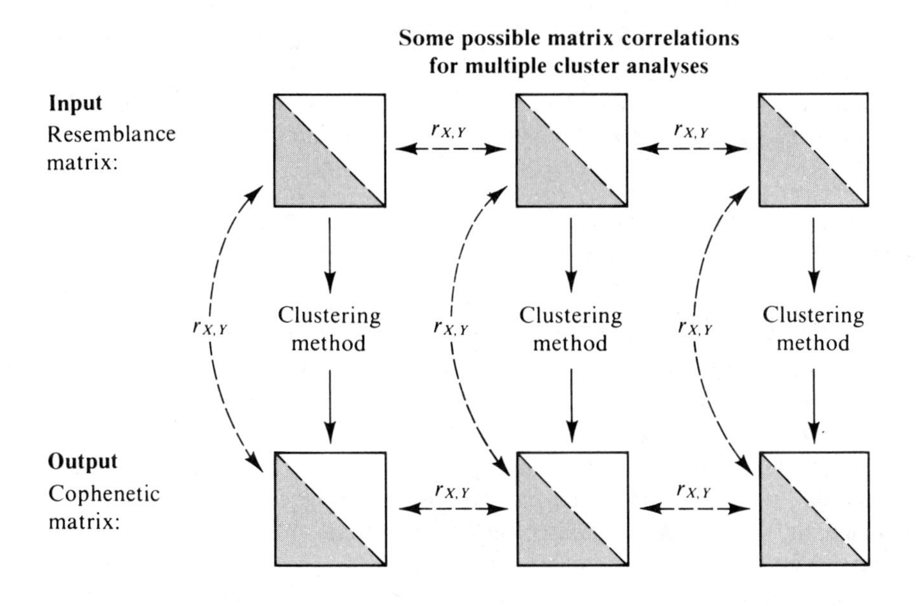

Figure 14.1. An illustration of various matrix correlations among and within multiple cluster analyses.

4. Suppose that we cluster-analyze a resemblance matrix separately with two clustering methods, say the SLINK method and the CLINK method. Then we unstring the two cophenetic matrices into lists, X and Y. It follows that $r_{X,Y}$ is an index of how well the two clustering methods give the same result when everything else in the problem is held constant.

These examples suggest some of the many reasons for comparing matrices and illustrate some of the possible comparisons. As often happens, a researcher will want to compare more than two resemblance matrices, or more than two cophenetic matrices. Figure 14.1 shows the variety of matrix correlations that are possible.

For example, at the input side of a cluster analysis, several resemblance matrices can arise from the same set of objects, based either on data matrices having different attributes or on different resemblance coefficients. For such cases, $r_{X,Y}$ can be used as a measure of the agreement between the different resemblance matrices. Sometimes a research goal can be achieved merely by computing $r_{X,Y}$ between pairs of resemblance matrices without executing a clustering method.

At the output side, several trees based on the same set of objects can arise, owing to any combination of (1) multiple data matrices based on different attributes, (2) multiple resemblance coefficients used with the same, or different, data matrices, (3) multiple clustering methods used with the same, or different, resemblance matrices. Some research goals can be achieved by computing $r_{X,Y}$ between pairs of cophenetic matrices.

14.1 RESEARCH GOALS INVOLVING THE CORRELATION OF RESEMBLANCE MATRICES

RESEARCH APPLICATION 14.1

For a study of human anxiety, Ekehammar and Magnusson (1978) used ten reactions to anxiety as objects, such as "hand begins to shake," "swallowing difficulties," or "heart beats faster." Their research goal was to find out if two different methods of measuring these reactions were the same. Each method produced a resemblance matrix of intensity ratings between pairs of reactions, and the correlation of the two resemblance matrices indexed the agreement of the two methods.

To compute the first resemblance matrix, 40 adolescents each gave an intensity rating to anxiety reactions based on their experiences in anxiety-producing situations. These ratings, measured on a five-point scale, ranged from "1 = not at all" to "5 = very much." Using the correlation coefficient, r_{jk}, for the resemblance coefficient, the researchers obtained a resemblance matrix. Next, the subjects were asked to produce a second resemblance matrix by using the sorting method described in Chapter 13. Based on their perceived frequencies of coinciding occurrences of the anxiety reactions, the subjects sorted the names of pairs of reactions into piles. When pooled over all subjects, a resemblance matrix of the frequencies of coinciding occurrences was obtained.

The unstrung X and Y lists of the two resemblance matrices approximated a linear relation between X and Y (a necessary condition for $r_{X,Y}$ to be valid), and the computed value of the matrix correlation coefficient was $r_{X,Y} = 0.56$. On this basis, the researchers felt that the strength of the relation between the two types of data, each purportedly measuring the same set of physiological responses, was "relatively high," and therefore of practical significance.

RESEARCH APPLICATION 14.2

In geographical research, Zonn (1979) studied patterns of housing among black workers in Milwaukee, Wisconsin, and then compared these with the patterns of housing among white workers in the same city for comparison. Zonn's research goal required the testing of the null hypothesis that housing for black and for white workers followed the same pattern. To do so, he constructed two resemblance matrices, one for blacks and one for whites, from 1970 census tract data. In each, the objects were workers' occupations —professional, managerial, sales, clerical, laborer, and so forth. The special dissimilarity index that Zonn used measured the degree to which workers in any two occupations resided within the same census tracts. The coefficient theoretically ranged from 0 to 100 (a value of 0 indicated that equal numbers of workers at any two occupations were housed within the same census tracts; a value of 100 indicated that workers in any two occupations never resided within the same census tracts). The actual values for both blacks and whites fell between these extremes.

Zonn visually compared the white housing pattern and the black housing pattern as revealed in the two resemblance matrices and concluded that housing by occupation was different for blacks than for whites, and that the hypothesis of similar housing patterns should be rejected. Zonn's findings support the contention that residential race segregation in America's large cities has restricted black populations to housing markets offering limited alternatives.

Zonn did not formally compute the correlation between the two resemblance matrices. But his analysis could have been put on a more objective footing had he done this. Using the two resemblance matrices given in his paper—one each for blacks and whites—the calculated matrix correlation coefficient is $r_{X,Y} = 0.52$. Inasmuch as a value of 1.0 would perfectly support the hypothesis of equal housing patterns, the correlation of $r_{X,Y} = 0.52$, reflecting Zonn's visual analysis, must be regarded as a numerical value too small to give support to the hypothesis.

RESEARCH APPLICATION 14.3

Another area of application of matrix correlation is in biogeography, a field that synthesizes aspects of biology and geography. It is concerned, in the broadest sense, with the life patterns of organisms on "islands" in space and time. For example, plants of a species living in a limited hospitable zone on a mountain find themselves bound to that zone as if to an island. If the plants are to extend their range they must find ways of leaping from one mountain to another. Countless forms of life give rise to a countless variety of ways in which islands of habitats can occur. (See Wilson and Bossert, 1971, for a good introduction to biogeography.)

In an application of matrix correlation in biogeography, Tobler et al. (1970) used New Zealand and ten of its neighboring islands as objects. For a pairwise measure of similarity between islands, they constructed what they called a geobotanical similarity coefficient. It indexed the number of plant species shared in common by pairs of islands, and this enabled them to compute a resemblance matrix for the 11 islands. Then they constructed a second resemblance matrix that contained the literal geographical distances between all pairs of the islands. They computed the correlation between the two matrices and found that the geobotanical resemblance matrix and the geographical resemblance matrix were not strongly related. That is, colonization of the islands by plants was not strongly related to the distances separating the islands.

14.2 CLUSTER-ANALYZING RESEMBLANCE AND COPHENETIC MATRICES

The comparison of two matrices generalizes to the pairwise comparison of any number of matrices. To make the comparison, we treat the unstrung matrices, either all resemblance matrices or all cophenetic matrices, as objects of a data matrix. This data matrix has $t(t-1)/2$ "attributes," one for each cell in the matrices being compared. After we construct the data matrix in this way, we cluster-analyze it, usually with r_{jk} as the resemblance coefficient. The resulting tree highlights the similarity structure among the matrices.

The need to compare more than two matrices can arise in numerous contexts. For example, Charlton et al. (1978), for their cluster analysis of obsidian artifacts and source mine material described in Research Application 4.4, generated 28 trees from the same data matrix by performing repeated cluster analyses for different combinations of resemblance coefficients and clustering methods. They compared the trees visually and found them all "surprisingly close and consistent." But their visual observation could have been put on firmer, more objective grounds by cluster-analyzing the 28 cophenetic matrices. If the analyses were really close and consistent, then the tree would show no well-defined dissimilarities among them.

14.3 ALTERNATIVES TO THE COPHENETIC CORRELATION COEFFICIENT

The cophenetic correlation coefficient is the measure that researchers have most often used to assess congruence between a tree and its resemblance matrix. But in the theoretical sense the correlation coefficient is based on axioms that are not entirely met in practice for this type of use. For instance, the correlation coefficient assumes there is a linear underlying relation between X and Y. However, the structure of the tree introduces constraints among the values in the cophenetic matrix, which causes this assumption to be not entirely valid. This can usually be seen in a scatter plot of X versus Y.

Methods that their inventors claim have superior properties have been proposed to overcome flaws in the cophenetic correlation coefficient.

Williams and Clifford (1971), Phipps (1971), Farris (1973), and Fowlkes and Mallows (1983) have all suggested methods for comparing trees based on their topological properties; Rohlf (1974) has reviewed more than a dozen alternatives to the cophenetic correlation coefficient. Virtually none of these have been used in practice, probably for two reasons. First, researchers are apparently tolerant of the typical degrees of nonlinearity between X and Y that they encounter in practice, and they regard the cophenetic correlation coefficient as able to cope with this despite its weaknesses. Second, the cophenetic correlation has become entrenched through habit, and researchers assume there are not better alternatives. The upshot is that nothing has surfaced in practice to replace the cophenetic correlation coefficient.

The following Research Application describes uses of matrix correlation methods in measuring genetic diversity.

RESEARCH APPLICATION 14.4

Special resemblance coefficients are used for measuring the pairwise genetic resemblance between subpopulations of animals of a species. These coefficients are used with attributes of blood groups, serum protein groups, red cell enzyme groups, and other chemical markers (see, for example, Balakrishnan et al., 1975). When used to compute a resemblance matrix for a set of t subpopulations existing in different places, the resulting genetic resemblance matrix can be compared with the geographical distance matrix between the subpopulations to see if the two matrices show the same patterns of similarities. If so, then genetic differences can be said to mirror geographic distances.

Three examples can be cited in which this procedure was used to study human genetic variation. In each, a research group computed a genetic resemblance matrix for selected groups of South American Indians. Each resemblance matrix was then compared with a matrix of geographical distance between the groups being studied: Ward and Neel (1970) studied seven Makiritare villages in southern Venezuela; Ward et al. (1975) tested several genetic hypotheses for 20 tribes; and Salzano et al. (1978) compared 22 tribes (and found that the genetic resemblance matrix was only weakly correlated with the matrix of geographical distance).

Four other examples show how the procedure has been applied to study genetic diversity among subpopulations of fish and of insects.

A study made by Imhof et al. (1980) of genetic diversity of lake whitefish in northern Lake Michigan revealed "much temporal and spatial overlap among population ranges."

Grant and Utter (1980) sampled subpopulations of walleye pollock from various locations in the Bering Sea and the Gulf of Alaska; then they computed pairwise measures of genetic resemblance among the sampled subpopulations, cluster-analyzed the resemblance matrix, and found that the subpopulations only weakly tended to form two groups (one corresponded to locations in the Bering Sea, the other to locations in the Gulf of Alaska). They concluded that "there are no distinct stocks within the southeastern Bering Sea nor within the Gulf of Alaska, but that there are minor genetic differences between fish in these two regions."

Grant et al. (1980) made a similar study of genetic diversity in sockeye salmon subpopulations at various locations in the central fishing district of Cook Inlet, Alaska.

Sluss et al. (1977) used matrix correlation to find that genetically described populations of mountain flies in the Colorado Rocky Mountains form two major groups. To reach this conclusion, the researchers performed a cluster analysis of genetic similarities among 11 populations of flies, and found "distinct northern and southern races, a distinction consistent with morphological data." From this observation they retroductively formulated the hypothesis that "selection due to host differences, temperature differences, or indirect selection on morphological traits may have combined with gene flow between semiisolated populations to produce the genetic differentiation which we have observed between regions and populations."

As a final example, interesting for its novelty, Lasker (1977, 1978) presented an alternative to using the composition of human blood as a measure of genetic resemblance, by proposing a similarity coefficient based on counts of human surnames. Lasker assumed that the number of surnames shared by two rural villages or communities could serve as a measure of their biological kinship. To compute this coefficient, he treated the villages as objects; the master surname list functioned as the attributes; and an entry in the ijth cell of the data matrix recorded the frequency with which the ith surname appeared in the jth village or community. For this, Lasker then computed a similarity coefficient he had developed. When feasible, Lasker's method may be cheaper than gauging genetic resemblance according to properties of blood, although it would seem less accurate.

SUMMARY

The need often arises in cluster analysis to compare matrices—for example, two resemblance matrices, two cophenetic matrices, or a resemblance matrix and a cophenetic matrix. Such matrices can be compared by unstringing them into lists, and then computing a pairwise measure of resemblance between them. Although the Pearson product-moment correlation coefficient is often used for the measure, other measures can be used—for example, any of the resemblance coefficients for quantitative attributes covered in Chapter 8.

How to Present the Results of a Cluster Analysis

OBJECTIVES

After you have performed cluster analysis you will want to convey your methods and findings to other researchers. You should tell them clearly what you have done, and convince them of what you have found.

This chapter discusses three topics central to this communication:

- First, key information that you must report. Without the key information your audience will be completely lost.
- Second, optional information. These are items that add to your audience's understanding but are not absolutely necessary. An example is the data matrix you used. Your audience will benefit from studying it, but you must balance this gain against the space required to present it. Thus, depending on its size, you may or may not want to present it.
- Third, alternative ways of displaying the tree from a cluster analysis. There are various ways to draw a tree, some of which are especially fitting for certain applications.

15.1 KEY INFORMATION

You must tell your audience how you framed and validated your application. Although we do not discuss in detail until Chapters 21 and 22 how to frame and validate applications, in this chapter we sketch the essence of these concepts.

Validation refers to how you have determined that you have attained your research goal. At the start of your presentation you must state your goal, and indicate what criteria you used to decide if the cluster analysis achieved it. This tells your audience what the purpose of your cluster analysis was. For example, the purpose might be to test a certain research hypothesis or create a specific-purpose classification—any of the six research goals discussed in Chapters 4, 5, and 6.

The **framing** of the cluster analysis refers to how you chose specific methods to address your research goal. For example, you must state the method of standardization, the resemblance coefficient, and the clustering method that you used.

Although when you performed the cluster analysis, you framed it first and validated it last, your audience wants to be told of these in reverse order. They want to be told where the analysis is headed before they are told what means you used to get there.

Validating a Research Application

1. State the research goal. For example, if the goal is to create a classification of forested plots based on environmental attributes and relate this classification to the geographic positions of the trees, then say this near the beginning of your presentation. Nothing annoys an audience so much as having to guess at the nature of your research problem.

2. Where they are not obvious, and where they can be operationally stated, specify any practical and statistical tolerances you were willing to accept in the design of your research. That is, tell your audience the criteria you used to decide whether or not the research goal was achieved. With some research goals, such as the generation of questions or hypotheses, these tolerances cannot usually be operationally stated, because subjective tolerances must be used to judge the worth of a question or hypothesis at the time they are created, and these tolerances can seldom be operationally defined. But for other research goals, such as assessing the strength of the relation of a classification of forested plots to their locations, the tolerances for a practical and statistically significant relation can be stated.

Framing a Research Application

1. Define the objects used in the cluster analysis. Say whether they are the plots of land, people, common stocks, or drug molecules. Explain why their nature is fitting for the research goal.

2. Describe how you selected the objects. If they are a sample from a known population, describe the population and explain how you drew the sample. Did you use a random sampling procedure? A systematic procedure? A haphazard procedure? Did you select them merely because they were convenient, or for other reasons? How did you determine the sample size?

3. Define the attributes. List other types of attributes you considered. Explain why those you chose are appropriate for the research goal.

4. Describe the scales of attribute measurement. State what other scales you considered and why you chose the ones you did.

5. Explain whether you performed a Q-analysis or an R-analysis, or both.

6. Was it necessary to standardize the data matrix? If so, tell how you chose the method of standardizing.

7. State the resemblance coefficient that you used, and explain why its selection was fitting for the research goal.

8. State the clustering method that you used. Of all the decisions required for a cluster analysis, the choice of the clustering method is usually the hardest to justify. Thus, it often suffices to state which method you used without giving reasons.

9. If you cut the tree into clusters to form a classification, explain why you cut the tree where you did.

10. State the value of the cophenetic correlation coefficient. State whether you believe it is or is not large enough—whether it does or does not show practical significance.

11. Reference the name and source of the computer program you used.

12. Cite the cluster analysis reference books you are using as background for your discussion.

Some of these steps will be difficult because they are subjective. You may feel that a given decision was the best one, but you may not be able to tell yourself or your audience why. Do the best you can, however.

15.2 OPTIONAL INFORMATION

Although in your written report of your research you may choose to include the data matrix, the resemblance matrix, and the tree, few journals accepting reports for publication will publish all of these owing to space limitations. Deciding which of these to include in a manuscript meant for publication requires an informal cost/benefit analysis. Here are some considerations.

1. If the initial data matrix is small, say 200 values of data or less, then including it will probably be worthwhile. Use a rearranged data matrix, however—it is easier for others to understand. Including initial data matrices larger than this will likely need to be decided case by case.

2. Often a large initial data matrix will translate into a small resemblance matrix, say 200 values of data or less. If so, including the rearranged resemblance matrix may be worthwhile. Juffs's (1973) article contains one of the largest resemblance matrices ever published: $t = 77$ objects, giving 2,926 values and requiring a special foldout to display it. The matrix is hardly informative enough to make the cost of such a treatment worthwhile.

3. Since a cluster analyst draws many conclusions directly from the tree, it usually makes sense to display it. If it is small, it can be shown in its entirety. Trees can be greatly reduced in size and still be informative. For example, Hodson et al. (1966) display a tree containing 107 objects in a journal illustration of only 3 by 5 inches. There are space-conserving ways to present the information larger trees contain (discussed in the next section).

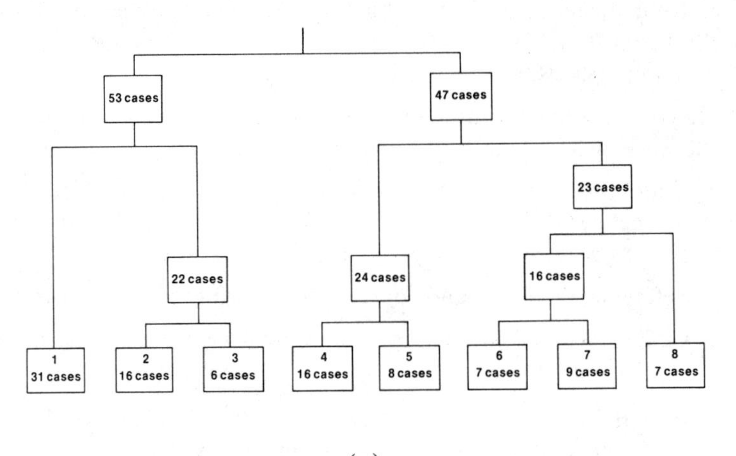

(*a*)

Figure 15.1. Examples showing various ways to present a tree. (*a*) A gross tree of psychiatric cases that summarizes its individual branches; there are eight major clusters of cases. (By permission, redrawn from Reid et al. (1978), Figure 1.) (*b*) A tree with shaded clusters. (By permission, redrawn from Scott et al. (1977), Figure 6a.) (*c*) A tree whose objects are related to the time and place of their collection. (By permission, reproduced from Sale et al. (1978), Figure 5.) (*d*) A tree showing similarities among birch trees according to their locations. (By permission, reproduced from Kirkpatrick (1981), Figure 6.)

4. When a research goal requires that a tree be cut into clusters, it is useful to devise a table showing the mean and standard deviation of the data for each attribute by cluster. For example, a problem with 100 objects and 10 attributes yielding six clusters can be summarized with a 6-by-10 table showing the means and standard deviations of the attributes by cluster.

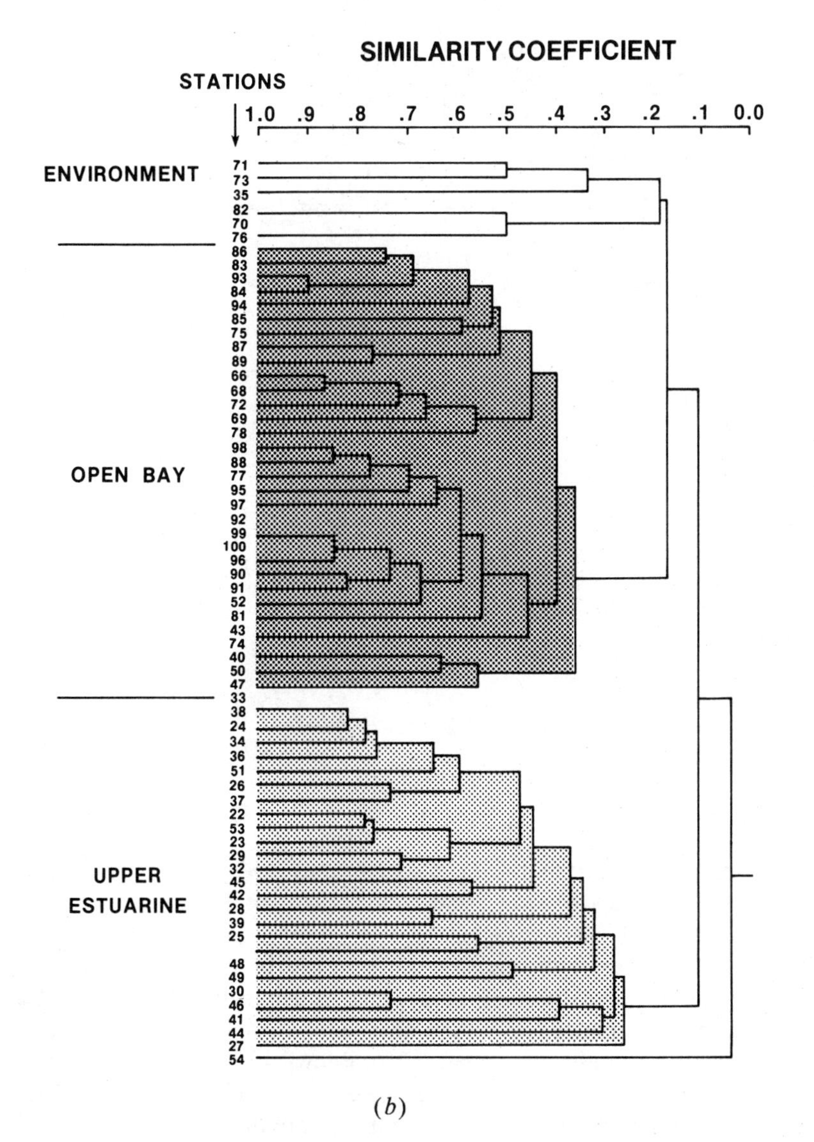

(*b*)

Figure 15.1. (*Continued*).

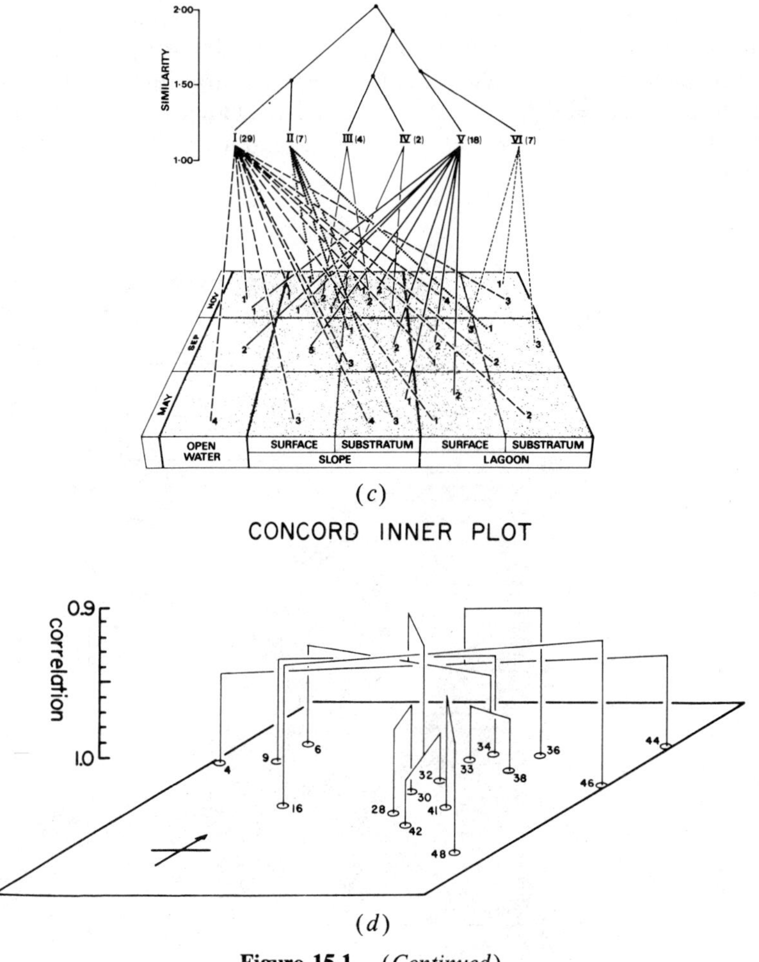

(*c*)

CONCORD INNER PLOT

(*d*)

Figure 15.1. (*Continued*).

15.3 WAYS OF PRESENTING THE TREE

Let us look at some alternative forms a tree may have.

Figure 15.1,*a* shows how Reid et al. (1978) presented a tree from a cluster analysis of 100 cases of severely retarded adults described by behavioral attributes. The original tree had 100 branches, but instead of including all of these in their report, the researchers identified 8 main branches and summarized the cases they contained. The resulting tree is probably easier to understand than the fullblown tree would be. Note that their tree does not have a vertical scale of resemblance because it is unnecessary: what matters is that the branches are proportionally correct in the vertical direction.

Figure 15.1,*b* shows how Scott et al. (1977) shaded the clusters in a tree to make them stand out. The objects are samples of Foraminifera from sediments collected in different locations of the Miramichi River estuary, New Brunswick, Canada. As the tree shows, the samples group in three major clusters corresponding to the three major regions of the estuary.

Visualizing a tree as a three-dimensional mobile makes it possible to show its relation to time and space. Figure 15.1,*c* shows how Sale et al. (1978) related a tree of samples of zooplankton, collected in Australian coastal waters, to the time and place of the sampling. Since the research goal was to investigate the relation between the classification of samples and the variables of time and space, drawing the tree this way was helpful.

Finally, Kirkpatrick (1981) displayed a tree in relation to a map on which the objects, which were birch trees, were sampled. His research depended on the fact that trees contain a record of their growth in the form of tree rings, and that local site factors of soil nutrients, soil moisture, available sunlight, and so forth determine yearly tree growth. Kirkpatrick's research goal was to determine if the similarity between growth patterns was related to where the trees grew. He cluster-analyzed a sample of birch trees described by their ring growth. He used 30 attributes, the tree ring widths for a recent 30 year period. Figure 15.1,*d* shows how he displayed the relation of the tree resulting from the cluster analysis to the locations of the birch trees. From this it is clear that the growth patterns of the birches are related to their locations.

SUMMARY

Because research is a communal enterprise, you must present your cluster analysis clearly and completely. You must give your audience key information that spells out how you framed and validated your application. They must understand what you did well enough to be able to do it themselves.

Space permitting, you may also give them optional information such as the data matrix. This will depend on the amount of data and how much it will add to the audience's understanding.

Finally, most reports of applications of cluster analysis include a tree diagram. It can be drawn in various ways that conserve space, that highlight certain features, or that help promote understanding.

PART III

How to Use Cluster Analysis to Make Classifications

> We take a handful of sand from the endless landscape of awareness around us and call that handful of sand the world. Once we have the handful of sand, the world of which we are conscious, a process of discrimination goes to work on it. We divide the sand into parts. This and that. Here and there. Black and white. Now and then. The handful of sand looks uniform at first, but the longer we look at it the more diverse we find it to be. Each grain of sand is different. No two are alike. Some are similar in one way, some are similar in another way, and we can form the sand into separate piles on the basis of this similarity and dissimilarity. Shades of color in different piles—sizes in different piles—grain shapes in different piles—grades of opacity in different piles—and so on, and on, and on.
>
> ROBERT M. PIRSIG
> *Zen and the Art of Motorcycle Maintenance*

Chapter 3 introduced you to the use of cluster analysis to make classifications. There, it was explained generally how a researcher must cut the branches of the tree in order to define a classification, and how the researcher can identify objects after defining a classification. Then, Chapter 5 described applications in which cluster analysis was used to make general-purpose and specific-purpose classifications.

Now, Part III explains important details omitted earlier.

Specifically, Chapter 16 offers a detailed treatment of how cluster analysis can be used to make classifications. Among topics treated there are: how to weight select attributes so that they contribute more or less to the similarities among objects; and how to determine the level of similarity at which to cut the tree to form a classification.

Chapter 17 shows in greater detail how to use a classification to identify objects. It explains how calculations of identification depend upon which resemblance coefficient and clustering method you used to create the classification. It also explains how, in situations where the calculations of cluster analysis are prohibitively expensive because of the large number of objects being cluster-analyzed, identification can be used to reduce the costs.

Finally, Chapter 18 presents a philosophy of classification and identification that you can use to frame the subjects.

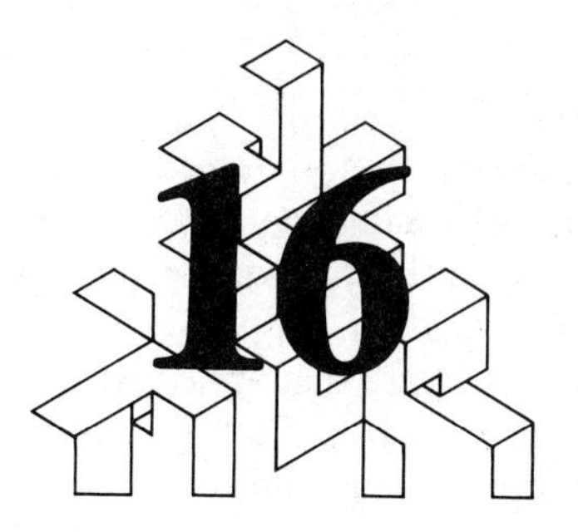

How to Make Classifications

OBJECTIVES

Our aim in this chapter is to discuss in some detail how to use cluster analysis to make classifications. Assuming a familiarity with the material on classifications covered in Chapters 3 and 5, our discussion will take us through the following topics:

- The steps required to make a classification.
- The kinds of classifications that arise in research.
- The weighting of attributes to change their contributions to the classification.
- How to decide the level of similarity at which to cut the tree to make the classification.

16.1 STEPS IN MAKING A CLASSIFICATION

For a general-purpose and a specific-purpose classification, the steps are the same. To illustrate them, we will enumerate the steps numerical taxonomists follow when making biological classifications. From there it will be easy to see how the steps generalize for any kind of classification.

Figure 16.1 shows the steps. It is taken from a figure that appeared in a paper by Sneath (1964*a*), and it retains the flavor of special terms used in numerical taxonomy. We will translate those terms as we go along.

1. Choose the Biological Specimens. The specimens are the objects of the data matrix. Normally they are a sample from a known taxon—for example, from a known genus that is to be partitioned into finer classes of

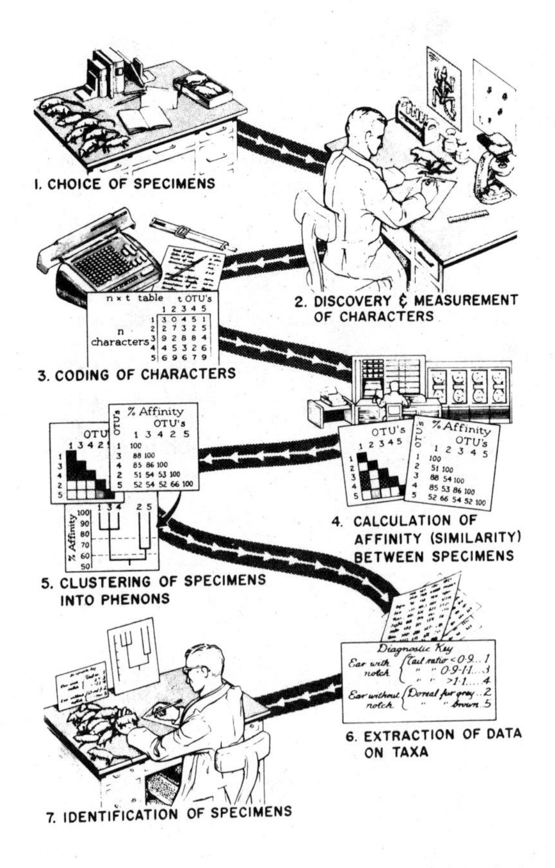

Figure 16.1. Steps for using cluster analysis to make a biological classification. (By permission, reproduced from Sneath (1964*a*, Figure 1.)

species, or from a known species that is to be partitioned into finer classes of subspecies. In the jargon of numerical taxonomy, a specimen is an **OTU**, which stands for **operational taxonomic unit**.

2. Choose and Measure the Characters. A **character** is another word for **attribute**. The numerical taxonomist decides what characters to measure, and their scales of measurement. Normally the characters are morphological measures describing the form and structure of the specimens. The guiding philosophy is that morphological differences in classes of life mirror biological differences.

3. Code the Characters. In preparation for data processing, the measurements of the n characters made on the t OTU's are entered into a data matrix. Optionally, the data matrix can be standardized.

4. Calculate the Similarity between Pairs of OTU's. The numerical taxonomist decides how the overall similarity between pairs of OTU's should be defined. For example, should size differences between data profiles contribute to similarity or should they be ignored? After deciding this, the taxonomist chooses the resemblance coefficient that makes the decision operational, and uses it with the data matrix to compute the resemblance matrix.

5. Choose and Execute the Clustering Method. This step transforms the resemblance matrix into a **phenogram** or **dendrogram**, two synonyms for what we have been calling a **tree**. To produce the classification, the numerical taxonomist cuts the phenogram into clusters of similar specimens (similar OTU's) called **phenons**. The phenogram is also used to rearrange the resemblance matrix, making the classes (clusters) stand out in the resemblance matrix. The cophenetic correlation coefficient is calculated to see if the phenogram faithfully portrays the similarities in the resemblance matrix.

6. Extract Key Data on Taxa. The numerical taxonomist finds and removes or "scrubs out" the inessential characters while retaining the essential ones. Essential characters are those that drive the clustering. Removal of the essential characters from the data matrix and performance of another cluster analysis should give a different set of clusters. On the other hand, inessential characters can be removed from the data matrix without greatly changing the set of clusters.

As a final check, the numerical taxonomist actually removes characters suspected of being inessential and cluster-analyzes the reduced data matrix to obtain a new phenogram. If the removed characters are really inessential, the same clusters should result. This requires a trial-and-error process of removing and reclustering until the final set of essential characters is found. The range of values of the essential characters for each cluster gives a diagnostic key for using the classification.

7. Use the Key to Identify New Specimens. By publishing the key to the classification, the taxonomist makes it possible for others to use it to identify (amateur naturalists often say "key out") specimens whose taxa are unknown.

Scrubbing out inessential characters eliminates errors due to the statistical "noise" that accompanies the inessential characters, and it reduces the costs of using the classification. For example, a classification of dogs and cats based on several hundred morphological characters will give clear identification. But it turns out that clearer identification can be had by finding one key character—let us say the number of teeth: dogs have 42 teeth, cats have 30. And instead of having to measure hundreds of characters to identify an animal, measuring one or a few will do the job better. However, the essential characters are only apparent in retrospect after a full set has been used and the inessential ones scrubbed out.

The steps for making a biological classification carry over to any kind of classification. For instance, instead of biological specimens, the objects could be drug substances, such as cocaine, heroin, phenobarbital, or others, with attributes from chemical analysis describing their infrared spectra. Cluster analysis will give a classification of drugs useful for identifying suspected drug substances. Moss et al. (1980) supply other details of drug analysis through clustering, and step for step they agree with Figure 16.1.

16.2 KINDS OF CLASSIFICATIONS

Let us make a fourfold categorization that will contain both general-purpose and specific-purpose classifications. To do this, we first note that **biological objects** are used to make biological classifications, and **nonbiological objects** are used to make nonbiological classifications. Biological objects are living entities—for example, specimens of plants or animals. Nonbiological objects are nonliving entities—samples of drugs or plots of land.

Likewise, attributes can be biological or nonbiological. A **biological attribute** is one with a biological basis. Biological attributes are of course used to describe biological objects, such as an animal's blood chemistry, hormone levels, respiration rates, or genetic makeup. Biological attributes can also be used to describe nonbiological objects. For example, plots of land (nonbiological objects) can be described by biological attributes that give the presence or absence, or the abundances, of the plant and animal species inhabiting the plots.

An example of a **nonbiological attribute** is a morphological attribute that describes some aspect of the form and structure of objects. Clearly, morphological attributes can be used to describe biological objects, or they can be used, as geologists and geographers use them in geomorphology, to describe nonbiological features of the earth.

While it is possible to use mixtures of biological and nonbiological attributes in a classification, such applications are rare.

The four categories of data matrices that we have described are shown in Figure 16.2.

The lefthand column denotes biological classification. Historically, biological classification has been dominated by the use of nonbiological attributes, specifically morphological attributes: the Linnaean system classifies the plant and animal kingdoms according to form and structure. More recently, technological advances have permitted classifications of life at the species level on the basis of biological attributes such as blood serum and plasma measurements.

The righthand column denotes nonbiological classification. The upper cell denotes a classification of nonbiological objects described by biological attributes. Classifications in this cell fall under the purview of ecology, where plant and animal ecologists make classifications—let us say, of plots of land or samples of water—described by the plant or animal life they contain. The objects (plots of land or samples of water) serve as a matrix that holds life.

The lower cell of the righthand column denotes classifications in which nonbiological objects are described by nonbiological attributes. For example, in recreation research a classification of park land could be made by using land form measurements. The research goal might be to relate this classification to users' perceptions of aesthetic quality.

		Objects	
		Biological	Nonbiological
Attributes	Biological	Example: Douglas fir trees described by their photosynthetic properties.	Example: Plots of land described by the species of plants and animals living on them.
	Nonbiological	Example: Finches described by their morphological or behavioral properties.	Example: Samples of ore described by their chemical properties.

Figure 16.2. A fourfold categorization of classifications. The objects can be biological or nonbiological, as can the attributes.

Being aware of these four types of classifications is important because the cluster analysis for each must be framed differently. Books and review articles on numerical taxonomy usually give framing criteria specifically for given subject areas. Though some of the framing criteria hold for all four types of classifications, in general they do not. There is always the danger of reading a book that has "numerical taxonomy" in its title, taking its advice, and ending up using the wrong framing criteria because its advice is meant for another subject area.

A book such as *Numerical Taxonomy* (Sneath and Sokal, 1973) is mostly concerned with biological classifications based on morphological attributes (such as the example in the lower lefthand cell in Figure 16.2). The framing criteria contained in Sneath and Sokal are normative principles for making biological classifications, principles that have evolved from thousands of numerical taxonomic studies. But a book such as *An Introduction to Numerical Classification* (Clifford and Stevenson, 1975) gives framing criteria for making classifications in plant ecology. Its advice is directed toward uses of the sort mentioned in the upper righthand cell of the matrix.

The framing criteria these two books give are somewhat different because each addresses separate aims, not because one is right and the other is wrong. For example, Clifford and Stephenson's book favors such resemblance coefficients as the Bray-Curtis coefficient and the Canberra metric coefficient, because the successful achievement of ecological research goals by means of these coefficients has reinforced their use. On the other hand, Sneath and Sokal downplay these two coefficients and put the average Euclidean distance coefficient and the correlation coefficient in a more prominent light. Similarly, the numerical taxonomy books of Mueller-Dambois and Ellenberg (1974) and Orlóci (1975) are for plant ecologists, and they reflect this in the framing criteria they recommend.

Sections of the book *Statistics and Data Analysis in Geology* (Davis, 1973) describe how cluster analysis is used to make classifications of samples of geological material based on chemical properties. Its advice is for geological research, which is of the sort indicated in the lower righthand cell in Figure 16.2. Other books, such as *Cluster Analysis for Applications* (Anderberg, 1973), do not distinguish these four types of classifications and instead compile extensive lists of mathematical options for using cluster analysis. What framing criteria they specify are based on desirable mathematical properties of the methods, not on knowledge of the subject matter to be classified. To date, no book has addressed the case in the top lefthand cell of Figure 16.2, in which biological objects are classified according to biological attributes.

In addition to the four categories of classifications so far explained, two other classification schemes are of interest. First, classifications can be built of **monothetic** or **polythetic** classes. Second, the classes can be **natural** or

artificial, which is the same as classifying them as general-purpose or specific-purpose.

Monothetic Classes. To illustrate a monothetic classification, let us consider a subpopulation of all people: those having red or blond hair, and blue or green eyes. We will invent a monothetic classification of these people, which will separate them into eight classes, its basis being three binary attributes: (1) color of hair, red/blond; (2) color of eyes, blue/green; and (3) height, shorter-than-6-feet/6-feet-and-taller. The three attributes, each with two states, give the following (2)(2)(2) = 8 classes.

Class	Defining properties
1	Red hair and blue eyes and shorter than 6 feet
2	Red hair and blue eyes and taller than 6 feet
3	Red hair and green eyes and shorter than 6 feet
4	Red hair and green eyes and taller than 6 feet
5	Blond hair and blue eyes and shorter than 6 feet
6	Blond hair and blue eyes and taller than 6 feet
7	Blond hair and green eyes and shorter than 6 feet
8	Blond hair and green eyes and taller than 6 feet

Were a fourth attribute to be added—say one with three states such as "age, young/middle-aged/old," how many classes do you think there would be? A little reflection shows the number of classes would triple, and there would be (2)(2)(2)(3) = 24 classes.

Supermarkets arrange produce according to monothetic classifications. An object that is yellow, oblong, and easily peeled belongs to the "banana" class, and no other fruit except bananas fits this class.

A problem of monothetic classifications is their tendency to have so many classes that they become useless. When many attributes are used, each with many states, the number of classes expands exponentially, easily exceeding the number of grains of sand necessary to fill up the visible universe—10^{100} by conservative estimate, or 1 followed by 100 zeros. Yet a monothetic classification using 100 attributes, each attribute described by 10 states, has exactly this many classes. Even with fewer attributes and states the number of classes quickly mounts. For example, a monothetic classification having 16 attributes, each described by 4 classes, has 4^{16} or 4,294,967,296 classes. Most of these classes will be empty—that is, nothing will exist to identify into them. For this reason, monothetic classifications cannot be used when the number of attributes is large.

Polythetic Classes. Polythetic classifications can have many attributes, yet still can be made to have a few classes. Their basis is a principle of overall shared similarity: for objects in a polythetic class to be regarded as similar, they must exhibit high overall similarity even though they can be dissimilar on a few attributes. Suppose that two objects are being compared on five attributes, and each attribute "votes" on the comparison: attribute #1 says, "There is close agreement on me, so I vote for high similarity"; attributes #2, #3, and #5 say the same; but attribute #4 says, "There isn't a good agreement on me, so I vote for a fair degree of dissimilarity." When the votes are tallied, four are for high similarity and one is for a fair degree of dissimilarity. The yeas win out, the two objects are therefore declared to be quite similar overall, and they are grouped in the same polythetic class.

Polythetic classes permit differences on a few attributes so long as the *average* difference over the whole set of defining attributes is tolerably small. From this description it is clear that the classifications made through cluster analysis are polythetic.

Polythetic and monothetic classifications are invented differently. Polythetic classifications are based on a sample of objects. First, a sample of objects from a population is selected. Second, those pairs of objects that show high overall similarity are placed in the same class. The average properties of the objects in each class, by attribute, define the classification. Monothetic classifications, however, are usually invented independently of sample objects. The attributes and the expected range of variation on each is first determined, and this defines and bounds the attribute space. This space is then partitioned into monothetic classes. Sneath and Sokal (1973, pp. 20–27) discuss the two types of classifications in greater depth.

Natural and Artificial Classes. Gilmour (1951, p. 401) distinguishes the concepts of natural and artificial classifications in this way: "classifications which serve a large number of purposes are called natural, while those serving a more limited number of purposes are called artificial." This means that a natural classification is another term for what we called a general-purpose classification in Chapter 5, whereas an artificial classification is what we called a specific-purpose classification. The reason for choosing these terms is that general-purpose and specific-purpose are self-descriptive.

Planners and managers are more often concerned with specific-purpose classifications than with general-purpose ones. Often a specific-purpose classification will predict what a planner needs to know. For example, Platts (1979) made a specific-purpose classification of sections of western trout streams. He chose physical attributes such as the type of stream bank and the sizes of the stream pools, specifically because he knew that the abundance of trout and salmon depends on these. Consequently his classification of stream sections could be used to predict fish abundance. This is useful information for environmental planners, and a reason for mapping stream sections according to this classification of fish habitat.

16.3 WEIGHTING ATTRIBUTES

Weighting attributes changes the contribution each makes towards the resemblance between objects. Moreover, weighting influences the entire cluster analysis and affects the similarities among objects in the tree. While in this section we examine the effects of weighting on those research goals that require a classification to be made, much of the discussion will carry over by analogy to the other research goals presented in Chapters 4 and 6.

There are four ways of weighting attributes. First, the researcher can deliberately choose to leave some attributes out of the original data matrix. Excluding attributes in effect gives them a zero weight. That is, the researcher distinguishes between a zero weight for what should be left out and a nonzero weight for what should be included.

A second way of weighting requires that an R-analysis of the data matrix be made before the Q-analysis is made. The R-analysis will find any sets of highly correlated attributes. Such correlated attributes are surrogates of each other, and the researcher may view their inclusion in the data matrix as a form of multiple counting of a single underlying attribute. If so, the researcher can select one attribute from each correlated set to represent the set, giving a new data matrix having a reduced set of nearly uncorrelated attributes. The attributes removed in this way receive zero weight and contribute nothing to the resemblance between objects.

For data matrices with quantitative attributes, a variation on this procedure uses principal component analysis for the R-analysis. It gives a new set of n attributes and their values: these are the principal components. Although the principal components are abstract and lack identifiable real-world meaning, they are nonetheless a basis for describing the objects. Moreover, they have the desirable feature of being pairwise uncorrelated with each other. Using the principal components in place of the original attributes is essentially a form of weighting.

Third, any of the methods described in Chapter 7 for standardizing the attributes changes the original attributes to standardized attributes. This alters how much each attribute contributes to overall resemblance.

Fourth, it is possible to introduce weights to make the attributes contribute desired amounts to the resemblance between objects. An attribute's weight can be increased by repeating its presence in the data matrix. As an explanation of this, suppose the original data matrix has only two attributes, and suppose we want the first attribute to contribute 60 percent toward overall resemblance and the second to contribute 40 percent, relative to what they contribute in their original forms. We would add two repeats of attribute 1 to the data matrix and one repeat of attribute 2. The new data matrix would contain five attributes, the first attribute of the original data matrix appearing three times and the second appearing two times. The first attribute would carry a weight of 60 percent, the second a weight of 40 percent.

It is easier to do this by leaving the original data matrix alone and instead simulating the weighting at the time the resemblance coefficient is computed. Here is how this works when the average Euclidean distance coefficient, d_{jk}, is used for illustration. We rewrite the coefficient, given by equation 8.2, as follows:

$$d_{jk} = \left[\frac{\sum_{i=1}^{n} W_i (X_{ij} - X_{ik})^2}{\sum_{i=1}^{n} W_i} \right]^{1/2}$$

The weights W_i are selected from the integers, $1, 2, 3, \ldots$. For the examples given above, we would make $W_1 = 3$, $W_2 = 2$ and $n = 2$. This would give 60 percent of the weight to attribute 1, 40 percent to attribute 2, and would not require that the original data matrix be rewritten with five attributes.

Because this procedure for introducing weights into resemblance coefficients is straightforward, we will not write the weighted form of each of the coefficients presented in Chapters 8 and 10: the procedure just outlined can be applied to these other coefficients.

The methods of weighting attributes can be combined to produce a variety of schemes. For example, we could first standardize the data matrix and then reweight the standardized data matrix to achieve a desired weighting. The effect would be to remove the natural weights first and then impose desired weights.

Even though the mechanics of weighting are concrete and easy to execute, deciding whether to weight the attributes and, if so, deciding in what ratios, is difficult. For weighting to make sense, we must concretely know the research goal we are addressing. The research goal and its context are necessary to give a scale of values against which the relative importance of individual attributes can be assessed. Putting this another way, decisions on weighting attributes are framing decisions, and they can only be made intelligently when the analysis is aimed at achieving a well-defined research goal.

Framing decisions are usually easier to decide when the research goal is "Goal 6: Facilitate Planning and Management" because planning must articulate the values against which good and bad plans can be distinguished. At the other extreme is "Goal 4: Make a General-Purpose Classification." Such classifications are meant to serve science in a general but unspecified way. Because the background values are vague, it is impossible to decide how to weight the attributes. It makes no sense to say that one attribute should count, say, five times as much as another when you have no specific use for your classification in mind.

On the other hand, Platts's (1979) classification of trout streams was made with the specific purpose of predicting the abundance of fish. For this purpose, it would make sense to weight some attributes more than others. The best set of weights could be found through trial and error, by making a series of classifications with different weights and choosing the one that gives a classification that predicts fish abundance the best.

16.4 HOW TO DETERMINE WHERE TO CUT THE TREE

The level of similarity at which we cut the tree determines the number of classes in a classification. Determining this level depends on whether the classification is general-purpose or specific-purpose.

To illustrate the general-purpose case, we will take eight of the animal dental formulas given in the data matrix shown in Figure 4.2,*a*, form a smaller data matrix, and then cluster-analyze it by using the average Euclidean distance coefficient and the UPGMA clustering method (we will use the smaller data matrix only because it makes the example shorter). Figure 16.3 shows the resulting tree, and it lists the dental formulas that define the classifications for each of three tentative cuts. For example, cut 1 at $d_{jk} = 0.4$ gives five clusters, and cluster 1 contains the raccoon and the weasel. But cut 2 at $d_{jk} = 0.6$, which relaxes the tolerance for cluster membership, gives three clusters, and cluster 1 contains not only the raccoon and weasel (of the old cluster 1) but also contains the pig and dog.

One strategy is to cut the tree at some point within a wide range of the resemblance coefficient for which the number of clusters remains constant, because a wide range indicates that the clusters are well separated in the attribute space. Let us explain what this entails. Looking at Figure 16.3 we see that in the range of $0.0 \le d_{jk} < 0.35$ there are eight clusters, each containing one object. At $d_{jk} = 0.35$, several clusters are simultaneously merged, leaving five clusters. In the range of $0.35 \le d_{jk} < 0.54$ there are five clusters. Similarly, in the range of $0.54 \le d_{jk} < 0.56$ there are four clusters, and in the range of $0.56 \le d_{jk} < 0.78$ there are three clusters—and so on until there is one cluster. Table 16.1 summarizes this, listing the number of clusters in order of the decreasing width of the ranges of d_{jk}. This means that the decision of where to cut the tree is least sensitive to error when the

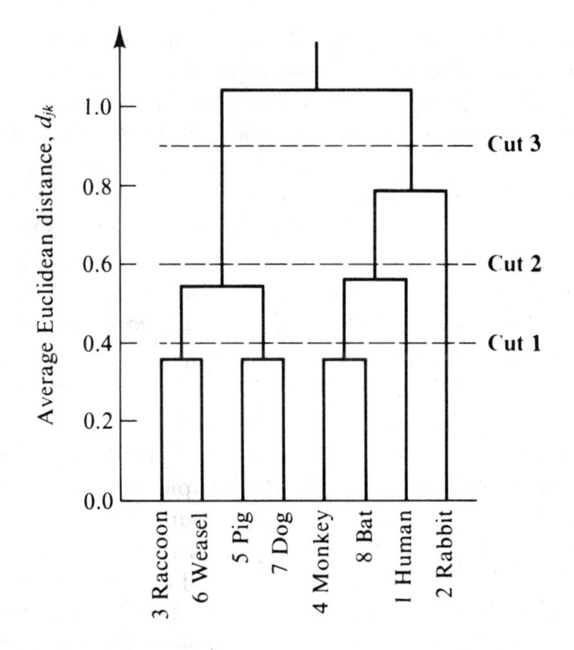

Cut 1: $d_{jk} = 0.4$

Cluster 1: (raccoon, weasel)
Cluster 2: (pig, dog)
Cluster 3: (monkey, bat)
Cluster 4: (human)
Cluster 5: (rabbit)

Cut 2: $d_{jk} = 0.6$

Cluster 1: (raccoon, weasel, pig, dog)
Cluster 2: (monkey, bat, human)
Cluster 3: (rabbit)

Cut 3: $d_{jk} = 0.9$

Cluster 1: (raccoon, weasel, pig, dog)
Cluster 2: (monkey, bat, human, rabbit)

Figure 16.3. Tree of animal dental formulas showing three levels of similarity for cutting the tree into clusters.

Table 16.1 Number of clusters obtained for different ranges of d_{jk}, for the tree shown in Figure 16.3.

Number of clusters	Range of d_{jk}	Width of range
8	$0.0 < d_{jk} < 0.35$	0.35
2	$0.78 < d_{jk} < 1.04$	0.26
3	$0.56 < d_{jk} < 0.78$	0.22
5	$0.35 < d_{jk} < 0.54$	0.19
4	$0.54 < d_{jk} < 0.56$	0.02
1	$1.04 < d_{jk} < \infty$	–

width of the range is the largest. For the example, this means that choosing two clusters is a good choice, while choosing four is a bad choice.

If we believe that nature is economic and therefore we should be too, then we should strive for a few clusters instead of many. For this example, cutting the tree in the range of $0.78 < d_{jk} < 1.04$ gives a relatively wide range of 0.26 units and two clusters, and this seems like a good compromise.

This strategy also holds for specific-purpose classifications, but for these another consideration enters the picture. The tree should be cut to produce classes that are maximally related to other specific variable(s) of interest. For many applications it will turn out that this agrees with what the economizing strategy just outlined dictates. Perhaps the reason for this agreement is that natural law operates somewhat the same on a set of objects across the several attribute spaces that we might choose to view them in. For example, natural law has made species of animals well defined in a morphological attribute space. But it has also defined them well in much the same way in a behavioral attribute space. Thus, knowledge of an animal's species allows us to predict much of its behavior. This is to say that the winds of evolution blow in much the same way in different attribute spaces, setting up correlations between animal form and behavior.

SUMMARY

The making of a classification follows the steps shown in Figure 16.1, which illustrates them for a biological classification.

Classifications themselves can be classified in several ways.

First, the objects can be biological or nonbiological, and the attributes can be biological or nonbiological, creating a fourfold classification. For example, the taxonomy of life is based on biological objects that are described by nonbiological (morphological) attributes. On the other hand, an ecologist's classification of land is based on nonbiological objects (plots of land) that are described by biological attributes (for instance, abundance of plant species).

Second, the classes can be polythetic or monothetic. Polythetic classes define similarity of the members of a class on the basis of their shared overall similarity. This allows members to be dissimilar on some attributes so long as they are similar when judged over all the attributes. Monothetic classes are those that contain members that closely agree on *all* their attributes. Cluster analysis produces polythetic classes.

Third, a classification can be natural (having a general purpose) or it can be artificial (having a specific purpose).

Attributes in the data matrix can be weighted to make certain ones contribute more or less to the similarities among the objects. Weighting only makes sense when the research goal is well defined and generates a scale of values that allows the researcher to estimate subjectively how much more one attribute should contribute over another.

The decision of where to cut the tree to make a classification depends on a combination of factors. We usually want the clusters to be few in number and to be well defined. However, the purpose of a classification also is a factor. Whether we make a general-purpose classification to serve science generally, or a specific-purpose classification to be related to other variables specifically, ultimately we want the classification to be judged valid —that is, we want it to work. Thus we should cut the tree at a level of similarity that makes it work the best. Often this will coincide with a few clusters that are well defined.

How to Identify Objects into a Classification

OBJECTIVES

Having seen how classifications are made, we now turn to how they are used.

- All uses require that objects be identified into their proper classes. This leads us naturally to
- The use of identification as a way of cluster-analyzing sets of objects that are too large for current computers.

17.1 HOW TO IDENTIFY OBJECTS

Suppose that you have cluster-analyzed a set of objects and cut the tree into a set of clusters to define a classification. Next you want to know which of these clusters (classes) an object U belongs to. How would you do this?

Chapter 3 contains an example that illustrates the calculations required to identify U. The classification in that example was made by using the Euclidean distance coefficient e_{jk} and the UPGMA clustering method. To identify an unknown object U into that classification, we first computed the mean data profile of each cluster. Then we computed the Euclidean distances between the unknown object U and each of the mean data profiles, and identified U into the cluster it was most similar to (that is to say, was closest in distance to). This procedure, however, only works well for the Euclidean distance coefficient and the UPGMA clustering method. For this reason we shall now examine a more general procedure for identification. This procedure depends on whether a dissimilarity coefficient or a similarity coefficient was used to make the calculation. Although we will explain it by using the Euclidean distance coefficient e_{jk}, a particular dissimilarity coefficient, and the correlation coefficient r_{jk}, a particular similarity coefficient, the explanation generalizes to any dissimilarity coefficient or any similarity coefficient.

To illustrate the procedure, consider a cluster analysis of five objects based on the Euclidean distance coefficient e_{jk}. Suppose the tree is cut into two clusters, (14) and (235), defining two classes of a classification. The procedure for identifying an unknown object U depends on the clustering method that was used to make the classification. For the UPGMA, SLINK, and CLINK clustering methods, the procedures are as follows:

UPGMA: Compute the average value of the resemblance coefficient e_{jk} between the unknown object U and each object member of each cluster. That is, compute $e_{U(235)} = 1/3(e_{U2} + e_{U3} + e_{U5})$, and $e_{U(14)} = 1/2(e_{U1} + e_{U4})$. Object U is identified with the cluster it is most similar to. Since e_{jk} is a dissimilarity coefficient, this means that (1) if $e_{U(235)}$ is the smallest, U belongs to (235); (2) but if $e_{U(14)}$ is the smallest, U belongs to (14).

SLINK: Compute the *minimum* value of e_{jk} between the unknown object U and each member of each cluster. That is, compute $e_{U(235)} = \min(e_{U2}, e_{U3}, e_{U5})$, and $e_{U(14)} = \min(e_{U1}, e_{U4})$. Object U is identified with the cluster it is most similar to. Thus, if $e_{U(235)}$ is the smallest, U belongs to (235); but if $e_{U(14)}$ is the smallest, U belongs to (14).

CLINK: Compute the *maximum* value of e_{jk} between the unknown object U and each member of each cluster. That is, compute $e_{U(235)} = \max(e_{U2}, e_{U3}, e_{U5})$, and $e_{U(14)} = \max(e_{U1}, e_{U4})$. Object U is identified with the cluster it is most similar to. Thus, if $e_{U(235)}$ is the smallest, U belongs to (235); but if $e_{U(14)}$ is the smallest, U belongs to (14).

This has illustrated calculations for e_{jk}, and the procedures hold for any other dissimilarity coefficient. When a similarity coefficient is used, such as the correlation coefficient r_{jk}, the procedures change. Suppose that instead of using e_{jk}, the cluster analysis that produced the classification had used r_{jk}. Depending on the clustering method used, the identification of objects would go as follows:

UPGMA: Compute the average value of the resemblance coefficient r_{jk} between the unknown object U and each member of each cluster. That is, compute $r_{U(235)} = 1/3(r_{U2} + r_{U3} + r_{U5})$, and $r_{U(14)} = 1/2(r_{U1} + r_{U4})$. Object U is identified with the cluster it is most similar to. Since r_{jk} is a similarity coefficient, this means that (1) if $r_{U(235)}$ is the largest, U belongs to (235); (2) but if $r_{U(14)}$ is the largest, U belongs to (14).

SLINK: Compute the *maximum* value of r_{jk} between the unknown object U and each member of each cluster. That is, compute $r_{U(235)} = \max(r_{U2}, r_{U3}, r_{U5})$, and $r_{U(14)} = \max(r_{U1}, r_{U4})$. Object U is identified with the cluster it is most similar to. Thus, if $r_{U(235)}$ is the largest, U belongs to (235); but if $r_{U(14)}$ is the largest, U belongs to (14).

CLINK: Compute the *minimum* value of r_{jk} between the unknown object U and each member of each cluster. That is, compute $r_{U(235)} = \min(r_{U2}, r_{U3}, r_{U5})$, and $r_{U(14)} = \min(r_{U1}, r_{U4})$. Object U is identified with the cluster it is most similar to. Thus, if $r_{U(235)}$ is the largest, U belongs to (235); but if $r_{U(14)}$ is the largest, U belongs to (14).

In summary, the steps of identification are consistent with the steps used to build the classification. To identify the class membership of an unknown object, the resemblance between the unknown object and each cluster is computed with (1) the same resemblance coefficient and (2) the same clustering method that were used to make the classification.

Identification in Ward's clustering method follows the same strategy of placing U in the cluster it is most similar to. This means placing U in the cluster that results in the smallest increase in the total sum of squares, E, computed for the t original objects plus the unknown object U. Tentatively, U is placed successively in each of the clusters, but U's tentative placement is not allowed to change the cluster means. Object U is identified into the cluster for which its inclusion gives the smallest increase in the value of E.

There are several other points to note. Identification can be done with either the full set of n attributes or a reduced set of key attributes. Regardless of which, the procedures are the same ones that have been outlined here. And if the classification was based on a standardized data matrix, then the attributes of the unknown object U should be standardized by using the standardizing function and parameters that were used to standardize the original data matrix. The standardizing function, with its parameters set by the original data matrix, should be applied to U before it is identified. As many unknown objects as you have can be identified into a classification. But each identification is treated as if it were the only case. That is, once an unknown object is identified, it is never included with the t original objects where it could affect the subsequent identification of other unknown objects.

Appendix 2 describes the computer program CLUSTID. For the resemblance coefficients and clustering methods covered in this book, it will identify objects into a classification.

Discriminant analysis, a method of multivariate analysis, is often used to identify objects. To do this, the objects' attribute measurements and their classes as determined by cluster analysis are entered as a set of data into a discriminant analysis. The fitted discriminant function can then be used to predict an unknown object's class membership by using its measured attributes.

17.2 STRATEGIES FOR CLUSTERING "TOO MANY" OBJECTS

When more objects are used than the cluster analysis computer program can handle, you can use one of the following strategies. Although only the first of these strategies depends on identification as a solution, this chapter is a convenient place to group and discuss other strategies.

The first strategy goes like this. From the available objects, pick at random as many objects as the computer can handle, say 400. Cluster-analyze them, cut the tree into clusters, and identify the remaining objects into the clusters. In the end, each object will belong to some cluster. With this strategy the randomly selected objects are used to define the clusters, and the other objects are identified into the clusters. A danger is that the randomly selected objects may not be typical of all the objects, and when identifications are made there will be atypical objects that do not fit any of the clusters well.

This strategy can be used even when the research goal does not require a classification to be made. Essentially, any tree can be regarded as a classification in which each of the t objects is the only member of its cluster; that is, the tree is cut at its bottom where similarity is maximal, giving t clusters. Identification will pair each unknown object with the object in the tree that it is most similar to.

While it cannot always be used, a second strategy is to average the properties of certain sets of objects before cluster-analyzing them. Each object in the data matrix then becomes an average object, and it is the average objects that are cluster-analyzed. For example, suppose the objects are specimens of various species, and the purpose of the cluster analysis is to produce a tree that shows the similarities among species. If there are too many specimens to handle in one cluster analysis, their attribute values can be averaged by species to produce an average data profile for each species. Then the average data profiles can be cluster-analyzed. Whether or not the details of the variation among specimens of each species can be sacrificed by averaging depends on the context of the research.

Another approach is to make a preliminary analysis aimed at finding a subpopulation of objects that is typical of the main population, and then cluster-analyze the subpopulation. Hodson et al. (1966) used this method for a cluster analysis of 350 Iron Age brooches, which at the time of their study proved to be too many to cluster as a set. They picked 100 brooches they felt were "reasonably representative of the variation" of the full set and cluster-analysed them. They based their research conclusions on this subpopulation, and they used analogy to scale their conclusions to the main population.

Still another approach, due to Sneath (1964*b*, p. 338) and called the **reference object method**, is as follows. When there are too many objects to cluster directly, a cluster analysis is carried out on a large random sample of them, and from each well-defined branch of the tree the analyst picks, say, three reference objects. Then a second cluster analysis is made that includes another sample of the remaining objects (those not in the first run) plus the reference objects. If any new well-defined branches emerge that do not contain any of the reference objects, the analyst picks three objects from each of these as additional reference objects. Then all of the reference objects from the first and second runs are added to another sample of the objects that remain to be cluster-analyzed, these are cluster-analyzed, and the analyst then looks for new well-defined branches of the tree that do not contain any reference objects.

When repeated until each object of the original set has gone through at least one cluster analysis, the final tree will contain all of the well-defined branches—each branch will contain reference objects. When this tree is cut to give the classification, the nonreference objects can be identified. This iterative procedure helps prevent an atypical sample of objects from being used to build the classification. Although the iterations could be automated with a specially designed computer program, it has usually been done by making a series of individual cluster analyses, as described.

SUMMARY

The way in which objects are identified into a classification should follow the way in which the classification was made. The method of standardization, the resemblance coefficient, and the clustering method used to make the classification determine how an object should be identified.

In instances when there are too many objects to cluster-analyze, a subset can be clustered first and the remainder identified into the classification. Moreover, there are other strategies, such as the reference object methods, for handling large numbers of objects.

Philosophy of Classification and Identification

OBJECTIVES

We have so far discussed classification and identification from the point of view of how to do these things. We now want to take a deeper view to illuminate what they are.

- We begin by presenting a philosophy in which classifications are viewed as islands of objects in attribute spaces.
- Building on this, we discuss the various uses of classifications in research.
- It will be seen that classifications are not static but must continually be remade; how they are used governs how they are remade.
- The chapter concludes with a discussion of how researchers should choose and sample the attribute spaces they base their classifications on.

18.1 CLASSIFICATIONS AS ISLANDS IN ATTRIBUTE SPACES

Imagine that we have a complete set of some kind of object that interests us, and that we have decided to describe this population of objects with a certain set of attributes. We plot all of the objects as points in their attribute space and notice that the objects group into separate classes to form what look like islands. The islands are rounded: in a two-dimensional attribute space they approximate circles; in a three-dimensional space, spheres; and in higher-dimensional spaces, hyperspheres. Within an island the differences between objects are much less than the differences between objects taken from different islands. That is, the islands are well separated in the gulf of the attribute space. The total "land area" of all islands is very much smaller than the area of the attribute space they are in.

If we want to be successful in using cluster analysis with sample objects to estimate the sizes and positions of these islands, it is more important that the islands be well separated than that they be rounded. For example, even if the islands are oddly shaped, if they are also well separated, then cluster analysis will usually work well. And in most domains of clustering applications the islands are well separated.

Let us look at the islands that represent "life forms," for example. At any randomly chosen point in the morphological attribute space, the chance of landing on such an island is close to zero. Putting this another way, the variety of life forms that we may imagine is much more diverse than the forms that we will actually find on our life-supporting islands. We will find no three-legged or six-legged humans, no two-legged horses, no creatures like centaurs, griffins, or unicorns. The islands of actual life are therefore well separated.

Similarly, we might explore "elemental composition of earth" on islands in an attribute space. The natural forces that stirred the matter of earth before it froze tended to segregate the light and heavy elements; consequently, when samples of ore are described by the attribute space that lists the abundances of the 103 known elements, they group in well-separated islands. Most of this attribute space is vacant, however, because very few of all known elements associate in quantity in the earth's geological crust.

As a more specialized, practical example, consider that plant ecologists study similarities among plots of land. The plots exist in an attribute space defined by the types of plant life that grows on them. Because of symbiotic affinities among some plant species and antibiotic affinities among others, and because of differential tolerances of species to environmental conditions, plots of land sampled from a region are usually sparsely distributed as islands in the biological attribute space. Much of the space will be empty.

Numerical taxonomists face two tasks in trying to discover the new, uncharted islands that exist in attribute spaces. First, they must select the attributes that define the attribute space that contains the islands. Second, they must sample objects from this attribute space and use numerical methods such as cluster analysis to estimate the islands. Each estimated island is a class, and a set of estimated islands that are close to each other, yet are also well separated, forms a classification.

The basic assumption of biological classification is that the islands are arranged hierarchically within the attribute space. For example, species of duck are separate islands. These islands form an archipelago corresponding to a genus. Sets of genera form a larger archipelago corresponding to a family, and so forth. Species are closer together in the attribute space than genera are to each other, the genera are closer together than families, and so on.

For scientists, the invention of classifications is only the beginning of their work. More important is the study of how different classifications are related. For example, scientists try to discover the relations and theories that span two or more classifications of the same objects described by different attribute spaces. As a case in point, the morphological-based classification of species of ducks is related in certain ways to the behavioral-based classification. In other words, the morphological differences from one species of duck to another are paralleled by behavioral differences. Both are a common result of evolution.

Scientists also seek out relations and theories that bridge classifications of *different* objects, each based on different attribute spaces. A case in point is the ecological niche theory that enables ecologists to link species of animals with classes of habitats.

From all of the examples used so far, we see that classifications are the building blocks of scientific theories and relations. They are not an end in themselves, but are the foundations for bridges of thought. The discovery of laws that bridge various classes of phenomena amounts to understanding and explaining nature.

Obviously a knowledge of these bridging relations, theories, and laws can be useful in practical applications of planning, management, and engineering, too—not only in science. For example, although ecological niche theory is being developed to aid in scientific investigation, it is useful also to the environmental management of wildlands; a wildlife manager considering alteration of a wildlife habitat needs to have a way of predicting the effects of this on animal populations.

18.2 THE USES OF CLASSIFICATIONS IN RESEARCH

Let us list and briefly describe six important uses of classifications in research.

1. Classifications are Used to Aid Memory and Thinking. Classifications bundle separately sensed data into class concepts. Detail about the objects is compromised to gain the economy of thinking about them with a single thought. For example, foresters do not rely on remembering details about individual trees they have seen, but they remember the species of trees and their average properties. A hundred thousand individual trees cannot be thought about in detail, nor is this necessary for the scientific and planning decisions foresters typically make. A forester can achieve a particular research goal just as well by thinking in terms of how many of each species of trees there are on a timbered tract, supplemented with statistics on their average ages, sizes, and locations.

2. Classifications are Used to Organize and Retrieve Objects. When correctly designed, classifications help locate objects quickly. Anyone who has used a letter file knows that the benefits of quickly being able to locate letters outweigh the costs of maintaining the file.

A study of how salvage yards organize used automobile parts would show a varied intolerance to disorganization. But even in the most disorganized yards some kind of classification is necessary. The larger the yard, the more important the classification becomes.

3. Classifications are Used as Building Blocks in Scientific Theories. As we noted earlier, classifications are like the pilings a pier sits on. And scientific theories are like a pier that spans the pilings. From this it follows that when a theory is experimentally tested and confirmed, this process determines the validity of the classifications, as well as their theoretic interrelations.

4. Classifications are Used to Predict the Values of Specific Variables. Such classifications have specific purposes, as discussed in Chapter 5. Arno and Pfister (1977) provide a good example of this. They quantitatively described plots of forested land in Montana by the composition of the vegetation growing beneath the trees, and cluster-analyzed the data matrix of plots, to produce a classification of land types. They went on to show that the classification of land types could be used for the specific purpose of predicting the land's capability for growing timber.

5. Classifications are Used to Discover Varied Inductive Generalizations. We say that a general-purpose classification is "richly inductive" when it yields many unexpected predictions and associations with other variables. Classifications made in the hope that they will be richly inductive are not designed to predict the values of specific variables. They are instead designed to lead to the prediction of many unspecified variables. They are designed with hopes that they will be catalysts for discovery.

Consider that a hypothetical classification of household pets, containing the classes "dogs" and "cats," turned out to be richly inductive. Suppose that this classification was originally based on morphological attributes that measure the animals' sizes and shapes. Once its ancient inventors had made the classification, they unexpectedly found certain behavioral attributes were also related to it. When replicated over many trials, they found the behavioral relations inductively generalized and that, for example, almost all dogs don't eat fish, whereas almost all cats do. Out of the original organization of the classification unexpectedly came another type of organization.

6. Classifications are Used to Improve Planning. A notable example of this is the use of cluster analysis to make classifications of psychiatric depression. In one application described by Blashfield and Morey (1979), a classification of depressed patients was related to the patients' responses to treatments with various drugs. This was of importance in planning the treatments of other patients through drug therapy.

Pratt (1972) describes uses of classifications in greater detail. For all uses, the primary measure of the validity of a classification is how well it achieves its intended use. When we can state the primary use of a classification, we automatically known how to assess its validity.

18.3 HOW IDENTIFICATION HELPS IMPROVE CLASSIFICATIONS

Ideally, a researcher setting out to make a classification of a part of the world would like to have all of the objects from that part at hand, and have the power to sort their numerical descriptions into similar classes. This would eliminate the possibility of basing the classification on an atypical sample of objects. When making a classification of species of oak trees, we should like to have all specimens of oak trees in the world to use for objects in one grand cluster analysis. We usually cannot have one millionth of our wish granted. Obviously, we could not begin to find every oak specimen; even if we could, we could not afford the data needed to describe them; and if we could collect all the data, no computer could process it.

Owing to these constraints, we have to use minute samples to estimate the classes that would result if we were able to use the whole population of objects. When they are cluster-analyzed, the classes based on sample objects become the archetypes for the true classes. Through experience, taxonomists have established principles for reliably estimating classifications. Usually the sample objects they choose are representative of all the objects actually present in the attribute space, despite the lack of a numerical index giving this assurance.

Numerical taxonomists sometimes make errors in their estimations, but it is by these errors that they improve their classifications. When an object is identified into a classification and found not to fit well, the taxonomist must acknowledge that the class the object belongs to was not represented by the sample objects used to make the classification. Such errors are the impetus for revising the classification—that is, remaking it with objects of the unrepresented type included. In this way, the act of identification fuels the trial-and-error evolution of classifications.

18.4 HOW ATTRIBUTE SPACES AFFECT CLASSIFICATIONS

To understand graphically how an attribute space determines a classification, let us consider the following I.Q.-type puzzle attributed by Gardner (1975) to puzzle designer Tom Ransom.

We are challenged to find out which of the five objects is the "most dissimilar" one. By analyzing the puzzle we can see how the answer depends on the attribute space we choose, and how, if we were to choose another one, we would get a quite different answer.

To begin, let us choose the four attributes that the puzzle designer had in mind. Using a value of "1" to mean that an object possesses the given property and a "0" to mean it does not, the data matrix appears as follows:

	Object				
Attribute	1	2	3	4	5
1. Overall triangular shape?	0	0	1	0	0
2. White outside border?	1	0	0	0	0
3. White intermediate area?	0	1	0	0	0
4. White inner circle?	0	0	0	1	0

The pattern of similarities in the data matrix clearly shows that object 5 is the correct answer because it is the "most dissimilar." It possesses none of the features, whereas each of the others uniquely possess one feature.

Note that had we abstracted differently by using other attributes, we would have gotten other solutions. In fact, there is a choice of attributes that makes object 3 (a more obvious choice when we compare the original objects) the "most dissimilar." How, then, do we know that the attributes we have used are the relevant ones?

A designer of an I.Q. puzzle makes framing decisions by choosing attributes that matter within an I.Q.-test context. Such a context may require finding a clear pattern in a relevant attribute space, perhaps by starting with a data matrix like the one shown above, and then transforming the data matrix to some graphical form that places the *relevant* pattern in the *background* and introduces an *irrelevant* pattern in the *foreground*. The designer will be able to validate the success of this approach if the test taker, accustomed to a world where perceptual "dirty tricks" are infrequent, takes the bait and assumes that what is prominent is relevant; through incorrect assumptions about the framing attributes, the test taker arrives at the wrong solution.

About this quality of I.Q. tests, Gardner (1975) says the following:

> Every intelligent, mathematically proficient person who ever took an I.Q. test has surely been annoyed by questions that involve a row of symbols and a request to identify the symbol that "does not belong" to the set. Sometimes there are so many different ways a symbol can be different from others that the brighter students are penalized by having to waste time deciding which symbol is "most obviously different" to the person who designed the test.

All told, I.Q. tests are designed according to a form of data processing that runs in a reverse direction—from clearly stated solutions to unclearly stated problems. Such tests are described by game theory, in which the puzzle designer and the puzzle taker attempt to second-guess each other's attribute space.

18.5 HOW TO CHOOSE AND SAMPLE AN ATTRIBUTE SPACE

An attribute space is not like a perceptual space that we see by simply opening our eyes. Seeing clusters of stars in the night sky is one thing. We automatically see the stars and the space they are in. All that remains is to find those stars that are close enough together for us to say they form clusters. But gazing into a mathematical attribute space and seeing clusters of objects is much more difficult. Before looking for clusters, we have to select the set of attributes (that is, the attribute space) in order to have a frame of reference for seeing the objects. Once we have decided on the attribute space, we will usually find that it has more than three dimensions and that our eyes are of little use in locating clusters. For this we need cluster analysis and other multivariate methods to make the clusters of objects apparent.

This brings us to an important question: how should we choose the attributes to be used to make a general-purpose classification? Numerical taxonomists who make biological classifications have already answered this. They choose attributes according to a principle called the **factor asymptote**. The procedure follows three steps.

First, a numerical taxonomist draws up a list of biologically meaningful attributes—attributes that enter into or reflect biological theory and law. Second, the taxonomist uses as many of the attributes from this list as possible to describe the specimens that are to be clustered. Third, a cluster analysis is performed and used to find essential attributes (the ones that help to discriminate clusters) and inessential attributes (the ones that do not help to discriminate clusters). The essential, or key, attributes become the attribute space; in a sense they are natural, because the objective procedures for choosing them always lead to essentially the same set of estimated classes. Numerical taxonomists have found that applying the principle of the factor asymptote to define an attribute space produces classifications that do indeed fulfill scientific functions, which in essence validates the principle.

The procedure is objective and will usually lead all taxonomists who follow it to the same classification. Sneath and Sokal (1973, p. 106–109) describe the principle of the factor asymptote in detail; in theory the principle should apply to general-purpose nonbiological classifications as well as to the biological.

Having seen how a natural set of attributes can be selected, let us now examine the kinds of errors which, if we were to make them when using cluster analysis, could prevent us from discovering the actual classes that exist as islands in the natural attribute space. For the discussion, we will assume an attribute space with only two attributes, a_1 and a_2, and with only two types of objects, A and B.

We wish to estimate the classes A and B by using three sample objects of each. As Figure 18.1,*a* shows, even if we were fortunate enough to obtain three sample objects from each class, we would fail if we used only attribute a_1 because we would see no well-defined clusters of objects projected on the axis that describes a_1. But we would succeed if we used a_2: we would then see two islands of objects in the two-dimensional attribute space. We would then also recognize that a_2 alone could have been used as the basis of the discrimination, as the projection of the objects on axis a_2 shows. The principle of the factor asymptote would lead us to regard a_1 as inessential and we would remove it when preparing the taxonomic key.

However, if the A's and B's existed in the two-dimensional attribute space shown in Figure 18.1,*b*, then both a_1 and a_2 would be required to

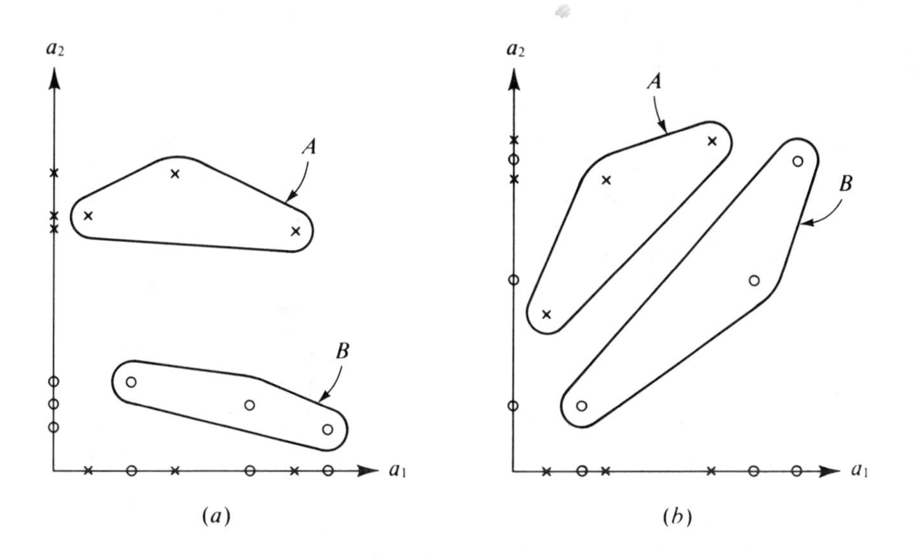

Figure 18.1. Attribute space containing two classes, *A* and *B*, described by two attributes, a_1 and a_2. (*a*) When three sample objects are selected from each of *A* and *B*, the sample objects define the classes by their projections on a_2, but not by their projections on a_1. (*b*) When the classes are oriented obliquely, neither the projection of the sample objects on a_1 or a_2 alone is sufficient to define the classes; both dimensions are required. Projections are shown on the axes.

discriminate the *A*'s from the *B*'s. Both attributes would have to be regarded as key attributes because neither the projection of the objects on a_1 nor that on a_2 gives a view that does not confound the distinction between the *A*'s and the *B*'s.

Now let us assume that we believe both a_1 and a_2 are key attributes. Even so, we can still err in several ways when we estimate the classification. First, it is possible that our sample consists of only *A* objects or only *B* objects. If this is so, since one of the classes is not represented in the sample, we will not be aware of the existence of this class. Second, even if we are lucky and have objects from *A* and from *B* in our sample, we may err if *A* and *B* are close together relative to their internal dispersions. For example, suppose the actual classes in the attribute space are as Figure 18.2,*a* shows, and suppose we select the eight objects shown for a cluster analysis. Figure 18.2,*b* idealizes how the tree would look. According to the tree, we will have erred in declaring that the attribute space contains the two classes of objects suggested by the two clusters. We may even go so far as to call these two classes *A* and *B*, but as can be seen, they will not agree with the natural classes *A* and *B*.

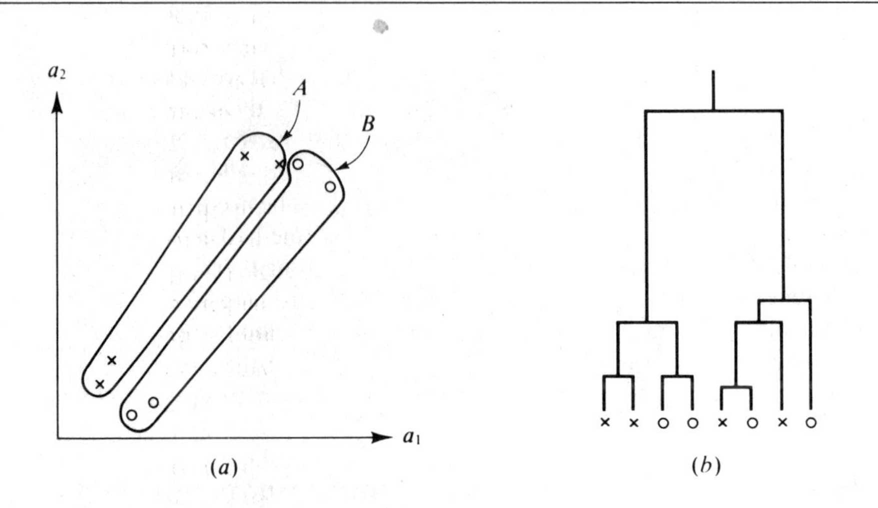

Figure 18.2. (*a*) Attribute space containing two classes, *A* and *B*, described by two attributes, a_1 and a_2. Because the classes are close together, and because the four samples in each come from the end of the sausagelike classes, the tree (*b*) from the cluster analysis fails to estimate the classes.

Figure 18.2.*a* also suggests why it is less important that the actual classes be rounded than that they be well separated. In the event the two classes are actually far apart, the ratio of the between-class dispersion among members of each class to the within-class dispersion will be large even when the classes are not rounded, and cluster analysis will be able to reflect actual distinctions between the classes. For example, if the sausagelike classes shown in Figure 18.2,*a* are moved far enough apart, thereby increasing the ratio of the between-class dispersion to the within-class dispersion, then cluster analysis will work when samples from both classes are present. But if the classes naturally exist close together, then only in the case where the sausagelike islands are arranged end to end will cluster analysis have some chance of being effective.

Using a large sample of objects, perhaps as many as several hundred, helps to guard against missing one of the naturally existent classes. And should a class be missed, or should one be poorly estimated because the sample has come from near one shore of its island, then this error will be discovered when the classification is used for identification and a hitherto unknown object is found not to fit any of the estimated classes well. If this happens, we must assume that we poorly estimated one or more natural classes.

If sausagelike islands naturally exist close together, in the way that Figure 18.2,*a* illustrates, then the methods of cluster analysis presented in this book may not be able to find them. The problem is that we do not know *beforehand* whether the actual classes are well separated—this is an assumption we must accept as true in order to use cluster analysis. To guard against the assumption being false, both Anderberg (1973, p. 188–194) and Everitt (1980, Chap. 4) suggest that other multivariate methods be used with cluster analysis on the same data matrix. Some of these methods assume the presence of oddly shaped classes and are helpful in finding them should they exist. No method is foolproof, however. All methods are based on assumptions that may not reflect reality. For example, if a method that can find oddly shaped classes assumes that all "odd" classes have the same kind of odd shape (as at least one method does), then it too may have trouble finding classes that are odd in another way.

While it is always possible to construct a hypothetical data set that will foil the use of any multivariate method, there is not strong evidence, in biological taxonomy at least, that natural data sets behave in this way. The successes of cluster analysis have made it the most-often-used method for helping taxonomists make biological classifications; there is much evidence that the classifications it helps produce are biologically sound and advance scientific thought.

While evolution seems to have made the islands of life at least roughly rounded and, more important, well separated by great gulfs in their attribute space, it is still good practice to peer into attribute spaces with as many kinds of "eyes" as possible, with as many different multivariate methods as possible, each providing a fresh perspective. For this reason, methods like principal component analysis and multidimensional scaling, and methods for detecting oddly shaped and nonlinear classes (as discussed by Everitt, 1980, for example) are useful adjuncts to cluster analysis.

SUMMARY

Numerical taxonomists use the principle of the factor asymptote to choose the attributes that define their general-purpose classifications. But even if the attribute space is correctly defined, errors can still be made in estimation of the actual classes of objects that are present: it is possible that the sample of objects to be cluster-analyzed does not represent all the actual classes present; it is also possible that the classes lie close together in their attribute space, and because of this the sample may be incapable of revealing their actual proximities.

For making classifications, the methods of cluster analysis presented in this book work best when the actual classes are rounded in shape and, more importantly, well separated in the attribute space. Because certain other multivariate methods are less sensitive to this condition, it is good practice to use them to complement cluster analysis.

Classifications are basic to research. They are used to aid memory and thinking, to organize and retrieve objects, to function as building blocks in scientific theories, to relate to other variables for specific and general purposes, and to improve planning and management. Their validity is determined by the extent to which they achieve their intended uses.

Classifications evolve through use. Identifying an object into a classification is a kind of feedback; either the object fits one of the classes well, and reconfirms the classification, or the object fails to fit any of the classes, making the classification suspect and suggesting that it be revised.

PART IV
Cluster Analysis—Philosophy

> [Consider] the question posed to a chess champion, "Tell me, Master, what is the best move in the world?" There simply is no such thing as the best or even a good move apart from a particular situation in a game and the particular personality of one's opponent.
>
> VIKTOR FRANKL
> *Man's Search for Meaning*

Research seldom fails because a researcher errs when collecting or transcribing data, or because of errors in the mathematical calculations: collecting data and making calculations are objective and can be perfected with practice.

More often research fails because the researcher errs when subjectively deciding what data to collect and what mathematical methods to use. The researcher errs by correctly solving, from the mathematical viewpoint, the wrong problem.

To do successful research, a researcher must get both its subjective and its objective parts right. Unlike the end-of-chapter exercises in mathematics and statistical texts, real research problems are born unstructured and unarticulated. Researchers must first subjectively structure and articulate their problems. Following this, they must objectively collect data and process it with their chosen methods, in effect solving the problem that they have articulated.

It is with this in mind that we now turn to the subjective aspects of cluster analysis. We want to learn how to make the subjective decisions that will correctly frame a cluster analysis. That is, how do we choose the correct objects and attributes, the correct method for standardizing the data matrix, the correct resemblance coefficient, and so on? And what are the criteria by which we can tell if our decisions are, at the time we make them, potentially the correct ones?

We also want to learn how to make the subjective decisions needed to validate the cluster analysis. That is, after having framed the cluster analysis, collected data, and performed the analysis, we find ourselves in our office looking at the tree on a computer printout. How can we tell if it is valid? How can we tell if potentially correct framing decisions were in fact correct?

Framing and validating are common to all mathematical analysis, not only to cluster analysis. So what you learn in this part of the book can help you in all of your research.

We begin with a discussion of research in general. Chapter 19 discusses six steps that are basic to all research analyses. These six steps connect the purpose of the research—what we have called the research goal—with the means by which the research goal can be accomplished. Chapter 20 defines the concepts of information, tolerances, and norms, and shows how they enter into subjective decision-making in research.

Reading these two chapters should help you understand that subjective decision-making in research is guided by unwritten and unspoken norms, and yet with experience we all come to learn what the guiding norms are. The norms set the values of the tolerances, and in turn these determine what data constitute information. The information, in turn, is necessary to decide if the research is valid.

In Chapter 21 this background is used to discuss the framing and validation of applications of cluster analysis; Chapter 22 illustrates this for each of the six research goals discussed in Chapters 4, 5, and 6. Then, Chapter 23 explains that the patterns of similarities that we can uncover with cluster analysis can be categorized according to a hierarchy of three orders. All research goals that can be pursued with cluster analysis depend upon finding one or more of these three orders of patterns in our data.

The Six Steps of Research

OBJECTIVES

- This chapter describes a model of research that has six steps.
- The steps are examined to determine the relative roles of objective and subjective decision-making in the research process.

19.1 A MODEL OF RESEARCH

Let the six steps in Figure 19.1 represent the chain of events and decisions followed by a researcher when solving a research problem.

Our concern is with research that uses quantitative methods, primarily cluster analysis, although the steps are general and apply to all research that uses qualitative methods.

Note that Figure 19.1 forms a feedback chain and that the six steps link the **ends** with the **means** of research. This feedback chain—from ends to means and back to ends again—is the basis for learning how to do better research.

The six steps comprised by the feedback chain are as follows:

Step 1. Identify a Problem. Let us define a problem in the way that Chadwick (1971) and Faludi (1973) do: a person senses existence of a problem when the gap between an ideal state and an actual state is intolerably large. The gap is often qualitative, but we can think of it as running on a scale. When the gap becomes large enough to exceed a threshold, the problem becomes perceptible. The size of the threshold depends on the importance the person attaches to the ideal state: if the importance is great, then a small gap will lead to awareness of the problem; if the importance is small, awareness will require a larger gap.

Applying this definition to problems in science, we see that the ideal state pertains to knowledge that scientists would like to have; their lack of knowledge, or ignorance, creates a gap. For instance, Sigleo's (1975) research on prehistoric Indian cultures in the American Southwest (as described in Research Application 4.3) began with the problem of not

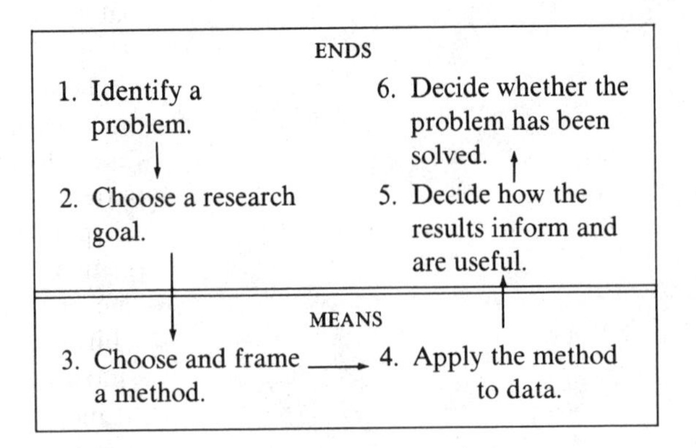

Figure 19.1. Six steps of research.

knowing the trading patterns among the cultures. For most people the gap between wanting to know of these trading patterns and the ignorance of not knowing would not constitute a problem—they wouldn't value this kind of knowledge dearly. The gap was great enough to create a problem in Sigleo's mind, and this was necessary to start her thinking about the design of a solution.

For problems in planning, the ideal state is the way a planner would like some aspect of the world to be. For instance, the manager of a factory would like a profit margin that equals, or exceeds, those normal for other factories within a particular industry. If the actual profit margin is below the norm, then at some point the manager becomes aware of a problem in the way the factory is operating. This is a necessary condition for beginning to think about the design of a solution.

In both science and planning, the researcher must identify a problem before beginning to plan its solution.

Step 2. Choose a Research Goal. The research goal is a statement of what the research should accomplish. For applications of cluster analysis in science, there are five types of research goals described in Chapters 4 and 5: (1) create a question, (2) create a hypothesis, (3) test a hypothesis, and make a (4) general-purpose or (5) specific-purpose classification. For example, Sigleo's research goal was to create a hypothesis that would answer the question about trading patterns among Indian cultures.

For planning applications, there is the general research goal of facilitating planning and management, discussed in Chapter 6. For example, the factory manager experiencing the problem of a low profit margin might state a research goal by specifying some change within the factory to alleviate the problem, such as finding a more efficient arrangement of machinery, thereby to reduce the costs of making products.

For both science and planning, the research goal is the aim of the analysis. Reaching the goal will solve the problem.

If we compare "identification of a problem" and "choice of a research goal," we find that usually the former is an ideal state and the latter is a specific state. For example, the impetus for Sigleo's research was an idealized desire to learn more about Indian trading patterns, whereas she formulated a more specific research goal—to create a hypothesis to answer a specific question. And in the example of the factory manager's problem, bettering the profit margin can be seen as an ideal; the research goal was more specific—to find an efficient arrangement of machinery.

Getting started with steps 1 and 2 is usually the most difficult part of research. For example, graduate students tend to spend inordinate amounts of time selecting a thesis topic. It takes them a long time to recognize what makes a good research problem and what its research goal should be. This happens because problem recognition depends on experience and background knowledge, and while graduate students are well trained in methods, they lack experience. Many of them put the cart before the horse and use methods they have been taught without first having a well-defined research problem and goal.

Step 3. Choose and Frame a Method. (a) The researcher frames the problem when designing an experiment. This should be done in two steps. First comes the choice of framing alternatives, which are a set of experiments each having high expectation of achieving the stated research goal. Second, the researcher makes framing decisions—that is, decides which of the alternatives seems best.

For example, the researcher might decide that similarities among certain objects must be determined in order to achieve the research goal. Determining similarities requires that various multivariate methods be examined—principal component analysis, factor analysis, cluster analysis, and so forth—and that the various kinds of data required by each method be examined too. The researcher might entertain several alternative uses of cluster analysis that would differ in the objects, attributes, standardizing functions, resemblance coefficients, and clustering methods they incorporate. From this list of possible framing alternatives the researcher would then select the one the researcher believed to be best.

(b) The researcher should also specify the validating criteria that will later be used (in step 5) to decide how informative the analysis is. For example, suppose the research goal is to test a certain hypothesis. Then the researcher must decide ahead of time how closely the actual and predicted trees of similarities must agree in order to confirm the hypothesis. Or suppose the research goal is to make a specific-purpose classification of land, in order to predict the number of waterfowl nesting sites that will be found there. In this case the researcher must form an idea of the weakest predictive relation that will be meaningful. By setting the validating criteria at this time, the researcher will be less tempted to adjust the criteria after the results are in hand.

Step 4. Apply the Method to Data. After the researcher selects a method (for our discussion we will assume that the method is cluster analysis), the data specified by the framing decisions in step 3 must be collected and processed with that method. Whether the data are processed by hand or with a computer, step 4 will produce numerical results. In cluster analysis, the main numerical results are the patterns of similarities between objects in the tree, and in the rearranged data and resemblance matrices.

Step 5. Decide How the Results Inform and are Useful. (a) First, the researcher must decide if the analysis is technically correct—that is, if it has been correctly executed so that the results bear directly on the research goal.

He must also decide if the value of the cophenetic correlation coefficient is large enough.

(b) Second, the researcher must compare the numerical results with the validating criteria (established in step 3) and decide what kind of information the results convey, and how they are useful. For example, if a scientific application, do the results strongly suggest a certain explanatory hypothesis? How important are the findings expected to be to other researchers and to the future course of science in the field? And if a planning application, do the results suggest a plan that will help achieve the stated planning objective?

Chapters 21 and 22 discuss and illustrate in detail how researchers decide whether their cluster analyses are valid.

Step 6. Decide if the Problem has been Solved. This will reconfirm that the research goal specified in step 2 was the right one for the problem specified in step 1, and that it was maintained throughout the analysis. Even with conscious effort, it is easy to misspecify the research goal or let it drift from its original statement.

19.2 OBJECTIVITY AND SUBJECTIVITY IN RESEARCH

The model of research we have outlined highlights the roles of objectivity and subjectivity in research. The only objective parts of the research process are the data processing methods themselves, in steps 3 and 4. By objective, we mean that a group of knowledgeable professionals will almost always agree on objective decisions such as the mathematical formulation of a resemblance coefficient and the computing of its value. The steps required to execute a clustering method are also objective, as is the Fortran code in the computer program used to get the results.

The other parts of the research process are subjective, because less rigid rules exist for identifying a problem and choosing a research goal, and for choosing the best objects, attributes, resemblance coefficients, clustering methods, and other details of these steps. Note, however, that a subjective decision is not necessarily a bad decision, nor a decision made without using intelligence.

When a group of professionals follows these less rigid rules, a consensus may result, but usually there will be differences of opinion as to which guides are best. The guiding rules for subjective decisions are the norms that professional fields sanction. The norms map out a range of what is and is not acceptable, rather than constraining the decision to one possibility.

Subjectivity is necessary for research to progress. If it were not permitted, then creativity could not flourish. While the mathematical and computing tools of research are objective, the use of these tools must be subjective to accommodate the vast variety of natural phenomena. Scientists must tailor the tools to the research problems at hand. And in the way that we don't expect a tailor to consult an objective set of rules when judging the fit of a suit of clothes, we shouldn't expect a scientist, planner, or applied mathematician to consult a set of rules when assigning real world content to symbols. What comes after the equal sign in, "Let $X = \ldots$," is properly a subjective decision. The researcher risks being wrong, but this is not worrisome because, in the end (as steps 5 and 6 indicate), the researcher will gain feedback indicating whether he was wrong or right.

There is a danger of forgetting about steps 1, 2, 5, and 6, and thinking that research only consists of steps 3 and 4. After all, data abound and are always easy to find and process with a computer. But of itself, this is not research. There are several reasons why reliance on method attracts us all; yet like Ulysses resisting the Sirens, we should resist such a mentality.

First, methods attract us because they are concrete, and we can get our hands on them and gauge our "progress" by measures such as how many computer runs we make in a day. When we work hard at collecting data and using a computer we have a tangible record of our productivity—numbers entered on data sheets or printed on computer printouts. The trouble is that productivity does not always measure progress.

Second, collecting data and working with computers can be so enjoyable to some people that it becomes an end in itself. In contrast, work on identifying or evaluating a research problem or a research goal is abstract and often difficult or frustrating.

Third, education fosters method-oriented research. It is natural to conceive the idea in college that subjects are mastered by mastering their methods. We come to think that when we master so many mathematical methods we will be able to do research.

Note that we have answered the question, without asking it, "Which is more basic, science or planning?" The answer is that planning is more basic. The six steps of research are a model for any type of planning. The model is characterized by having a goal, having values implied by the goal, and having a set of alternative decisions, each with various expectations for achieving the goal. It is also characterized by the making of the decision whose expectation is best, and by the evaluation of how well this decision achieves the goal. Clearly, science fits the mold of planning. Science is only a special kind of planning with peculiar values, and with its peculiar subjective and objective norms that constrain the kinds of alternative decisions a scientist is permitted to consider.

SUMMARY

Whether in science or in planning and management, all research follows six steps. It begins with our identifying a problem (step 1) and then choosing a research goal (step 2), which, if attained, can be expected to solve the problem. Next, we choose methods (step 3) and use them with data to produce numerical results (step 4). We evaluate the numerical results to decide how they inform and are useful (step 5). Finally, we decide whether the original problem has in fact been solved (step 6).

This sequence of steps forms a feedback chain. In essence, we propose what we want to accomplish, choose the methods that seem best suited to accomplishing it, and then decide at the very end whether we accomplished what we set out to do. If we succeed, this reinforces our choice of methods. If we fail, this causes us to restudy the methods we chose and the reasons we chose them, so that we may learn and do better next time.

While these six steps are well known, they are not always easy to follow. There is a strong temptation to substitute the means of research for the ends of research, to spend all of one's time choosing and using mathematical methods to process data while losing sight of the purpose. None of the research applications given in this book do it that way. Had they, they would not have been published.

The Roles of Information, Tolerances, and Norms in Research

OBJECTIVES

Now we discuss how researchers draw conclusions from their processed data.

- How do researchers study the tree from a cluster analysis and then, for example, decide that it confirms or falsifies the hypothesis they were testing?
- To understand this subjective decision-making, we must see how the concepts of information and tolerance are crucial to decision-making.
- Then, we consider how norms within a research community set the range of acceptable tolerance, thereby determining how data inform us.

20.1 INFORMATION

We define information as Bateson (1980, p. 250) defined it: *information is any difference that makes a difference*. The first use of the word "difference" applies to patterns in data. The second use applies to our beliefs. Patterns of data must show certain types of differences in order to affect beliefs in certain ways.

Let us illustrate this with what is called an **information function.** It is an x–y graph. On its vertical y-axis are n classes of information labeled $I_1, I_2, \ldots, I_n$, and on its x-axis is a scale of data. The data are such that we may make a mental relation of them to the n classes of information.

The following hypothetical example illustrates an information function. Suppose it is a cold day and we want to decide whether we should go outside. The two classes relevant to this decision are: I_1, "should stay inside"; I_2, "needn't stay inside." The relevant data are outside temperatures, in degrees Fahrenheit. This is shown in Figure 20.1.

The temperature of $T_T = 20°F$ is a threshold that prescribes the boundary between classes I_1 and I_2. (The subscript "T" stands for "threshold.") Once this information function is specified, it acts as a policy for transforming data into beliefs. Suppose we measure the outside temperature and it is $T = -5°F$. We mentally enter this value on the horizontal axis of the information function above and read I_1, "should stay inside."

The idea of a tolerance is implicit in the threshold of $T_T = 20°F$. We have a tolerance for any pattern of temperatures of less than $T_T = 20°F$. If we look at temperatures throughout the day and they are $T = 10°F, 12°F, 8°F, 16°F$, then for any of these values our belief goes unchanged: we

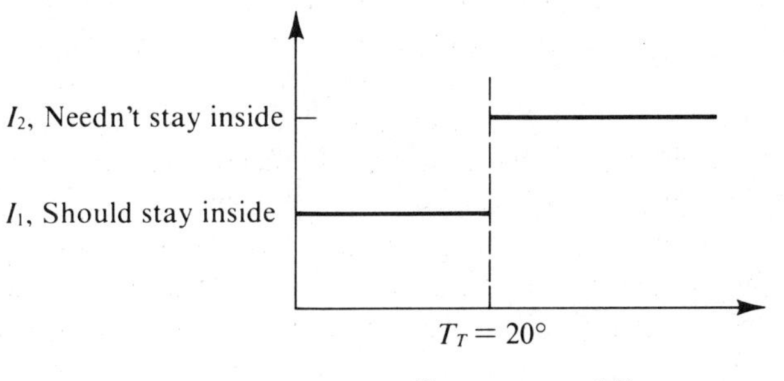

Figure 20.1. Hypothetical information function relating the temperature outside to two classes of information: I_1, "should stay inside," and I_2, "needn't stay inside." The temperature $T_T = 20°F$ is the threshold that separates the classes.

continue to think I_1, "should stay inside." Similarly, when temperatures are $T = 45°F$, $38°F$, $52°F$—these all reinforce the single belief I_2, "needn't stay inside." It is only when the pattern crosses the threshold $T_T = 20°F$, as it would if the temperature rose from $T = 16°F$ to $T = 24°F$, that we change our belief.

The idea of a tolerance eases the burden of thinking. Were the information function to identify each separate value of data as a separate belief, we would need an infinite number of classes of information. There would be an impossible number of separate beliefs to deal with. And we would constantly be making errors because even the smallest errors in data would result in erroneous beliefs.

The threshold of $T_T = 20°F$ shows what Bateson meant: it shows how data determine what we believe.

20.2 TOLERANCES AND NORMS

The thresholds of information functions are set by trial and error from past experience. In the example of the previous section, setting $T_T = 20°F$ signified that through experience we learned we were content outside when the temperature exceeded 20°F and uncomfortable when it did not. The threshold value of 20°F differs, of course, from person to person within a group, and especially from group to group (people living in the Tropics might have one range of threshold values, and people living in the Arctic might have another). Each group has a separate norm.

Now let us see how the concept of a norm that sets the policy for decisions applies to research. In order for researchers practicing in the same field to reach essentially the same conclusions, and to be able to communicate with each other, they must hold essentially the same information functions. There will, however, be minor differences between their information functions in the threshold values they hold—just enough to allow some disagreement on the conclusions that should be drawn from the data while still generating a consensus belief. Should colleagues within one field not be able to reach a consensus, they might decide to split into new fields, each reestablishing its own consensus.

The following examples from research and everyday life illustrate how norms set the thresholds, and hence the policies, for research decisions.

Example 1. Consider the question often asked in applications of regression analysis, "How good is my value of R^2?" Or in other words, "Does my value of the coefficient of determination, R^2, show **practical significance?**" To answer this, we will assume that the observed value is $R^2 = 0.66$, and we will use the information function shown in Figure 20.2, *a*. Its horizontal axis scales R^2 and its vertical axis gives the classes of information: (1) nontrivial, and (2) trivial. The threshold is drawn at $R_T^2 = 0.6$, and values in the range

$0.0 \leq R^2 < 0.6$ are trivial, while values in the range $0.6 \leq R^2 < 1.0$ are nontrivial. Once the information function is specified, the observed value of $R^2 = 0.66$ signifies that the regression is nontrivial—that it is of practical significance.

Different fields of science draw the line for R_T^2 at different places. For chemistry the threshold is placed higher than for biology. The reason is that the "hard sciences" study phenomena having smaller variability and higher regularity, and accordingly "hard scientists" set higher threshold values. Conversely, the "soft sciences" are beset by variability and so their scientists must necessarily set lower threshold values. Likewise, the planning and management professions have their own normative thresholds.

How does a profession decide what it will tolerate? That is, how are the normative threshold values set? When a researcher uses a given value such as $R_T^2 = 0.6$ in a published paper, its being accepted during the peer review

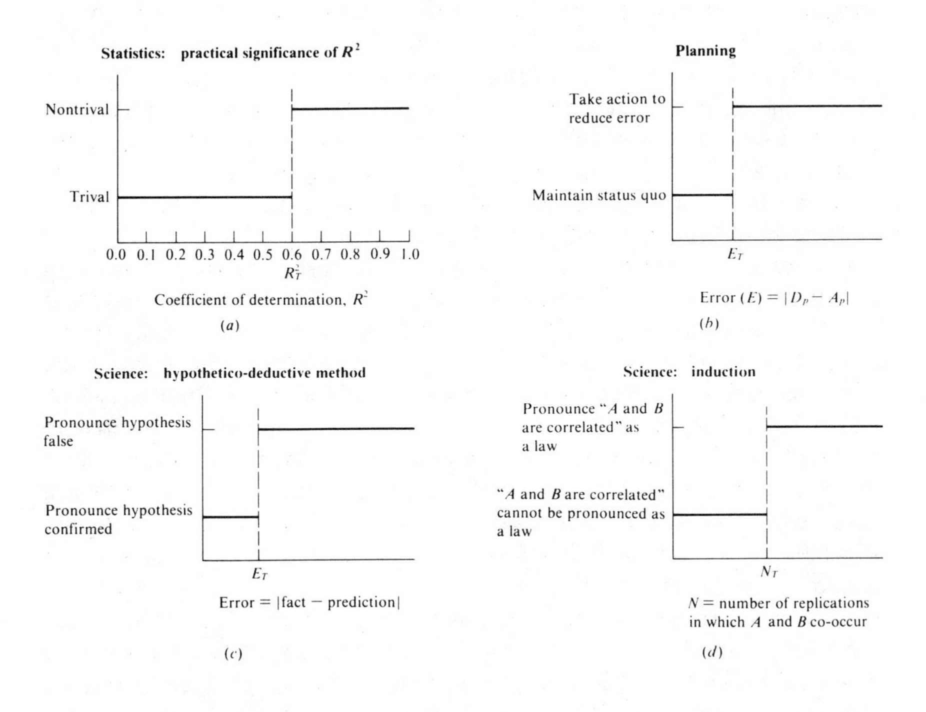

Figure 20.2. Hypothetical information functions for (*a*) deciding the practical significance of R^2 in regression analysis; (*b*) planning (for example, to control a rocket's path); (*c*) deciding whether a research hypothesis being tested should be declared false, or confirmed; (*d*) deciding whether events are inductively related as a law.

the paper undergoes while being published as available literature helps to reinforce, or change, the norm. The matter is simply that what gets published sets the standard. Most researchers will not nitpick with each other over different threshold values so long as they are in the normal range, but they will argue about values they feel are outside of it.

Journals do not directly state norms. For example, they do not take the articles they have published over the past several years and make histograms of the threshold R_T^2 values that published researchers have used. What prevents this is that researchers seldom state R_T^2 in their papers, and instead simply state whether the value they observed was nontrivial or not. So each reader must infer the normative range of thresholds through his own reading.

Example 2. The tolerance for **statistical significance** is normative in the same way. Generally, biologists will accept $\alpha = 0.1$, or smaller, while drug companies testing a new drug to decide whether to market it usually do not go above $\alpha = 0.01$. Although deciding the proper value of α is a matter of weighing the short-term benefits of making correct decisions against the short-term costs of making incorrect decisions, it is also a long-term matter of keeping each field self-sustaining. A profession will set its tolerance, α, in a range where it causes neither too few, nor too many, statistically significant results.

To explain why both "too few" and "too many" are undesirable, imagine for a moment two researchers in the same profession. The first always uses $\alpha = 0.00001$, and consequently almost never rejects the null hypothesis being tested. The second always uses a large value, $\alpha = 0.8$, and almost always rejects the null hypothesis. Their actions are virtually predetermined, and it is a waste of their time to collect data to decide the outcome! Using $\alpha = 0.05$ is a compromise between these extremes, which allows most sciences to progress—to afford the sample sizes required to reject a few null hypotheses, but not too many null hypotheses. Researchers in one field might sanction a tolerable range of, say $0.0 < \alpha \leq 0.1$, in which case fewer researchers would use $\alpha = 0.01$ or $\alpha = 0.10$ than would use $\alpha = 0.05$, yet all values in the range would be permitted. However, $\alpha = 0.2$ would not be permitted.

This is a sociological explanation. Professions set ranges of tolerance for both practical and statistical significance that will allow them to make continual progress in their quest for knowledge. Looking to the long term, they set standards that allow themselves to stay in business, standards that compromise rates of error. Using different tolerances makes the quality of knowledge each profession produces different and not comparable by another's standard. A profession will sanction tolerable ranges for practical significance to accommodate the typical degrees of irregularity in the patterns of data it studies. It will also sanction tolerable ranges for statistical significance to allow it to cope with the inherent variability, or noise, that obscures the patterns it studies.

Example 3. Berger and Luckman (1967) give the following example of the process through which people come to realize a norm exists. As a child, one first becomes aware of a specific standard, such as "Mummy is angry with me *now*." This progresses with other incidents, to something such as "Mummy is angry with me *whenever* I spill soup." When everyone in the family expresses anger over soup spilling, the child thinks "*My family* gets angry when I spill soup." Then a neighbor chastises the child who spills soup, and the child thinks "*Everyone* gets angry at me when I spill soup." Eventually the child comes to believe the norm, "*No one* should spill soup." In this way a norm generalizes out of specific instances.

Example 4. A rocket's feedback control system shows why a tolerance is necessary to balance opposing errors. Take the sending of a rocket to the moon, for instance. The goal is for the rocket to stay on its desired path, D_p, to the moon. But its actual path, A_p, does not always agree with its desired path, the error between the two being $E = |D_p - A_p|$, where the vertical lines mean the absolute or positive value.

Figure 20.2, *b* shows an idealized information function that transforms the error E into a policy for controlling the rocket. The threshold, E_T, separates the controlling actions into the following two classes: (1) take corrective action to reduce the error and bring the rocket closer to its desired path; and (2) maintain the status quo and do nothing. Mathematically stated, the policy is:

1. When $E \leq E_T$, do not take action to reduce the error E.
2. When $E > E_T$, take action to reduce the error E.

It is easy to see that either extreme of $E_T = 0$ or $E_T = \infty$ is undesirable. If a zero tolerance is used, any drift from the desired path, no matter how small, will change "take no action" into "take action." The rocket will continually home in a constant state of fine tuning, and as a result its fuel supply will soon be used up. It is possible to get to the moon more efficiently by allowing a value of E_T that results in fewer midcourse corrections and in a corresponding savings of fuel. But the tolerance cannot be too large because the efficiency decreases with long and protracted midcourse maneuvers. Thus some finite threshold value E_T signifies a compromise.

Each engineer in a group that has been given the problem of finding the optimal value of E_T will find nearly the same solution because all will use nearly the same assumptions, theory, and data as a result of their uniform, objective educations. But there will be minor tolerable differences among them. There has to be this tolerance, because no two engineers ever agree exactly. By allowing a tolerance, they can communicate with each other.

Example 5. Evolution equips animals with innate feedback control systems, which they use to cope with the circumstances of their lives. With one exception, an animal's control system is like a rocket's. Both an animal and a rocket have a desired path (a goal) and an actual path. The exception is that evolution has influenced an animal's desired path and the optimal value of its threshold for detecting error. For example, think of the tolerance that an organism uses in responding to a certain prey. The tolerance cannot be set so narrowly that an animal responds to every prey stimulus, no matter how slight the stimulus might be, nor so wide that an animal does not respond at all. Between these extremes is an optimal value that balances the cost of wasting too much energy responding unnecessarily and the cost of not responding at all.

Yet all animals of a species do not respond in exactly the same way when shown the same stimulus, just as all people do not respond to a noise in the night in exactly the same way. But as a species, the animals' behaviors comprise a norm, a range that evolution has made tolerable. This range of behavior gives animals—especially humans—the appearance of what we call subjectivity. It is this subjectivity that makes survival possible. If all animals of a species acted exactly the same, they would all be subject to the same risk of a catastrophe, which could eliminate the entire species. By giving animals a little bit of say in the matter, nature has made them uniform and objective enough to communicate and breed, while at the same time increasing the chances that the species can survive natural disasters.

Example 6. As Chapter 4 discusses, according to the hypothetico-deductive method of science, a hypothesis H is declared "confirmed" when a prediction P made from it agrees with an experimental fact F. Otherwise, it is declared false. Clearly, for science to be possible it must tolerate some amount of error between P and F, because if it insisted that they match without error, then no hypothesis could ever be confirmed and knowledge about natural processes would be impossible to get. Figure 20.1,c shows an idealized information function for testing a research hypothesis. The threshold value of error, E_T, transforms the error, $E = |F - P|$, into two classes of belief; one where H is called "confirmed," and one where it is called "false."

It follows that scientists in a single field will have different ideas of what the value of E_T should be in a given situation, but were the scientists able to plot their values of E_T in a histogram, this would show a normative tendency—something like a normal curve. The acceptable range of E_T would be bounded by the limits of the curve; it would be the range that captures what most practicing and publishing scientists in that field are allowing each other to use. The problem that an individual scientist faces is that it is not possible to look at the normal curve. For applications, the scientist must state a value of E_T felt to be within the limits, and the editors of the journal to which the scientist submits an article will be the final arbitrators.

Example 7. The scientific method of induction also relies on a normative threshold. Imagine that an experiment is run under controlled conditions, and that on the first trial A and B co-occur. This repeats on the second, third, and fourth trials. If this keeps repeating, there will come a trial after which a scientist will proclaim (1) "A and B are correlated" and (2) the scientist will then stop replicating the experiment. All scientists would not make this proclamation on the same trial. But the behaviors of many scientists, independently performing the same experiment without knowledge of each other, would show a normative central tendency. Few would proclaim a law based on one trial. No one would insist on 10,000,000 trials. There would be an acceptable range, and most would fall in the central region of the range.

Figure 20.1,d shows an idealized information function for scientific induction. When the number of continually successful replications, N, is less than the threshold, N_T, then the idea "A and B are correlated" is not yet accepted as law. In other words, a scientist refrains from declaring a law when $N < N_T$. But when $N = N_T$, this changes the scientist's mind; the scientist stops the experiment, and declares the belief that the observed regularity is a law. Although these information functions differ from scientist to scientist, the values of N_T that each uses will be enough alike to force a consensus.

Russell (1974) notes that lower animals perform induction to establish laws about environmental correlations, as when a cat experiments with a new owner's lap and after a sufficient number of trials decides it is safe to sleep there. The number of trials required for the generalization varies from cat to cat, but a curve made over a histogram of the threshold values would doubtlessly approach the shape of normality. So to this extent the cat is a scientist, without realizing it.

Example 8. Consider that soldiers in battle experience cognitive dissonance brought on by the difference between their personal information functions and those imposed by the army. On a personal level, the enemy is usually more tolerable than on the army's level. The soldier is caught between two conflicting standards. Often a conflict like this appears in research. A researcher, well aware of the normative standards for statistical and practical significance that a profession sanctions, may find their use leads to personally intolerable results. That is, the researcher may feel something is true even though the data do not say it according to the sanctioned standards—may, in fact, try to get away with passing off statistical significance at the $\alpha = 0.15$, or greater, level.

The purpose of army training is to mold soldiers to the army's norm. The purpose of science training is, similarly, to mold scientists to each science's norm. All organizations must establish sanctioned norms among their members; otherwise, as we've said, their members could not communicate and act together to achieve a common goal.

Let us conclude by showing how these concepts of information, tolerance, and norms bear upon cluster analysis. When coming to step 5 of the research method given in the previous chapter, the researcher must decide about conclusions—what to believe. The researcher looks at patterns of data in the tree (how similar and dissimilar certain pairs of objects are, for example), and must decide whether the patterns sway belief one way or another. For example, if testing a hypothesis, the researcher must decide whether the patterns are strong enough to favor confirmation or falsification. This requires that the researcher hold something like an information function in mind, with a threshold value that is based on knowledge of the norms in his field. In this way a decision, while necessarily subjective, will not be arbitrary.

SUMMARY

Reaching conclusions in research is necessarily subjective but is guided by professional norms. This chapter has shown how the concepts of information, tolerances, and norms illuminate subjective decision making.

States of information are a finite set of beliefs. The role of data is to "turn on" the proper belief. Since patterns of data are continuous in form, an information function with thresholds is required to relate data to beliefs. There must be regions of tolerance in which patterns of data can be different yet still prompt the same belief. These regions are defined by the thresholds, and the thresholds are set by the norms of the professions.

As a result, no two researchers from the same field will carry in his mind exactly the same information function, yet there will be enough agreement among all researchers in the field to generate a consensus about what conclusions can be drawn from any given analysis.

Framing and Validating in Cluster Analysis

OBJECTIVES

Our interests now turn to how applications of cluster analysis are framed and validated.

- We first discuss the nature of framing and validating.
- Then we discuss measures of primary and secondary validity.

21.1 THE NATURE OF FRAMING AND VALIDATING

While planning to make an application of cluster analysis, a researcher must make several decisions that will condition the analysis. These are subjective, and include (1) the choice of objects, (2) the choice of attributes, (3) the choice of scales of measurement for the attributes, (4) the choice of whether or not to standardize the data matrix and, if so, how, (5) the choice of a resemblance coefficient, (6) the choice of a clustering method, and (7) the choice of the level of similarity at which to cut the tree (if the research goal is to make a classification). These choices are interdependent and cannot be made without regard for each other.

The actual choices the researcher makes are called **framing decisions** because they frame, shape, or channel the data from its raw state (the input) to the tree (the output). If these decisions are made differently, the tree will be different.

For each choice listed above there are alternatives. These are called **framing alternatives**. The problem for the researcher is to consider the various framing alternatives with respect to their expected efficacy for achieving the research goal being addressed, and then to choose the best ones—the best objects, the best attributes, the best way to standardize the data, the best resemblance coefficient, and so on. These framing decisions are entered into a cluster-analysis computer program where they determine how the data matrix is processed.

Having a research goal is essential to making the framing decisions. Each of the research goals discussed in Chapters 4, 5, and 6, when set in context, generates values that make the framing process meaningful. And if you don't believe this, find a table of data in a scientific journal and cluster-analyze it out of context. You will have an impossible time deciding which resemblance coefficient to use, for example, because having no research goal strips the values from the problem, and without values you cannot decide what the best choice is. "Best" is meaningless without the sense of direction a research goal gives.

Framing the cluster analysis takes us through step 4 of the six steps of research presented in Chapter 19. The final two steps evaluate how well the research goal was achieved—that is, how valid and useful the conclusions are; against a background of **validating criteria**, in steps 5 and 6 the researcher must decide the worth of his analysis.

The standards used to frame and validate a research problem depend on each other, as the following illustration shows. Suppose we have the goal of firing a rifle shot close to the center of a target. We draw up a mental list of the framing alternatives—a tolerable range of choices necessary to fire a rifle shot: choices of rifle, choices of bullets, methods of adjusting for slope and windage, positions for aiming, and so on. Then we make the framing decisions—we make the choices that will give us the greatest expectation of hitting the target's center.

The validating criteria, on the other hand, describe a tolerable region around the target's center, inside which our shot must fall in order to achieve our goal. This tolerance cannot be set at zero, for this would give virtually no chance of success. But neither can it be set a mile wide, for success would always result.

Then comes the moment of truth. We implement the framing decisions and make the shot: when the smoke clears, we judge whether or not the bullet hit within the region of tolerance around the target's center. From this outcome, we learn how to make better framing decisions the next time we try. Out of the analysis of why the shot failed to achieve the goal, or why it succeeded, we learn. In this way we become better marksmen—and it is in the same way that we become better researchers.

We learn by repeating the steps of research again and again over many cluster analyses. This is because the steps form a feedback loop. At the start of the loop we make the framing decisions that optimize our expectations of achieving valid conclusions. At the end we judge whether we have succeeded. The trial and error of making framing decisions and then comparing them with feedback to see if they are valid modifies how we make framing decisions in the future. We learn and improve with experience.

Much of the learning is metaphorical, a point Bateson (1980) makes throughout his book *Mind and Nature*. Consider for a moment four people who learn to fire a shotgun by using four different feedback systems. Person *A* has always shot quail. Person *B* has always shot clay pigeons at a trap shooting range. Person *C* has only fired at noises in the night, and has never seen a target of any kind. Person *D* has never shot a gun.

Suppose the four decide to hunt live pigeons. Who will do the best? Person *A* will, and could do even better by training to shoot pigeons rather than quail. Person *B* has never shot at live birds, but will do better than person *C*. Person *C* has never seen a target of any kind, but has learned something about shooting and will do better than person *D*.

Learning can be transferred from the pursuit of one goal, where we have experience, to the pursuit of another where we have little experience. The amount of transfer depends on how closely the two systems are metaphors of each other.

As well as from their own experience, researchers learn from the experiences of others, by reading books and applications articles and talking to other researchers, all of which help the beginner to acquire some of the experiences of others.

The framing alternatives and validating criteria evolve hand in hand, each endlessly affecting the other. When repeated success at achieving research goals is met, researchers tighten the tolerances of the validating criteria, thereby reducing the probability of future success when the same framing decisions are made. This in turn forces researchers to reexamine, change, and expand the list of framing alternatives in order to increase their success rate. As the cycle of trial and error is repeated with new applications, it is as if researchers were gradually zeroing in on absolute truth—never getting there but always improving.

Research conclusions are not absolute. Today's regression analyses that give $R^2 = 0.65$ may be judged nontrivial, but that same R^2 may be judged trivial at some time in the future. The progression of tightened standards is a result of researchers learning to make better framing decisions.

Of multivariate methods, cluster analysis offers the greatest number of framing alternatives. You will not find many options in a computer program that does, say, principal component analysis or factor analysis. That's because these techniques are the results of deductive mathematics, rather than an algorithmic sequence of steps such as cluster analysis is based on. In order to derive these techniques, most of the framing decisions have already been made by mathematicians, and the researcher cannot change them. On the other hand, the more framing alternatives there are to choose from, the more flexibility there is to tailor the mathematics to reality.

21.2 PRIMARY AND SECONDARY VALIDITY

We have seen that how well a cluster analysis achieves its research goal and generates interesting and useful conclusions is a measure of its **primary validity**. However, there are certain features that we would like the cluster analysis to have, and these are measures of its **secondary validity**.

The following measures of secondary validity have appeared in the literature. Some of them apply only to making classifications. We will list them before commenting.

1. Obtaining Well-structured Clusters. To make a classification of ticks, Herrin and Oliver (1974) performed two cluster analyses of the same data matrix by using two resemblance coefficients, the average Euclidean distance coefficient, and the correlation coefficient. In assessing secondary validity they stated that "in the formation of well-structured clusters the distance phenogram [tree] produces much more pleasing results than the correlation phenogram." In other words, they compared each tree to their subjective referent of what an ideal tree with "well-structured clusters" should look like and found the tree based on the average Euclidean distance coefficient to be in better agreement.

2. Agreement with Existing Classifications. Alexander (1976) used cluster analysis to make a classification of the skeletal remains of some Sioux

Indians. He used an existing classification of the skeletal remains (based on traditional methods) as the referent for secondary validity. By trial and error, he even changed the attributes in a series of cluster analyses to arrive at a classification that closely approximated the existing one. He put more faith in the existing classification because he chose it as the referent, rather than the other way around.

3. Agreement with Expert Intuition. Hodson et al. (1966) and Sneath (1968) found that their cluster analyses of Iron Age brooches agreed with an intuitive classification of the brooches made by expert archaeologists. In another example, Gupta and Huefner (1972) used expert intuition as the referent for secondary validity in making a classification of industries. To make the classification, they cluster-analyzed apparel, tobacco, textile, paper, machinery, and other industries described by five financial ratios. Commenting on their criteria of validity, they noted: "we surveyed a group of individuals knowledgeable in the area of industry economics and asked them to classify industries according to several financial ratios, based on economic characteristics of the industry. We recognize that such a procedure has certain inherent limitations... but this appeared to offer the best available quantifiable standard against which we might compare our results." The agreement, it turned out, was reasonably close.

In another example, from medical research, Zajicek et al. (1977) used cluster analysis to make a classification of histopathological sections comprised of various glomerulopathies. They concluded that "the fact that clustering matches the pathologists' classification indicates that all necessary observable variables [attributes] have been taken into account and thus unique diagnosis is possible."

4. Agreement of the Classification with One Derived by Using a Different Data Matrix. Stretton et al. (1976) cluster-analyzed species of fungi of the genus *Aspergillus* by using attributes that described their long-chain fatty acids. The resulting classification agreed well with one produced by another research method based on DNA hybridization data and techniques. In another example, Fujii (1969) cluster-analyzed eight strains of bean weevils based on characteristics of their population dynamics such as development time of males and of females; adult longevity of males and of females. The structure of the tree agreed well with another tree from an independent cluster analysis of a data matrix that was based on the weevils' morphological characteristics.

5. Agreement of Different Multivariate Methods. As a measure of secondary validity some researchers have used the agreement of classifications produced from the same data matrix processed by different multivariate methods—for example, methods such as cluster analysis, principal component analysis, multidimensional scaling, and factor analysis. For example, Hicks and Burch (1977) cluster-analyzed specimens of 20 southern red oak varieties that they described by 15 twig, bud, and floral characteristics, and made a classification of oaks having two classes. They then analyzed the same data matrix with factor analysis and found that the oaks separated into the same two groups.

Other researchers have coupled the use of cluster analysis with discriminant analysis, in a procedure that goes as follows. First, they use cluster analysis with a data matrix to produce a classification of objects. Then, using the objects' attribute measurements given in the data matrix as independent variables and their class memberships given by the cluster analysis as the dependent variable, they fit a discriminant function to the data. They use the fitted discriminant function and the objects' independent variable measurements to see how correctly the discriminant function identifies each object's class. They take the percentage of correct predictions as a measure of the secondary validity of the classification. This procedure makes a circle, first using cluster analysis to make a classification and then, with the same data, seeing whether another method, such as discriminant analysis, will identify the objects into their known classes. This procedure has been used by Bradfield and Orlóci (1975) and by Green and Vascotto (1978).

6. Agreement of Classifications Based on Split Samples of Data. Sinha (1977) has suggested that the objects of the original data matrix be randomly split into two subsamples and separate cluster analyses be run on each to produce two classifications. To be judged valid, the two classifications should agree: the number of classes should be the same and their defining attributes should be the same.

7. Demonstration of Stability and Robustness. Williams (1967) describes these two properties this way: "A classification should be truly *stable* in that it is not disturbed by the addition of further information [for example, adding a few new objects or attributes to the data matrix]; it should be *robust* in that alterations to a small number of items of information [for example, removing one or two objects or attributes from the original data matrix] should not produce major changes in the classification."

8. Agreement with a Researcher's Prior Expectations. Sinha (1977) states, "Before the analysis the... researcher should develop, in as much detail as possible, the structure of the solution he expects. Then he should compare it to the results of the analysis. *A priori* specification of a model gives a more objective benchmark than is provided by *a posteriori* rationalization or appeal." His comments apply to *a priori* prediction of a tree's gross or detailed features, or prediction of the structure of a classification obtained from a tree.

Most of these secondary goals are worthwhile. For example, suppose that a cluster analysis gives a classification that fails to agree with one obtained from expert opinion. Searching for the source of the disagreement —which could be either in the cluster analysis, the expert opinions, or in both—could lead to better understanding and ultimately to a classification that has a better chance of achieving primary validation. Or consider that a classification that is demonstrated to be robust, stable, and insensitive to random splitting of the data matrix is probably reliable, and not an artifact of the sampling process.

But in the end the assessment of primary validity is also crucial. For it is possible to have a valid classification with respect to several measures of secondary validity yet still miss achieving primary validity.

For example, suppose the research goal is to create a research hypothesis. A cluster analysis is performed and it achieves secondary validity when judged on the above criteria. But if the scientist ultimately fails to create an interesting research hypothesis, one that leads to further testing and proves useful to the advance of science, then despite scoring high in matters of secondary validity the research will be undistinguished.

The statement of the research goal, and the assessment of whether or not the cluster analysis achieved the research goal and is useful, measures the primary validity of the whole research process from start to finish. This overall test completes the feedback chain that is necessary to make science, and planning and management, self-correcting endeavors.

SUMMARY

For all types of data analyses, the processed data are framed by the researcher's choices of the kind of data to be collected and the mathematical equations to be used to process the data. These are framing decisions.

The researcher must choose the primary validating criteria with respect to his research goal, and use these as a standard against which he judges the overall validity and worth of his analysis. This is true of all types of data analyses, not only cluster analysis.

Framing and validating a cluster analysis requires subjective decision-making. The framing alternatives and the primary validating criteria are set by the norms of a research community. In general they depend on the research goal and its context. Thus, a cluster analysis that aims to generate an interesting scientific question will be framed and validated differently than one that aims to help solve a problem in planning.

Apart from primary validity, which measures how valid the cluster analysis is overall, there are several measures of secondary validity. They assess how well the cluster analysis has certain desirable properties. For example, we would like the cluster analysis to be stable—that it not be altered by the addition of a few new objects or attributes to the data matrix.

When secondary validity occurs, it increases the chances that primary validity will occur but it does not guarantee it.

Examples Illustrating How to Frame and Validate Applications of Cluster Analysis

OBJECTIVES

- For each of the research goals presented in Chapters 4, 5 and 6, we now illustrate how researchers frame and validate their applications.
- Studying these examples can help you to understand how researchers make their subjective framing and validating decisions.
- You will then be better able to make such decisions yourself.

22.1 RESEARCH GOAL 1: CREATE A QUESTION

Of all research goals, the decision-making required for this one is the most subjective. Why? Because with other goals such as testing a hypothesis, the hypothesis is known and serves as a background that helps show how the cluster analysis should be framed. But in creating questions, what kind of knowledge may exist as a background is less obvious and this makes the process of framing more open ended.

Even so, there is always some background knowledge available to interject a scale of values into the problem. For one thing, the framing alternatives for selecting objects and attributes are circumscribed by the accumulated knowledge within each field; for fields are defined by what they study—by the properties and concepts that enter into their theories and laws—and this focuses the direction of investigation.

Take the example of the two anthropologists who cluster-analyzed the skins of species of primates in Research Application 4.1 to generate new questions. They chose anthropologically meaningful objects—skins—to discriminate among primates. Furthermore, they described the skins with attributes commonly used in dermatology. This was appropriate because each field has its milieu, its precedent of concepts and theory and knowledge, which defines the types of objects and attributes that are meaningful and therefore suitable for study. By focusing on meaningful objects and attributes, meaningful questions often arise.

Once the data matrix is framed with these meaningful objects and attributes and the data are collected, it should be cluster-analyzed in many ways. That is, it should be standardized in a variety of ways, and with a variety of resemblance coefficients and clustering methods. Any combination of framing decisions is permissible as long as the researcher understands the mathematical consequences of the choices. Each cluster analysis will show the similarities among objects from a different framing perspective, and this will increase the chances of finding interesting patterns in the tree from which interesting questions can result.

The importance of understanding the mathematical consequences of the framing decisions should not be minimized. For example, two cluster analyses, one using the average Euclidean distance coefficient and the other using the correlation coefficient, are apt to give different trees. But for meaningful questions to follow, the researcher must know that the first coefficient is sensitive to size displacements among data profiles whereas the second one ignores them (as Chapter 8 discusses). Otherwise the researcher will not be able to interpret the trees correctly.

Now for validation. By what criteria are questions validated? How can we know how good a question is? Naturally, when the cluster analysis fails to raise questions, the research is a failure. But suppose questions are created. It is then necessary to evaluate their quality, and to do this we must know what the properties of good scientific questions are.

One property is that they challenge existing knowledge. For instance, when a researcher looks at a tree and thinks, "The tree suggests *this*, but before I ran the cluster analysis I thought *that*—now I'm perplexed," he has created a good question. The more fundamental and dogmatic the challenged knowledge is, the more important the challenging question is.

Good questions help direct future scientific research so that knowledge grows fast, and so that the knowledge finds use in applied science—in planning, management, engineering, and technology. The payoff from a good question is always in the future. History ultimately judges how good a question was, although at the time the question is raised the researcher may already have suspected its ultimate importance.

If history shows that a question led to the creation of a research hypothesis, which then led to its testing, and that the outcome was vital knowledge that could be linked with existing knowledge, thus causing the field to advance and thus eventually benefiting society, then in retrospect the question must be judged as good. Some researchers clearly have a knack for anticipating what these ultimate outcomes will be and what history will say.

Success in creating a good scientific question also depends on what the researcher already knows. Two researchers who set out independently are not likely to create the same questions. For this reason, brainstorming done as team research, using a variety of cluster analysis and other multivariate methods, is more likely to be fruitful than individual research.

22.2 RESEARCH GOAL 2: CREATE A HYPOTHESIS

For an example we will use Sigleo's (1975) research, as described in Research Application 4.3. Recall that she cluster-analyzed turquoise artifacts and source mine material to create a hypothesis that answered this question: geographically, how wide-ranging were the trading patterns among prehistoric Indian cultures that lived in the American southwest? It became apparent to her that if she could find out the similarities and dissimilarities among artifacts and source mine material, this might show patterns that would allow her to create a meaningful hypothesis. This led her to the class of methods that can show patterns of similarity among objects, and from these methods she selected cluster analysis.

To frame the analysis, Sigleo searched for attributes that would unambiguously discriminate among the turquoise samples. She could not use attributes of color and beauty, for they are not unique "birthmarks." However, she reasoned that elemental compositions would be reliably unique for distinguishing among turquoise mined in different areas and thus would be suitable. How should she measure the attributes? Not on a presence/absence qualitative scale because the elements that make up turquoise would be present in all samples. To meet the criteria of discrimination, she determined that the quantitative abundances of the elements would have to be measured.

Should she standardize the data? She knew some elements would be much more abundant than others. If not standardized, they would contribute unequally to overall resemblance, shadowing the rare elements and possibly reducing the uniqueness among the data profiles. Therefore, standardizing would be in order.

Should she use the correlation coefficient r_{jk} to measure the similarity between data profiles? No—for the reason that size displacements among data profiles would differentiate samples and endow each with a distinguishing birthmark; and to ignore these would ignore a critical feature of the samples' origins.

Sigleo could have standardized the data matrix by using equation 7.1, and then have used the average Euclidean distance coefficient, d_{jk}. Both choices are consistent with her framing criteria. Instead, she chose a self-standardizing similarity coefficient, g_{jk}, which is written

$$g_{jk} = \left(\frac{1}{n}\right)\sum_{i=1}^{n}\frac{\min(X_{ij}, X_{ik})}{\max(X_{ij}, X_{ik})}, \qquad 0.0 \le g_{jk} \le 1.0$$

where X_{ij} and X_{ik} are the elemental concentrations of the ith element for the jth and kth turquoise samples, and "min" and "max" denote the minimum and maximum, respectively, of the values between the parentheses. Note that g_{jk} is the average of n fractional terms.

To illustrate the calculation of g_{jk}, consider these data:

	Object	
Attribute	1	2
1	100	120
2	10	12
3	10	5

It follows that for objects 1 and 2, for example,

$$g_{12} = \left(\frac{1}{3}\right)\left(\frac{100}{120} + \frac{10}{12} + \frac{5}{10}\right) = 0.722.$$

Clearly g_{jk} is self-standardizing: attribute 1 is ten times attribute 2, yet both contribute the same to overall resemblance.

We can imagine Sigleo's thinking as she framed the cluster analysis. Most likely she mentally played out the consequences of alternative cluster analyses done with various objects, attributes, and resemblance coefficients, imagining how the resulting trees might suggest hypotheses. By this process of performing "thought experiments," she eliminated what her imagination suggested would not work, leaving what might work.

Such thinking draws heavily on mathematical knowledge and on experience in using cluster analysis. It is not possible to imagine the consequences of alternative analyses without understanding the mathematical natures of resemblance coefficients and cluster methods, and so forth. Nor would Sigleo be able to think what objects and attributes might work without having gained experience from actual practice, from reading journal articles of studies that dealt with similar problems, and from talking with other researchers.

Before Sigleo performed the cluster analysis she had framed, she formulated possible outcomes and their meanings. Some of these are the following:

Possible Outcome 1: The artifacts would cluster by themselves, and the source mines would cluster by themselves. If this happened, the only plausible hypotheses would be that the turquoise came from an unknown mine whose location has been lost, or that the actual source mines were much more distant than current knowledge suggests.

Possible Outcome 2: All source mines and artifacts would be quite similar and fall into one large cluster. If this happened, no hypothesis could be formed and the research goal would not be achieved.

Possible Outcome 3: All of the artifacts would cluster with samples from one mine. This outcome did occur, allowing Sigleo to form a hypothesis about the extent over which Snaketown residents traded. In this way she achieved her research goal.

Success also depends on how good the generated hypothesis is. How do we judge the goodness of a hypothesis? One measure is the researcher's expectations about whether or not further testing will confirm it. A second measure is how it stands in relation to existing knowledge. Does it stand as a potential link between two or more highly-valued bodies of knowledge? If so, and if the hypothesis can be confirmed, then the bodies of knowledge can be connected into a new whole. This means that a researcher, when deciding if a hypothesis is good, must base that decision on two uncertainties—the uncertainty that future testing will confirm the hypothesis, and the uncertainty that the research community will value the hypothesis if it is confirmed.

As with questions, history is the judge of how good a hypothesis is. A researcher will have hunches about what history may ultimately say, but only time will tell.

Note that whereas creating a question and creating a hypothesis are both acts of creation, more definitive background values exist for creating a hypothesis. Thus, it is easier to frame a cluster analysis to create a hypothesis than it is to create a question—the former will be less of a shot in the dark. The reason is that to create a hypothesis you must already have a question in mind, and this question acts as a focus that makes the framing of the cluster analysis more apparent.

22.3 RESEARCH GOAL 3: TEST A HYPOTHESIS

To use cluster analysis to test a research hypothesis using the hypothetico-deductive method, the hypothesis must allow a prediction about the pattern of similarities and dissimilarities among a set of objects; the assumed truth of the hypothesis must entail a prediction of how the tree would look—perhaps not the detailed structure of the whole tree, but certainly some features. Later, when the tree's predicted features are compared to its actual features as revealed by the cluster analysis, they will either agree or disagree, according to a tolerance specified by the validating criteria. Agreement will confirm the hypothesis, and disagreement will disprove it.

The framing alternatives that bound the selection of objects, attributes, methods of standardization, resemblance coefficients, and so on are predicated on the nature of the hypothesis. To cite Research Application 4.4 as an example, the researchers tested a hypothesis about the extent of pre-Hispanic trading patterns among Indians in central Mexico. The hypothesis was general, applying to any of the things that the Indians might have traded. But it was known that they made obsidian artifacts, that their most likely source of raw obsidian would have been through trade, and that chemical properties of obisidan uniquely mark where it was mined. So obsidian artifacts from different living sites and samples from various mines could be used for objects. And any method that would uniquely discriminate the objects' chemical properties could be used to measure the attributes. Of the possible methods, the researchers chose neutron activation analyses, possibly because this method would not deface the artifacts.

The hypothesized trading patterns led the researchers to predict that virtually none of a sample of artifacts from a site called Chalcatzingo would cluster with the source mine material from the Paredón mine, 150 kilometers away. Here, the phrase "virtually none" is a validating criteria that most scientists would certainly translate into "a few percent or less." When the cluster analysis was run, 32 percent of the Chalcatzingo artifacts were most similar to the Paredón obsidian, disproving the hypothesis.

Part of the validating criteria for hypothesis-testing is that the types of error that statistics books label as Type I and Type II should be minimized. A **Type I error** is committed when the researcher declares the hypothesis false when really it is true, and a **Type II error** is one whereby a researcher calls true a hypothesis that really is false. In planning-research, both types of error carry social costs when technological products or environmental decisions are incorrectly based on them. Both are costly to scientific research too—but the Type II error is the more costly.

If a scientist makes a Type I error, he simply fails to discover a law of nature. Even if he publishes his proclaimed falseness of a true idea, other scientists can still do research; they will simply choose not to base their research on the falsely nullified hypothesis. The situation is similar to not being able to use yet-to-be-discovered truths as a basis for scientific research, a situation we all must work within.

Type II errors, however, can have disastrous consequence. When a scientist falsely proclaims a truth, other scientists will likely build their work on the false knowledge. Many other scientific results may have to be retracted once the mistake is eventually found. Mainly to avoid Type II errors, science insists on the independent validation of all important knowledge claims.

22.4 RESEARCH GOAL 4: MAKE A GENERAL-PURPOSE CLASSIFICATION

Using two Research Applications, let us illustrate how to frame and validate a general-purpose classification. The first concerns a classification of species of blueberries; the second, a classification of the cells of organisms. While both use biological objects, the attributes that frame the two classifications are different. The classification of blueberries is based on nonbiological attributes describing the morphology of blueberries, whereas the classification of cells is based on biological attributes that describe the chemistry of the cells' molecular structures.

RESEARCH APPLICATION 22.1

Using cluster analysis, Vander Kloet (1978) made a classification of species of blueberries. From 43 sites in eastern North America he collected 378 specimens as objects for the cluster analysis. On them he measured 15 attributes: (1) plant height, (2) growth habit (open colony, compact colony, or crown-forming), (3) leaf persistence (evergreen, deciduous, or intermediate), (4) mean leaf width, (5) mean leaf length, (6) type of abaxial leaf surface (glaucous, glabrous, pubescent, or intermediate), and so forth.

Vander Kloet dichotomized the quantitative attributes in this list, so that all were qualitative. Then, because the 378 specimens proved to be too many for one cluster analysis run, he used the reference object method, described in Chapter 17, with the simple matching coefficient and the UPGMA clustering method. The specimens clustered in the tree according to their existing species designations, and this reconfirmed the status of *Vaccinium pallidum* as a well-defined species in its own right.

Vander Kloet found that three key attributes (plant height, leaf width and shape, and leaf margin) were sufficient to separate *V. pallidum* from *V. angustifolium*; that three attributes (plant height, growth habit, and leaf width) would separate *V. pallidum* from *V. corymbosum*; and that three attributes (leaf size and shape, glandular indumentum, and calyx pubescence) would separate *V. pallidum* from *V. tenellum*. In addition, he found only a few natural hybrids among the specimens.

RESEARCH APPLICATION 22.2

Let us next examine Woese's (1981) research on the classification of the cells of organisms. Until this research, biologists described cells by morphology, by what they saw when they looked through their microscopes. And from what they saw, they thought all cells could be classified with two classes: the *eukaryotes*, cells with a well-formed nucleus; and the *prokaryotes*, cells without a nucleus. Woese, however, used a relatively new technology to measure cell nucleotides, which has made it possible to analyze the molecular structure of cells.

To make a molecular-based classification, he formed a data matrix having 16 species of bacteria for its objects, with attributes describing their nucleotides. Thus, each object signed itself with a unique molecular signature. Then, using the Bray-Curtis coefficient, Woese cluster-analyzed this data matrix, and he found that in addition to the two recognized classes, a new one emerged—the *archaebacteria*. He concluded that from a molecular viewpoint, cells of this new class are as different from the prokaryotes as they are from the eukaryotes.

In the future the validity of this classification will be determined by stages. Woese's research has initiated the process of validation: first, he has shown that when cells are described molecularly they group into three classes, a finding contrary to those based on the traditional morphological viewpoint (and this raises the unanswered question of why morphological differences do not fully mirror molecular differences); second, he used his findings retroductively to derive a hypothesis about the evolution of the archaebacteria that implies they are at least as old as the two previously recognized classes. As Woese's classification works its way into new theories, these new theories will be tested, and this will be the basis of ultimate validity.

The framing of general-purpose classifications is guided by the precepts of the disciplines that practice numerical taxonomy. As for validating criteria, a classification must eventually integrate itself into scientific theory or prove to be related to other variables. Hence, history is the ultimate judge here, too.

A newly proposed classification has an **expected efficacy** for making a variety of predictions and functioning in theories and laws. How well the new classification is expected to work must be subjectively judged against how well the existing classification (if there is one) does work. When expectations are high, scientists set out to see if the new classification does in fact produce better theories and laws; if it does, it comes to supersede the old classification.

22.5 RESEARCH GOAL 5: MAKE A SPECIFIC-PURPOSE CLASSIFICATION

Let us give two Research Applications that illustrate the framing and validating of specific-purpose classifications. Each requires that a classification be made in order to test a research hypothesis. Thus, both also fulfill Research Goal 3 (test a hypothesis).

RESEARCH APPLICATION 22.3

Gregor (1979) offers evidence in support of the hypothesis that short men in primitive tribes are accorded lower social status. He formed the data matrix shown in Table 22.1. The objects are 15 Mehinaku Indians, men from a tribe in central Brazil. These he described with five attributes reflective of social status (number 5 measures political influence in terms of being a chief or nonchief). The lowermost row of the table is not a social attribute and therefore is not part of the data matrix. The height of the men is the dependent variable that the hypothesis suggests is related to social status.

Table 22.1 Data matrix of 15 Mehinaku Indian men described by five attributes. Data are from Gregor (1979). "N" denotes missing values of data.

	Mehinaku men														
Attribute	1	2	3	4	5	6	7	8	9	10	11	12	13	14	15
Number of girlfriends	6	10	7	7	6	3	3	4	3	2	4	2	3	6	3
Rich(1) or poor(0)	1	1	1	1	N	N	0	N	0	0	N	0	0	0	0
Rituals sponsored	1	1	1	2	1	0	1	0	1	0	0	0	0	0	0
Rituals participated in	1	3	3	1	2	2	0	0	0	0	1	1	0	1	0
Chief(1) or nonchief(0)	1	1	0	1	0	0	0	0	0	0	1	0	0	0	0
Height in inches	69.25	65.00	64.25	64.25	64.25	64.25	64.25	63.75	63.75	63.75	63.00	62.75	60.75	60.25	59.75

Note how the hypothesis directs the framing decisions. The hypothesis is that short men in primitive tribes (Mehinaku Indians in particular) are accorded lower social status than tall men. Suppose we were to test this hypothesis by using cluster analysis; to do so requires that we assess the relation between two variables: (1) physical height, and (2) social class. Physical height is a quantitative variable, a straightforward measurement. But social class is a qualitative variable, with no universally accepted

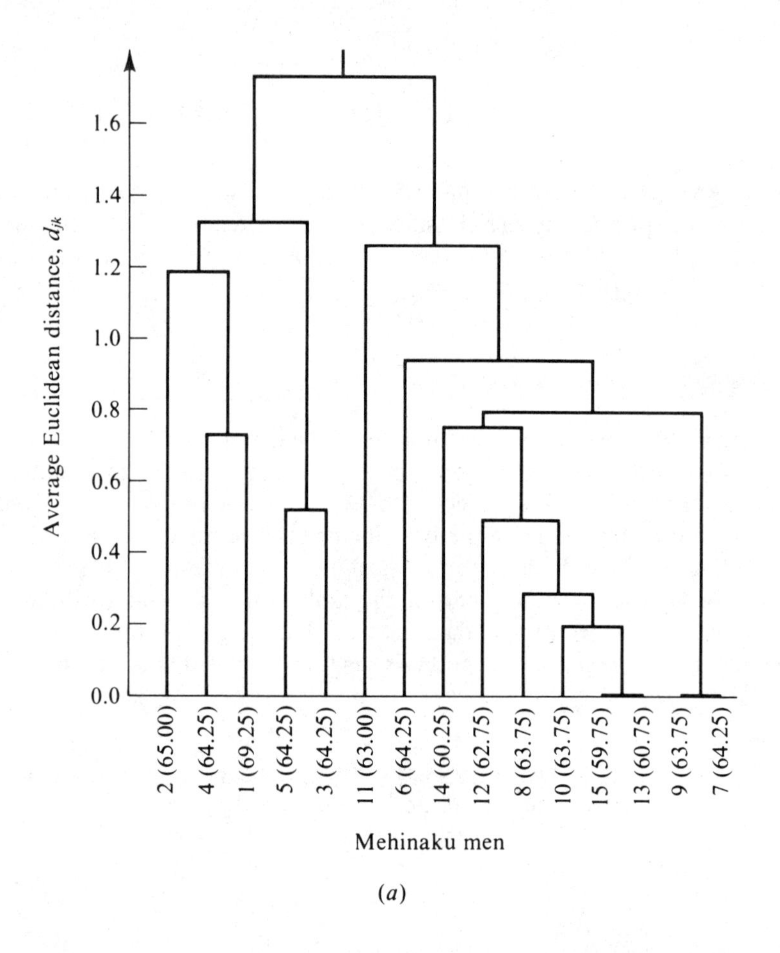

Figure 22.1. (*a*) Tree of 15 Mehinaku Indian men produced by using the average Euclidean distance coefficient and the UPGMA clustering method. Their heights in inches are shown in parentheses on the branch tips. (*b*) Contour diagram showing the relation of the clusters of Mehinaku men to their heights.

objective definition for it. Thus we must manufacture it for our analysis. This we do by measuring those attributes that anthropologists believe are indicators of status among the primitive tribes. This frames the data matrix, and the role of our use of cluster analysis is to manufacture the qualitative variable of social status—the classification.

Gregor confirmed the hypothesis subjectively by visually examining the data matrix and observing that most of the attributes seemed to conform to the hypothesis. Our approach would be to cluster-analyze the data matrix; in our approximation we would neglect the fact that attributes 2 and 5 are nominal-scaled, and treat all five attributes as if they were quantitative attributes. But to equalize the contribution of each to the overall resem-

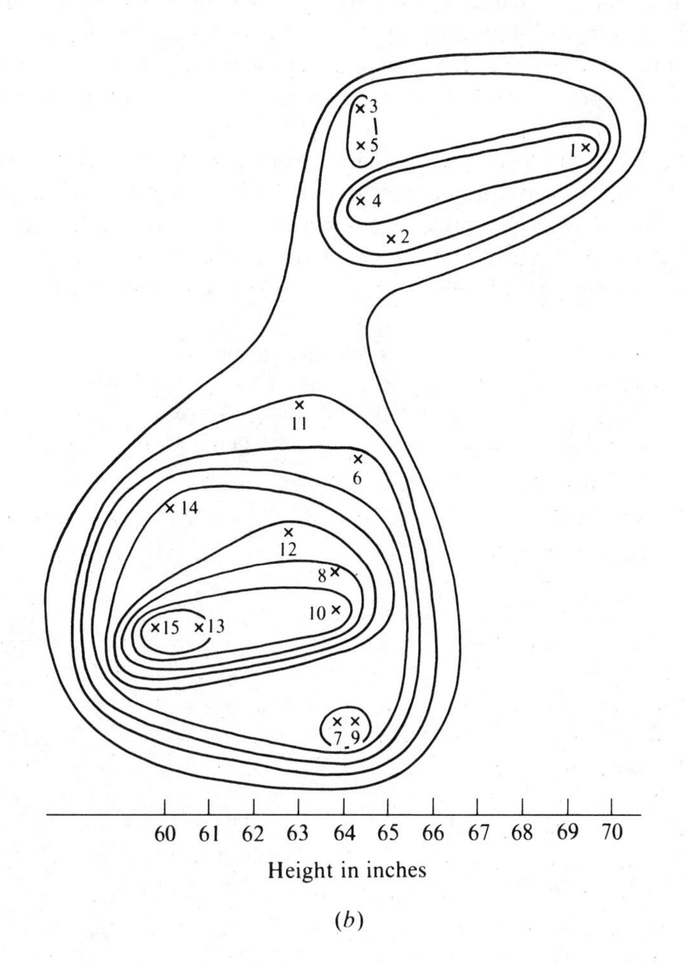

(*b*)

Figure 22.1. (*Continued*).

blance between the men, we would standardize the data matrix with equation 7.1 (the standardized data matrix will not be shown).

Having standardized the data matrix, next we compute the resemblance coefficient. Clearly, size displacements in data profiles matter, for we are measuring social status in terms of amount, for instance the number of girlfriends. For this reason we choose to frame the analysis by using the average Euclidean distance coefficient, d_{jk} (the missing data matrix values prevent us from using e_{jk}). Cluster analysis by use of the UPGMA clustering method then produces the tree shown in Figure 22.1,*a*.

Cutting the tree near $d_{jk} = 1.4$ gives two clusters of Mehinaku men; the first contains five men; the second, ten men. Table 22.2 summarizes the mean values of the attributes by cluster. For all attributes, the five men in cluster 1 score higher; that is to say, they have more status. Furthermore, on average they are taller, which supports the hypothesis. Since the standard deviations are mostly small relative to the differences in means, the results are clear cut.

Figure 22.1,*b*, a contour diagram, gives another perspective. Its horizontal axis is the height variable and its vertical axis is unscaled. The men are plotted and contours are subjectively drawn around them consistent with the order of their clustering in the tree. The contour diagram shows considerable overlap in height among the two main social clusters, and it

Table 22.2 Mean values of the five attributes, by cluster, for the cluster analysis of Mehinaku Indian men shown in Figure 22.1,*a*. Mean height of the men is also shown. Standard deviations are shown in parentheses.

	Cluster	
Attribute	1	2
Number of girlfriends	7.2 (1.6)	3.3 (1.2)
Rich (1) or poor (0)	1.0 (0.0)	0.0 (0.0)
Rituals sponsored	1.2 (0.04)	0.2 (0.4)
Rituals participated in	2.0 (1.0)	0.5 (0.7)
Chief (1) or nonchief (0)	0.6 (0.5)	0.1 (0.3)
Height in inches	65.4 (2.2)	62.6 (1.7)

suggests that were it not for the inclusion of man #1 the hypothesis would not be as strongly supported.

Overall, the cluster analysis is more objective and more insightful than merely looking at the data matrix and forming conclusions. It allows other views of the data, and is more convincing than Gregor's original subjective analysis.

Note from Table 22.2 that all five attributes play a role in driving the clustering. That is to say, if there were an attribute whose mean was about the same across all clusters, it would be an inessential attribute and could be removed from the analysis. Conversely, an attribute that drives the clustering must show a large difference in its mean value (relative to the standard deviations) across two or more clusters. The question of statistical significance cannot be answered with an inferential statistical analysis (for example, with an analysis of variance). The men are a deliberately chosen sample, not a random sample, and this invalidates statistical inferences.

RESEARCH ILLUSTRATION 22.4

According to Galtung's (1971) theory of imperialism, there are two classes of nations of the world, called *central* and *peripheral*. Central nations dominate peripheral nations in a feudal sense. For instance, central nations are primarily exporters of finished goods while peripheral nations are primarily exporters of raw materials. Moreover, volume of trade is large between central nations, less between central and peripheral nations, and small between peripheral nations. One of the main consequences of Galtung's theory is that central nations and peripheral nations should form two discernible classes within an overall classification based on attributes describing their economies.

Gidengil (1978) used cluster analysis to test Galtung's theory, to see if the nations would separate into the two clusters as the theory predicts. She used 68 nations, categorized as either central or peripheral, as objects; and she used seven economic indicies as attributes to describe each nation. Note that Galtung's theory dictated that the analysis be framed with countries as objects described by economic attributes.

For the resemblance coefficient, Gidengil considered using the correlation coefficient, r_{jk}, but she rejected it because it ignores additive translations of data profiles. Instead, she framed the analysis by using the cosine coefficient, c_{jk}, because it is sensitive to the additive translations she wanted to detect.

Gidengil cut the tree resulting from the cluster analysis into five clusters, computed the means of the seven attributes for each cluster, and compared them. The analysis confirmed Galtung's theory: two of the clusters were composed of central nations; the other three were of peripheral nations.

Gidengil also learned that the central nations differ among themselves and divide into two camps, and that there are three types of peripheral nations. Her analysis of the data profiles suggested why this is so.

22.6 RESEARCH GOAL 6: FACILITATE PLANNING AND MANAGEMENT

Planning and management cannot be practiced without well-defined objectives. These objectives give well-defined values, so that planning and management applications of cluster analysis almost frame themselves. Let us give three examples to illustrate this point.

RESEARCH APPLICATION 22.5

We begin with an application from plant genetics. There, specimens of plants play the role of experimental units. Usually a geneticist does not apply a treatment to the specimens but instead crossbreeds dissimilar specimens to produce new varieties having desirable properties.

Davis's (1975) research in producing new kidney bean hybrids is an example. His research goal was to produce new hybrids by crossing members from a stock of 50 hybrids. Then he could examine the new hybrids and find those having the most desirable properties. But this would have required $(50)(50 - 1)/2 = 1{,}225$ experimental crossings, and to run all these experiments would have been extremely costly and inefficient: first, there would be the cost of 1,225 experiments; second, there would be the cost of assessing the properties of 1,225 new hybrids. And finally, many of the crossings would likely result in new hybrids with similar properties. It was therefore necessary to plan a smaller experiment.

What Davis did was to reduce both the cost and the redundancy in experimental outcomes. He described the 50 hybrids with a data matrix whose 50 columns were the hybrids, and whose attributes measured their morphological features, their rates of phasic development, and their yield components. He framed the analysis with these attributes because all were of commercial importance. Following the cluster analysis, he selected representative hybrids from each branch of the tree, 26 in all, in order to have a smaller set whose properties mirrored the variety found in the larger set of 50. This reduced the number of possible pairings from 1,225 to $(26)(26 - 1)/2 = 325$.

Davis felt that hybrids resulting from pairings among the smaller stock would not compromise discovery; that is, nearly all of the commercially important hybrids that could be discovered in the 1,225 experiments could also be achieved in the 325 experiments; other, similar hybrids could be eliminated beforehand, without needing to test them. Cost and redundance were reduced, thereby validating the procedure.

RESEARCH APPLICATION 22.6

Another application that requires a diverse set of objects mirroring the range of dissimilarity in a larger population is the construction of food tasting panels. That is, how do you select people to form a typical panel?

As Golovnja et al. (1981) describe, a human population is randomly sampled for candidates who will become the objects for a cluster analysis. The research goal of selecting a food tasting panel dictates that the attributes frame the selection process in terms of the candidates' tastes. Thus, these researchers used as attributes the candidates' scores on a series of taste tests made with graded concentrations of different flavors in water. The candidates rated the taste properties of each flavor on a ten-point scale, giving a data matrix.

Obviously, size displacements between candidates' data profiles indicate ability to detect differences in tastes, and so a resemblance coefficient such as the average Euclidean distance coefficient, d_{jk}, which is sensitive to size displacements, was used. Furthermore, because the data matrix was based on scores that covered the range of ten points, it did not have to be standardized (however, in related research problems there might be considerations we have not noted here).

The cluster analysis showed clusters of people whose tastes were similar, and by taking a few people from each cluster a representative taste panel was assembled.

RESEARCH APPLICATION 22.7

Datapro Research Corporation (Computerworld, 1980) had the research goal of organizing proprietary computer programs so that potential buyers could easily compare their features. This objective automatically frames the analysis. The objects were the 301 computer programs of interest. Each was rated by users according to a one-to-four point scale (1 = poor, 4 = excellent) on the attributes that potential buyers care about: reliability, efficiency, ease of installation, ease of use, vendor's technical support, and so forth. Following the cluster analysis, the Datapro analysts cut the tree into four clusters and listed the programs and their features by cluster.

The need to end up with a few clusters was dictated by the need to make comparison-shopping easier. The analysts could have used as many as 301 clusters or as few as one cluster, both extremes having decided disadvantages. They framed the cutting of the tree into four clusters because they felt that four would prove the most useful to their customers.

Datapro's ability to satisfy their customers, to keep their customers subscribing to their service, and to gain new customers, measures the validity of the research.

At its simplest, the criterion for validity in achieving planning and management research goals is how well you succeed in staying out of the poorhouse. If you work for a company and make a "wonderful" research discovery by using cluster analysis, you will know exactly how wonderful it was by its effect on the financial health of you and your company. As we saw earlier, Davis (1975) reduced the number of experiments required from 1,225 to 325 in Research Application 22.5. Assuming no technical errors in his work, his research was valid simply because it resulted in a plan that saved his employer money.

SUMMARY

This chapter has given examples of how to frame and validate applications of cluster analysis for six research goals. The examples are meant to be illustrative rather than prescriptive.

The examples emphasize that (1) framing alternatives should be considered, that (2) framing decisions are made based on their expected efficacy for achieving a research goal, and (3) that validating criteria are set by considering the research goal in its context. These subjective decisions are based on (1) an understanding of the mathematical tools of cluster analysis, and on (2) a reading of professional norms within each field. What constitutes validity differs from field to field, and from one research goal to another, and this must be felt and understood.

Research problems must be framed so as to attain valid and useful conclusions. When attained, these reinforce the framing decisions. When not attained, they force the researcher to modify the framing decisions on subsequent analysis.

This is how all trial-and-error learning goes. Unfortunately, it is easier said than done.

The Orders of Patterns of Similarity

OBJECTIVES

Let us conclude this book by discussing patterns of data.

- First we define what a pattern is.
- Next, we examine the variety and orders of patterns.
- Then, we see how patterns gain significance and become information.
- Finally, we illustrate patterns of data by using examples from cluster analysis.

23.1 WHAT A PATTERN OF DATA IS

A **pattern of data** is a design made by the relations among a set of objects. Only when all the relations are the same is there no pattern. There must be contrast among some of the individual relations to produce a pattern. When the contrast is great enough, the pattern becomes information that informs the researcher.

Of the many kinds of relations in science—relations of prediction, or of association, or of one object acting on another, or relations of similarity—this last one is the focus of this book.

Scientists study patterns of similarity among data on different levels, and these levels have been catalogued by the biologist Gregory Bateson (1980, p. 11) into a hierarchy of three orders. The first level, the similarities made by the relations among a set of objects is a **first-order pattern** of similarities, the most basic kind. But scientists also study the pattern of similarities among two or more first-order patterns, this making a **second-order pattern** of similarities. Finally, scientists study the pattern of similarities among two or more second-order patterns of similarities, this making a **third-order pattern** of similarities. Bateson concluded that biologists only look for these three orders of similarities when they study and contrast biological forms.

23.2 THREE ORDERS OF PATTERNS OF SIMILARITIES

First-order Pattern of Similarities. Bateson (1980) explains the first-order pattern by focusing on the anatomy of the crab, which he describes as "...repetitive and rhythmical... like music, repetitive with modulation." Bateson asks us to imagine a crab, head upward, with an imaginary axis from head to tail bisecting its body.

From this vantage point we see the bilateral symmetry shown in Figure 23.1,*a*. Each side has five numbered "legs" (one a pincer). Taking these 10 legs as the entities of a pattern, and taking the $(10)(10 - 1)/2 = 45$ pairwise comparisons of similarity among the legs as the relations, we see a pattern of similarity between the legs on the left and right sides. Bateson calls this kind of pattern within an individual creature a set of "first-order connections." We have renamed it a first-order pattern of similarities.

We can use cluster analysis to recast the crab's symmetry. To do this, let us number the five major segments of each leg from the tip segment inward to the body, as Figure 23.1,*a* also shows. Measuring the length of the segments and using these as five attributes to describe each leg, we get the data matrix shown in Figure 23.1,*b* (the values are relative to the picture of the crab). Next we use the Euclidean distance coefficient, e_{jk}, to compute the 45 pairwise values of resemblance between legs, as Figure 23.1,*c* shows. And finally, we use the UPGMA clustering method with this resemblance matrix to obtain the tree shown in Figure 23.1,*d*.

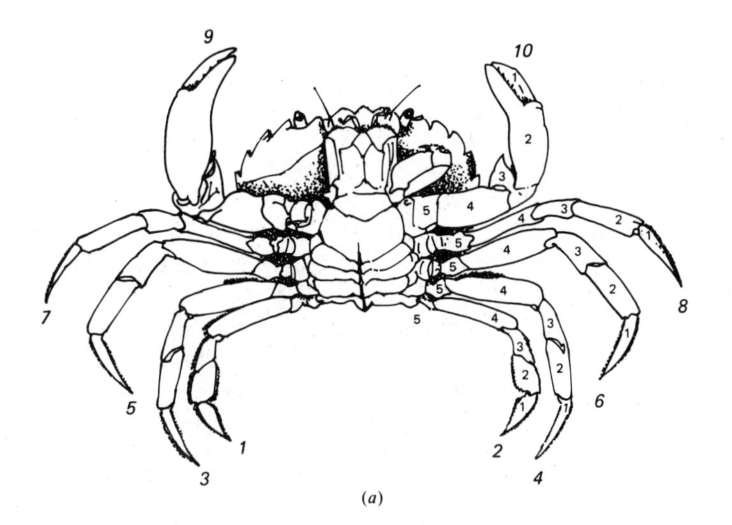

(a)

Leg number

	1	2	3	4	5	6	7	8	9	10
1. Dactylus	10	10	14	14	14	14	13	13	14	11
2. Propodus	8	8	11	11	13	13	15	15	34	26
3. Carpus	7	7	8	8	9	9	7	7	5	4
4. Merus	18	18	20	20	21	21	13	13	15	14
5. Ischium	3	3	4	4	5	5	6	6	6	6

(b)

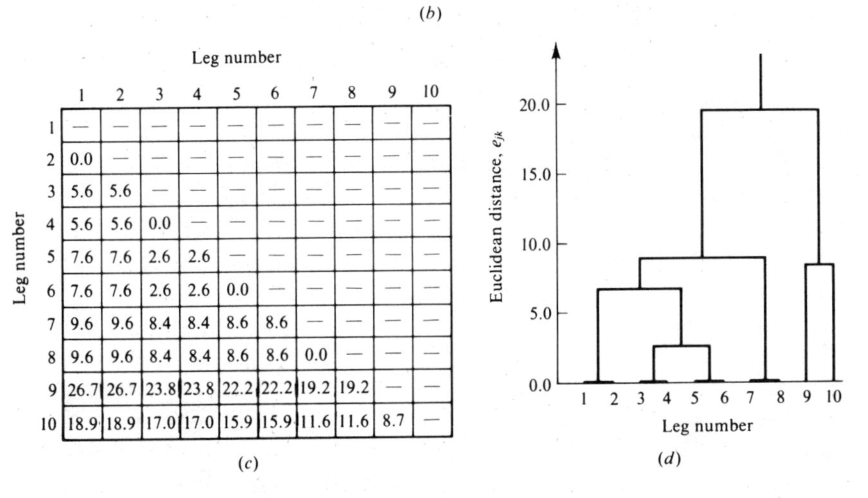

Leg number

Leg number	1	2	3	4	5	6	7	8	9	10
1	—	—	—	—	—	—	—	—	—	—
2	0.0	—	—	—	—	—	—	—	—	—
3	5.6	5.6	—	—	—	—	—	—	—	—
4	5.6	5.6	0.0	—	—	—	—	—	—	—
5	7.6	7.6	2.6	2.6	—	—	—	—	—	—
6	7.6	7.6	2.6	2.6	0.0	—	—	—	—	—
7	9.6	9.6	8.4	8.4	8.6	8.6	—	—	—	—
8	9.6	9.6	8.4	8.4	8.6	8.6	0.0	—	—	—
9	26.7	26.7	23.8	23.8	22.2	22.2	19.2	19.2	—	—
10	18.9	18.9	17.0	17.0	15.9	15.9	11.6	11.6	8.7	—

(c)

(d)

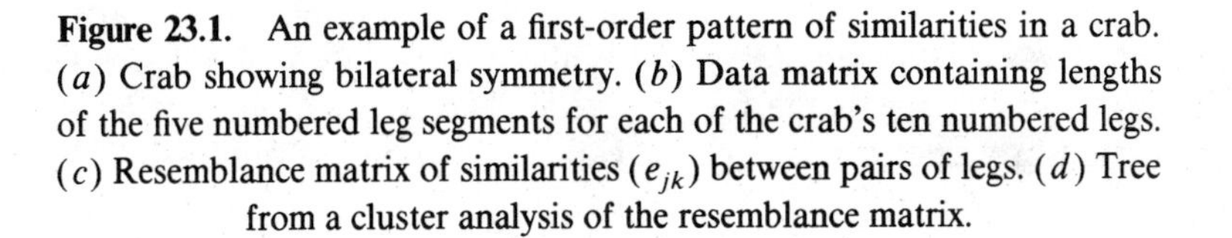

Figure 23.1. An example of a first-order pattern of similarities in a crab. (a) Crab showing bilateral symmetry. (b) Data matrix containing lengths of the five numbered leg segments for each of the crab's ten numbered legs. (c) Resemblance matrix of similarities (e_{jk}) between pairs of legs. (d) Tree from a cluster analysis of the resemblance matrix.

This tree captures the first-order pattern of similarities. The pattern has the special property of **bilateral symmetry**—symmetry between the relations of similarities on one side of the crab's body and those on the other side.

Sigleo's (1975) research, described in Research Application 4.3, is a search for an asymmetric first-order pattern of similarities. The entities she cluster-analyzed were two kinds of turquoise, artifacts and source mine material. However, Sigleo was not looking for a symmetric first-order pattern of similarities of the sort that existed in the left and right legs of the crab, for in science a pattern doesn't have to be symmetric to be useful. Instead, she was looking for one of the following sets of conceivable asymmetric first-order patterns: (1) the artifacts would cluster alone; (2) the artifacts would cluster with the source turquoise material from one, and only one, mine; (3) the artifacts would cluster with the turquoise from several mines. She defined beforehand how the patterns of similarities that could conceivably result would inform her. As we will see in the next section, she had an idea of the information function that relates the conceivable patterns of similarities to information.

Second-order Pattern of Similarities. Suppose we have two first-order patterns of similarities, each of which can be studied independently. Suppose further that they are in some sense comparable—for example, we may want to see if they are similar. Their comparison makes a second-order pattern of similarities that draws our interest.

As examples of second-order patterns, Bateson uses the similarity of (1) the first-order pattern formed by the parts of crabs, and (2) the first-order pattern formed by the parts of lobsters; and (1) the first-order pattern formed by the parts of men, and (2) the first-order pattern formed by the parts of horses.

For an example from cluster analysis, suppose we compute the cophenetic correlation coefficient, which measures the similarity between a resemblance matrix and a cophenetic matrix. The resemblance matrix describes the similarities between the objects before they are cluster-analyzed, and the cophenetic matrix describes the similarities between the same objects after they are cluster-analyzed. Both are separate first-order patterns of similarities. By comparing them, we are investigating a second-order pattern. In general, when we compare any two resemblance matrices, or any two trees, for the same set of objects, we are investigating a second-order pattern of similarities.

For an example of a second-order pattern in which the two first-order patterns are not patterns of similarity, but rather are patterns of association, let us reconsider Research Application 5.6. There, a second-order pattern of similarity was used to test a hypothesis about the location of a hydrochemical front running across the Indian Ocean. The diagram on page 62 shows the second-order pattern. The classification of samples of the fossil *Orbulina universa*, C_1, forms a first-order pattern of association with the locations at which sample specimens were collected, Y. That is, there is a pattern of association between classes of samples and the sampled locations. This we denote as $C_1 \leftrightarrow Y$, a first-order pattern. Similarly, the classification of samples of living *O. universa*, C_2, forms a first-order pattern with the sampled locations, Y. This we denote as $C_2 \leftrightarrow Y$. The hypothesis test hinges on the prediction that the two first-order patterns of association must be similar if the hypothesis is true, that $(C_1 \leftrightarrow Y) \leftrightarrow (C_2 \leftrightarrow Y)$, as diagrammed on page 62. Thus, the prediction involves a second-order pattern of similarities.

Note that a second-order pattern has the structure of an analogy. For example, take the generalized analogy written as A : B : : C : D, which reads "A is to B, as C is to D." Clearly, "A is to B" is a set of relations, "C is to D" is a second set, and the two sets are themselves related.

Third-order Pattern of Similarities. Scientists discover third-order patterns by taking two second-order patterns and examining the relations between them. Bateson gives as an example of a third-order pattern the similarity between the second-order pattern of crabs and lobsters and the second-order pattern of men and horses.

Although in science third-order patterns occur less frequently than first- and second-order patterns, Research Application 5.6 also contains an example of a third-order pattern. As stated, the scientists held a hypothesis about the hydrochemical front that led to a predicted second-order pattern. To see if the prediction were true, they obtained the facts of the second-order pattern by performing two cluster analyses to get the classifications C_1 and C_2. Then, they had the predicted second-order pattern and the actual second-order pattern. When they compared them, they were looking at a third-order pattern of similarities.

As a summary of the three orders of patterns of similarity, in Table 23.1 we give Bateson's example that compares the parts of crabs, lobsters,

Table 23.1 Three orders of patterns of similarities illustrated with Bateson's (1980) example of crabs, lobsters, men, and horses. For each order of pattern, the comparison between the arrows takes place. Parentheses indicate that a comparison has already been made between the enclosed items at a lower level.

Pattern	Comparison			
First-order	→ Crab parts ←	→ Lobster parts ←	→ Man parts ←	→ Horse Parts ←
Second-order	→ (Crab parts)	(Lobster parts) ←	→ (Man parts)	(Horse parts) ←
Third-order	→ ((Crab parts)	(Lobster parts))	((Man parts)	(Horse parts)) ←

men, and horses. For each order of pattern, the comparison between the arrows takes place. Parentheses indicate that a comparison has already been made between the enclosed items at a lower level.

23.3 HOW PATTERNS BECOME INFORMATION

Like individual values of data, patterns of data can also inform. We can think of this process as an input/output device, in which patterns of data are input through the senses and information is output from the mind:

Pattern → | Mind | → Information

The diagram, like an engineer's black box diagram, draws our attention to the mind as a coupling device that stands between factual patterns and states of belief.

With one exception, this is the concept of an information function we studied in Chapter 20. There, single values of data were used to label the x-axis of the information function. Here, the x-axis is labeled with the abstract idea of a pattern. That is, instead of entering the x-axis with a single value of data as before, we enter it with a pattern. We can, in fact, imagine an index of "patternness" that would assign to any first-, second-, or third-order pattern a numerical value summarizing how it is patterned, and the value of this index would label the x-axis of the information function.

To understand more about how patterns produced by cluster analysis become information, let us diagram what happens from step 3 through step 5 of the six steps of research given in Chapter 19—the steps that span between the gathering of data and the generation of information needed to draw conclusions. Here is how it looks:

Multivariate data → | Cluster analysis | → Tree → Pattern → | Information function | → Information

Beginning on the left, the researcher frames the problem, deciding what multivariate data need to be collected and how to tailor the cluster analysis (or other multivariate analysis method). This requires that a vantage point be chosen that can frame the pattern that will eventually be seen. Next, data are collected, the cluster analysis is made, and a pattern of similarities is found in the tree. The researcher studies the tree, and from this further frames the pattern, again deliberately selecting and abstracting its essential properties. Then, using this twice-framed pattern, the researcher mentally passes it through an information function to decide what should be believed. In this chain, raw data become reduced data, reduced data are interpreted as patterns, and the interpreted patterns become information—a serial metamorphosis of data into thought.

From this chain, it is also clear that there are two points in an analysis where the pattern of data is framed. The first is when the researcher chooses objects, attributes, scales of measurement, a method of standardization, a resemblance coefficient, and a clustering method—all in accordance with the dictates of the research goal. The second point is when the researcher examines the tree and abstracts the relevant features from the irrelevant ones, consciously or unconsciously.

The idea that a logical information function can transform patterns into beliefs is parallel to the idea that an *emotional* information function can transform patterns into feelings. For example, symmetric patterns are a special class of all patterns, and in some instances they cross our innate thresholds for beauty.

23.4 SOME EXAMPLES OF THE ORDERS OF PATTERNS IN CLUSTER ANALYSIS

Most applications of cluster analysis address research goals that require first-order patterns of similarity be studied. To give some examples, first let us compare two applications that depend on first-order patterns of similarity—one is from paleontology, one from art history.

For the paleontology application, Harris and Porter (1980) wanted to know whether the fossil remains of what appeared to be a new species of horse found in Dry Cave, near Carlsbad, New Mexico, fell along lines of descent morphologically similar to known species in the *Equus* genus. As a start at forming a hypothesis, they cluster-analyzed a fossilized limb bone of the species in question, which they named *E. scotti*, along with the fossilized limb bones of 19 other species collected from locations around the world. For attributes, they used average measurements made on specimens of the bones, and they used the correlation coefficient, r_{jk}, for the resemblance coefficient because it ignored arbitrary size displacements possibly owing to arbitrary age and sex differences among the data profiles. This produced the tree shown in Figure 23.2.

From this, they decided that the relation of similarities between *E. scotti* and the other species is a pronounced first-order pattern. *E. scotti* is weakly similar to the African zebras, but overall it stands apart from the others, like a bump on an otherwise undistinguished first-order pattern. Because of this, they retroductively proposed the hypothesis that *E. scotti* does not fit well the accepted line of descent of the *Equus* genus, and they propose further study to resolve its position.

For the art history study, Guralnick (1976) used cluster analysis retroductively to propose a hypothesis that some early Greek sculptors copied their styles from Egyptian sculptors. To do this, she selected a set of Greek and Egyptian figurative statues as objects and described them with attributes measuring forms such as distance from top of the head to the eyes, distance from the eyes to the chin, and width of the head. After standardizing to adjust for the effects of arbitrary size displacements among the statues, she used cluster analysis to find the first-order pattern of grouping. As it turned out, some of the Greek and Egyptian statues clustered together, and using this significant feature of the pattern plus other information, she concluded that "a few archaic Greek sculptors certainly used grids based on the Egyptian Second Canon grid in order to proportion their statues."

We saw earlier in Research Application 4.3 how an anthropologist used cluster analysis to group turquoise artifacts and source mine material, making a first-order pattern of similarities that led her to a specific hypothe-

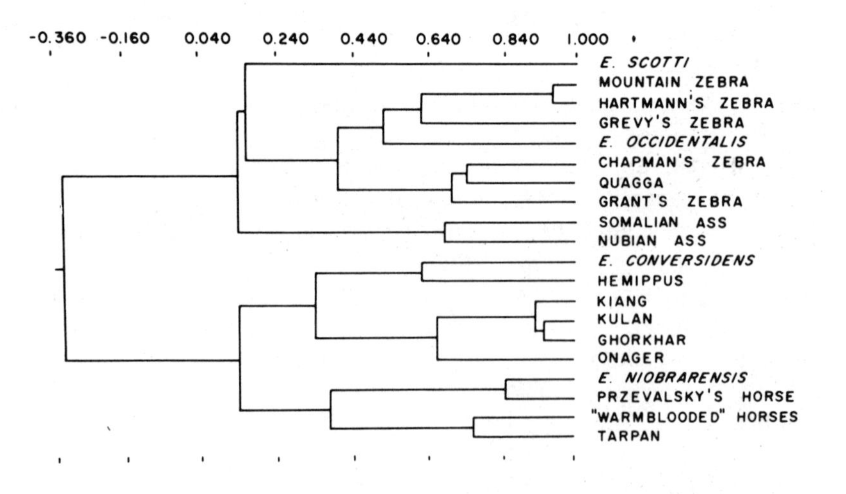

Figure 23.2. Tree of horse bones obtained by using the correlation coefficient r_{jk} as a measure of similarity. The relations of similarities between *E. scotti* and each of the other species is a pronounced first-order pattern. (By permission, reproduced from Harris and Porter (1980), Figure 8.)

sis about the origin of the artifacts. Likewise, the paleontologists used cluster analysis to group fossil bones, one of uncertain origin, to make a first-order pattern of similarities that led to information about where the unknown fossil stands in a known line of descent; and the art historian used cluster analysis to group Greek and Egyptian statues, making a first-order pattern of similarities that led her to conclude that some Greek sculptors were influenced by Egyptian style. Only the contexts were different: these researchers were all thinking alike.

Now for a final example that contains a second-order pattern. Clover (1979) used cluster analysis to test a hypothesis about the ecological diversity of populations of lizards that evolved in isolation from each other on islands. The hypothesis predicted three separate first-order patterns of association involving similarities, so Clover set out to obtain actual first-order patterns, thereby to test the hypothesis.

For the cluster analysis, he selected between 4 and 38 lizards from each of 31 islands and used these lizards as objects. He described the populations by averaging the sample values for each of 15 morphological attributes, such as width of occipital cap, internal width, head length, and snout-vent length, which are all properties that evolution has influenced. Using Ward's method of cluster analysis, he obtained a tree and cut it into a classification of populations.

Clover then determined the relation of association between the classification and each of three variables— the three relations that would be found if the hypothesis were valid. These were between the classification and (1) the islands' geographic locations, (2) their land areas, and (3) the species of lizards living on the islands. In this way the classification was used in three separate first-order patterns of association.

He concluded that indeed the classification was related (1) to the geographic location of the populations, (2) to island areas, and (3) to species. That is, by comparing these actual first-order patterns to the ones predicted earlier—a comparison that generated three second-order patterns of similarities—Clover concluded that another hypothesis held previously must be false. He stated that "these [second-order] patterns appear to contradict similarities implied by previous subspecific designations based largely on color and pattern differences."

Clover ended his article by suggesting that an expanded analysis along the same lines, but one that would include lizards from the mainland populations, would test what ecologists call the *time-divergence theory of insular evolution*—a theory that states that evolutionary rates of island-bound populations are inversely proportional to island size, the smaller islands with their unique environments causing strong selective pressure and a corresponding erosion of genetic variation. The theory leads to the test prediction that "populations on large islands, young small islands, and old small islands should be increasingly dissimilar to the mainland populations."

SUMMARY

Patterns of similarities can be classified according to a hierarchy. A first-order pattern is the most basic. The tree from a cluster analysis shows a first-order of pattern of similarities among the objects.

A second-order pattern is made by the relations between two first-order patterns. The similarity between two trees (that is, two cophenetic matrices) is a second-order pattern of similarities that is measured by a matrix correlation coefficient.

A third-order pattern is made by the relations between two second-order patterns. If we wish to test a hypothesis that predicts the form of a certain second-order pattern, to perform such a test, we must obtain the actual second-order pattern and compare it to the predicted one. This comparison involves a third-order pattern.

Patterns of similarities inform us, and the concept of an information function is useful when thinking about how patterns prompt beliefs, and about the thresholds that determine how beliefs are formulated.

EPILOGUE

One can love science even though it is not perfect, just as one can love one's wife even though she is not perfect. And, fortunately, for just a moment and as an unexpected and undeserved reward they sometimes do *become perfect and take our breath away.*

ABRAHAM MASLOW

The Psychology of Science

APPENDIX 1

Books and Articles on Cluster Analysis and Other Multivariate Methods

As an introductory book, this one is merely a starting point: only through additional study will you be able to shape your own views of how multivariate analysis, and particularly cluster analysis, can be used in science and planning.

The following books and articles will begin to take you beyond this book. Some of them cover primarily the theory of cluster analysis; and some are about applications. Others cover multivariate methods such as principal component analysis and discriminant analysis. Except where noted, they are mathematically simple. Some of these books were published a decade or more ago; all were chosen for this list because of their excellence, not because of how recently they were written.

A1.1 BOOKS

Numerical Taxonomy, P. H. A. Sneath and R. R. Sokal (1973). Although dated, this book is widely regarded as the authoritative reference to the philosophy and application of cluster analysis in numerical taxonomy. With a dictionary in hand for defining a few unfamiliar terms, the nonbiologist can also reap rewards. Inasmuch as its advice is geared for numerical taxonomy—specifically for the making of biological classifications—it does not always offer good advice for studies with other research goals. This book is a complete revision and updating of an earlier book, *Principles of Numerical Taxonomy*, R. R. Sokal and P. H. A. Sneath (1963).

Numerical Ecology, L. Legendre and P. Legendre (1983). Although set in an ecological context, this is a practical handbook that gives thorough treatment to cluster analysis and several other multivariate methods. Particularly noteworthy are several chapters covering resemblance coefficients that can be used with qualitative and quantitative data. The authors have assembled a most complete list of resemblance coefficients, many of which are not well known outside of ecology.

An Introduction to Numerical Taxonomy, H. T. Clifford and W. Stephenson (1975). This book explains the methods and applications of cluster analysis and of principal component analysis as used in plant ecology. It contains a detailed discussion of resemblance coefficients.

Cluster Analysis, B. S. Everitt (1980). This book is a short and readable introduction to cluster analysis that features the centroid clustering method.

Cluster Analysis for Applications, M. R. Anderberg (1973). A comprehensive, context-free introduction to hierarchical and nonhierarchical cluster analysis, this book has probably been widely read because researchers often cite it when describing their applications.

Clustering Algorithms, J. A. Hartigan (1975). A collection of various clustering algorithms, this book is concerned more with the mathematical aspects of cluster analysis than with research applications, although a variety of entertaining examples are given (such as clustering hardware items, clustering automobiles described by their frequency of repair records, and clustering wines described by connoisseurs' ratings).

Cluster Analysis, B. S. Duran and P. L. Odell (1974). Quite mathematical, this book reviews theories of cluster analysis; it contains no applications.

Cluster Analysis Algorithms, H. Späth (1980). Written by a German mathematician, this book discusses some hierarchical and nonhierarchical methods of cluster analysis. Späth's perspective, and the book's treatment of topics neglected in many American books, as in the chapter on the optimal presentation of data profiles, make the book worth reading.

Cluster Analysis, E. J. Bijnen (1973). This Dutch author describes methods of cluster analysis that are popular in sociological research. These methods, some strange (method of Beum and Brundage, method of Coleman and MacRae) and others familiar but not well covered in the literature (method of Tyron, McQuitty's syndrome analysis), are described with almost no use of mathematics.

An Introduction to Systems Analysis: with Ecological Applications, J. M. R. Jeffers (1978). Jeffers introduces the methods of multivariate analysis, including cluster analysis, principal component analysis, discriminant analysis, as well as several methods seldom used outside ecology, (for example, reciprocal averaging). He gives a good discussion of minimal spanning trees, an alternative to the tree for displaying the results of a cluster analysis.

Sampling Design and Statistical Methods for Environmental Biologists, R. H. Green (1979). Using the philosophy of experimental design as a framework, this book covers applications of multivariate analysis, including cluster analysis, in environmental research.

Mathematics and Computers in Archaeology, J. E. Doran and F. R. Hodson (1975). A seminal work, this book lays the foundations for a quantitative archaeology. As a scholarly work it is on a par with Sneath and Sokal's *Numerical Taxonomy*. Using very little mathematics, it covers concepts and applications of principal coordinate analysis, principal component analysis, discriminant analysis, multidimensional scaling, and cluster analysis.

Quantitative and Numerical Methods in Soil Classification and Survey, R. Webster (1977). Although primarily a primer for soil scientists that explains the uses of cluster analysis and multivariate analysis in soil classification, this book can be easily read by anyone.

Statistics and Data Analysis in Geology, J. C. Davis (1973). This book clearly explains a variety of multivariate methods, including cluster analysis. It explains the necessary matrix algebra in a simple way that nonmathematicians will appreciate, and it can be read by those knowing little about geology. It includes listings of computer programs.

Multivariate Statistical Methodologies Used in the International Pilot Study of Schizophrenia, J. J. Bartko and W. Gulbinat (1980). The authors give their view of how cluster analysis and other methods of multivariate analysis should be used in human behaviorial research.

Applied Multivariate Analysis, J. E. Overall and C. J. Klett (1972). This book presents the theory and application of multivariate analysis, including cluster analysis, in the context of research into mental disorders. It includes listings of computer programs.

Data Analysis Strategies and Designs for Substance Abuse Research, P. M. Bentler et al. (1976). This introduction to multivariate methods includes a discussion of cluster analysis. It also gives good introductions to multivariate analysis of variance, and to path analysis—two topics infrequently described in concise and clear language.

Statistical Analysis in Geography, L. J. King (1969). Primarily an introduction to the methods of statistical geography, this book includes a well-written introduction to principal component analysis.

Computation Methods for Multivariate Analysis in Physical Geography, P. M. Mather (1976). In detail this book describes cluster analysis and other multivariate methods, including multidimensional scaling. Its context is geography, but its advice has a wider range of application. It contains listings of computer programs.

Application of Numerical Classification in Ecological Investigation of Water Pollution, D. F. Boesch (1977). An introduction to cluster analysis, this report was published by U. S. Environmental Protection Agency.

Quantitative Ethology, P. W. Colgan (1978). This book has chapters on cluster analysis, multidimensional scaling, principal component analysis, factor analysis, multivariate analysis of variance, and discriminant analysis—each written by researchers in animal behavior.

Multivariate Techniques in Human Communication Research, P. R. Monge and J. N. Cappella (Eds.) (1980). Containing chapters written by different experts in human communication, this book describes discriminant analysis, canonical correlation analysis, factor analysis, cluster analysis, nonmetric multidimensional scaling, network analysis, and time-series analysis.

Introduction to Multivariate Analysis. C. Chatfield and A. J. Collins (1980). This book is for statisticians and other research workers who are familiar with the basics of probability, statistical inference, and matrix algebra. It is about as easy to understand as a book can be that still retains a mathematical basis. It has chapters on principal component analysis, factor analysis, multivariate analysis of variance, multivariate analysis of covariance, multidimensional scaling, and cluster analysis.

Methods for Statistical Data Analysis of Multivariate Observations. R. Gnanadesikan (1977). Written mainly for statisticians, but suitable for other researchers familiar with distribution theory and matrix theory, this book presents principal component analysis, factor analysis, cluster analysis, and multidimensional scaling. It quite thoroughly discusses data transformations and the detection and removal of outliers, to allow some of the methods to be used for statistical inference.

A1.2 ARTICLES

"Numerical Taxonomy," R. R. Sokal (1966). Written by one of the founders of numerical taxonomy, this article explains what numerical taxonomy is and what numerical taxonomists do.

"Classification: Purposes, Principles, Progress, Prospects," R. R. Sokal (1974). This article explains the role of computers in classification, defines the main terms used in classification, and gives the purposes and principles of biological classification.

"Spiders & Snails & Statistical Tales: Application of Multivariate Analyses to Diverse Ethological Data," W. P. Aspey and J. E. Blankenship (1977). The authors explain the uses of multivariate analysis, including cluster analysis, in animal behavior research.

"Fundamental Problems in Numerical Taxonomy," W. T. Williams and M. B. Dale (1965). This article contains philosophical discussions of the foundations of numerical taxonomy. It presents the viewpoints on monothetic versus polythetic classifications, hierarchical versus nonhierarchical classifications, and metric (quantitative) versus nonmetric (qualitative) resemblance coefficients.

"Multivariate Approaches in Ecology," R. H. Green (1980). This article describes the uses of multivariate analyses, including cluster analysis, in ecology.

"On the Application of Numerical Taxonomy in Soil Classification for Land Evaluation," R. L. Wright (1977). Wright presents a case study of an application of cluster analysis to develop a soil classification.

"Statistical Methods in Soil Classification," R. J. Arkley (1976). The author gives opinions on how numerical taxonomy should be applied in making soil classifications.

"Cluster Analysis," K. D. Bailey (1975). An introduction to cluster analysis for sociologists, this article gives comprehensive coverage to the methods that have found favor in sociological applications, (for example, the different cluster analysis methods of McQuitty).

"Pattern Seeking Methods in Vegetation Studies," M. B. Dale (1978). The author suggests strategies for using cluster analysis to make grassland classifications.

"Numerical Methods of Classification," D. W. Goodall (1973). This leading ecologist reviews the methods of numerical taxonomy typically used in making classifications of vegetation.

"Interpretation of Research Data: Other Multivariate Methods," R. N. Zelnio and S. A. Simmons (1981). The authors discuss some of the advantages and disadvantages of four methods of multivariate analysis: discriminant analysis, factor analyses, multidimensional scaling, and cluster analysis.

"Cluster Analysis," B. S. Everitt (1977*b*). This tutorial paper on cluster analysis contains examples of applications.

"Unresolved Problems in Cluster Analysis," B. S. Everitt (1979). Writing for professional statisticians, Everitt discusses philosophical problems of cluster analysis, such as how to choose the best clustering method, how to determine where the tree should be cut, and so forth. He also discusses statistical problems, such as how to make cluster analysis give clusters that are in some sense optimal.

APPENDIX 2

Computer Programs for Cluster Analysis

In this chapter we will describe the most widely used computer programs for cluster analysis. Three of these are programs within the commercially available statistical analysis packages SAS, BMDP, and CLUSTAN. (The SPSS package does not at present contain any programs for cluster analysis.) One advantage of using a statistical analysis package is that in addition to hierarchical cluster analysis you can use the package's general procedures to manipulate and display the data, and use its other programs to perform other kinds of multivariate analyses. For example, once you have read your data matrix into the package you can also perform nonhierarchical cluster analysis, principal component analysis, discriminant analysis, as well as other statistical analyses.

In addition to these packages, we will also describe the stand-alone computer programs NTSYS, CLUSTAR, and CLUSTID. These are programs written in Fortran, and are available to individuals at moderate costs. For example, the latter two programs will do all the calculations in this book. They are published on magnetic tape as *Clustar/Clustid: Computer Programs for Hierarchical Cluster Analysis*, and the Users manual is published as *User's Manual for Clustar/Clustid Computer Programs for Hierarchical Cluster Analysis*, both by Lifetime Learning Publications, Ten Davis Drive, Belmont, California 94002.

All of the programs have advantages and disadvantages, and the ones best suited for your needs will depend on your situation. The SAS, BMDP, and CLUSTAN programs are widely available at university, corporate, and government agency computer centers. They will do most, but not all, of the methods described in this book. For example, SAS will do Ward's clustering method but BMDP will not. Neither provides the option of using the Bray-Curtis resemblance coefficient. They generally take longer to learn how to use than the stand-alone programs. On the other hand, they offer considerable advantages. Among them are a wide variety of subprograms for alternative multivariate methods, allowing you to analyze your data matrix from other points of view. Furthermore, the computer centers that support the packages usually have well-informed consultants to help you and to answer your questions.

It is a good idea to know how to use several of the programs. For example, if you can use either the SAS or BMDP statistical packages, and can use one of the stand-alone programs, you will have a powerful set of methods available for most any conceivable analysis.

Table A2.1 A comparison of features in the SAS, BMDP, CLUSTAN, NTSYS, and CLUSTAR/CLUSTID computer programs for cluster analysis "Y" denotes the feature is present and "N" denotes the feature is not present.

	SAS			BMDP						
	CLUSTER	FASTCLUS	VARCLUS	P1M	P2M	P3M	PKM	CLUSTAN	NTSYS	CLUSTAR/ CLUSTID
No. of quantitative resemblance coefficients	1	1	2	4	4	1	1	13	9	10
No. of qualitative resemblance coefficients	0	0	0	0	0	1	0	27	16	14
Cluster Methods:										
WPGMA	Y	N	N	Y	N	N	N	Y	Y	Y
SLINK	N	N	N	Y	N	N	N	Y	Y	Y
CLINK	N	N	N	Y	N	N	N	Y	Y	Y
UPGMA	N	N	N	N	N	N	N	Y	Y	N
Ward's	Y	N	N	N	N	N	N	Y	N	Y
Centroid	Y	N	N	N	Y	N	N	Y	Y	N
Flexible	N	N	N	N	N	N	N	Y	Y	N
Other	N	Y	Y	N	Y	Y	Y	Y	Y	N
Rearranged data matrix	N	N	N	N	N	Y	N	N	Y	Y

Table A2.1 (*Continued*)

	SAS			BMDP						
	CLUSTER	FASTCLUS	VARCLUS	P1M	P2M	P3M	PKM	CLUSTAN	NTSYS	CLUSTAR/ CLUSTID
Rearranged resemblance matrix	N	N	N	Y	Y	N	N	N	Y	Y
Cophenetic correlation coefficient (or other measure)	N	N	N	N	N	N	N	Y	Y	Y
Matrix correlation	N	N	N	N	N	N	N	Y	Y	Y
Identification	N	N	N	N	N	N	N	N	N	Y
Other methods of multivariate analysis:		SAS			BMDP					
Multivariate										
Analysis of variance		Y			Y			Y	Y	N
Factor analysis		Y			Y			Y	Y	N
Canonical correlation		Y			Y			Y	Y	N
Discriminant analysis		Y			Y			Y	Y	N
Multidimensional scaling		Y			N			Y	Y	N
Principal component analysis		Y			Y			Y	Y	N
Principal coordinates analysis		N			N			N	Y	N

We next briefly describe each program. Table A2.1 contains additional information about the programs. Furthermore, all of the programs allow you to standardize the data matrix in a variety of ways, so we shall not discuss standardizing further. All of the programs are designed to be run from computer terminals.

A2.1 SAS PROGRAMS

SAS is an abbreviation for Statistical Analysis System. It is a collection of programs for data display, univariate analysis, and multivariate analysis, bound together by common procedures for entering and manipulating data. Because of this, once you learn how to input data and run several SAS programs, your learning will transfer to running other programs in the package. SAS is supported by SAS Institute Inc. Until recently the programs were limited to use with a few makes of computers, principally large IBM computers. SAS is widely used by universities, government agencies, and corporations. The SAS Institute leases SAS to subscribers, updates it periodically, and issues documentation manuals.

Among the SAS programs are three for cluster analysis: (1) CLUSTER is a hierarchical cluster analysis program for Q-analysis; (2) FASTCLUS is a nonhierarchical cluster analysis program for Q-analysis; (3) VARCLUS, which can be used for either hierarchical or nonhierarchical cluster analysis, is for the R-analysis of attributes.

CLUSTER, the program for hierarchical cluster analysis of objects, is restricted to quantitative data. If qualitative data are used, they are treated as if they were quantitative. It provides for three clustering methods: UPGMA, centroid, and Ward's method. The first two of these must be used with the Euclidean distance coefficient as the measure of the resemblance between objects, while Ward's method uses the sum of squares index described in Chapter 9.

FASTCLUS performs a variation of what is called the **k-means method** for nonhierarchical cluster analysis. With the k-means method, k objects are initially used as cluster seeds and the remaining objects are tentatively assigned to the cluster seed each is nearest in distance ("most similar to"). What sets the variants of the k-means methods apart (such as the k-means subprogram PMK in the BMDP package, to be described in the next section) is how the initial k objects are selected to be seeds, and how the process iterates from the initial set of k seeds to the final set and the clusters surrounding them.

In particular, FASTCLUS executes the following four steps:

1. To begin the first iteration, the program selects k cluster seeds so as to have a large Euclidean distance between them. That is, the k seeds cover all regions of the attribute space in which objects are expected to reside.

2. It then forms temporary clusters by sequentially assigning the remaining objects to the cluster seed each object is nearest. As objects are assigned, the cluster seed is recomputed and made to be equal to the mean of the data profiles of the objects that are in the cluster. As a consequence, the cluster seed (usually) changes when an object is tentatively assigned to a cluster.

3. When the first iteration is completed, the final set of cluster seeds are taken as the k initial seeds to start the second iteration. The process is repeated, sequentially assigning the objects to their nearest cluster seed, and updating the seeds as the process moves along.

4. After a number of iterations, when the change in the positions of the cluster seeds is tolerably small from one iteration to the next, the program terminates. The k final clusters are composed of the objects surrounding the k cluster seeds from the final iteration.

The SAS manual claims that FASTCLUS is good at finding initial cluster seeds, so that the cluster-finding process typically converges within a few iterations. Because of this, FASTCLUS can be used to cluster-analyze upwards to 100,000 objects.

VARCLUS is a program that optionally can be used for either hierarchical or nonhierarchical cluster analysis of attributes (variables). It is restricted to using either the correlation coefficient or a covariance measure as an index of resemblance between attributes, and the data must be quantitative. VARCLUS is based on a method that closely resembles principal component analysis. Its objective is to split up the set of n attributes into nonoverlapping clusters such that each cluster can be described by an equivalent, but abstract attribute—that is, by a principal component or a centroid component.

These programs are described in the SAS manual (*SAS User's Guide*, 1982). In addition, there is a specialized cluster analysis program (*SAS Supplemental Library User's Guide*, 1980) called IPFPHC. It is used to cluster-analyze transaction flow tables, which are typically found in social science research. These tables are a square matrix with the same set of objects labeling the rows and columns. The ij-cell entry is a measure of the effect of the ith object on the jth. For examples, research on migration, social dominance, occupational mobility, and input-output analysis, produce tables of this sort.

For more information on SAS, see the user's manuals or write to: SAS Institute Inc., Box 8000, Cary, North Carolina 27511, U.S.A.

A2.2 BMDP PROGRAMS

BMDP, standing for Biomedical Computer Programs P-Series, is a set of programs for data display and univariate and multivariate analyses. The programs share procedures for manipulating and displaying data. The result is a system for analyzing data with different methods once it is entered into the computer. A comprehensive user's manual that describes the programs is available (Dixon, 1983).

BMDP has four programs for cluster analysis: P1M, P2M, P3M, and PKM. P1M is specialized for R-analyses of data matrices (cluster analysis of attributes). P2M is for Q-analyses (cluster analyses of objects). The BMDP manual refers to attributes as "variables," and to objects as "cases." P3M is for the simultaneous execution of R-analysis and Q-analysis (the clustering of attributes and objects at the same time; this is called *block clustering*). Finally, PKM is a program for the nonhierarchical cluster analysis of objects using the k-means method.

The reason BMDP and SAS have separate programs for R-analyses and Q-analyses is as follows. Under the view that the profiles of two attributes are comparable because they cut across the same set of objects (R-analysis), yet the profiles of two objects are not strictly comparable in the same way because they cut across attributes of a different nature (Q-analysis), the argument can be made that these differences demand different resemblance coefficients.

The view that R-analyses and Q-analyses should be restricted to certain types of resemblance coefficients is common (see Anderberg, 1973, Chaps. 4 and 5). For a number of reasons, however, the treatments of cluster analysis made in Sneath and Sokal (1973) and in the present book do not sustain the distinction. To start with, in some problems it is hard to decide which things should be the objects of the data matrix and which should be the attributes. For instance, suppose questions on a given topic are answered by a set of people. Should the people be made the objects and the questions be made the attributes, or vice versa? Are people the same stuff while questions are separate stuff? Or vice versa? It depends on the philosophical frame that one adopts, and different people will frame the problem in different ways.

A second reason for not observing the distinction of using separate resemblance coefficients for R-analyses and Q-analyses is that the research goal may dictate that one of the "forbidden" coefficients be used. For instance, the Euclidean distance coefficient, e_{jk}, is not recommended for the R-analysis of attributes, yet there may be research problems in which the researcher wants to compare data profiles of attributes across objects so that size displacements in data profiles do affect the similarity between profiles.

In any event, the P1M program is intended for the R-analysis of attributes. It has four resemblance coefficients for quantitative data but makes no provision for qualitative data. It makes available three clustering methods: UPGMA, SLINK, and CLINK. The program rearranges the resemblance matrix.

P2M is intended for Q-analyses that cluster-analyze objects. It has four resemblance coefficients for quantitative data, one of which is the Euclidean distance coefficient, which is not available in P1M. Qualitative data are not permitted, except if the user is willing to analyze them by using resemblance coefficients intended for analyzing quantitative data. The user has the option of two clustering methods. P2M also rearranges the resemblance matrix.

P3M does block clustering (the simultaneous clustering of objects and attributes). In one run, it does what required two cluster analyses to do in Research Application 10.2. There, the objects of the data matrix were machine components and the attributes were the machines used to make the components. A value of "1" was scored for each component that had to pass through a given machine; a score of "0" meant otherwise. Two independent cluster analyses were performed, a Q-analysis of the components and an R-analysis of the machines. When the columns and rows of the data matrix were rearranged, blocks of 1's and 0's appeared in the data matrix. This block structure could be useful for designing the layout of a factory so that components could be efficiently routed through machine areas.

For block clustering, P3M should find clusters of blocks that are better separated than can be found by running independent Q- and R-analyses. The reason is that the Q- and R-analyses should be made dependent on each other, since clusters of objects affect clusters of attributes, and vice versa. P3M works off an algorithm that incorporates this dependence. In addition, P3M will handle block-clustering of both qualitative and quantitative data.

PKM is a nonhierarchical cluster analysis program that performs a variant of k-means clustering for Q-analysis. The user specifies an integer value, k, which is the number of clusters desired. The method, which is restricted to the Euclidean distance coefficient, is programmed to perform clustering by division. It begins with all objects in one grand cluster. Then stepwise, the number of clusters is increased by one at each step by splitting one of the existing clusters into two. The program terminates when the objects are partitioned into *k* clusters, and it prints out the objects belonging to each cluster, the distance between cluster centers, and other descriptive statistics. PKM, and FASTCLUS in the SAS statistical package, are alternative variants of *k*-means nonhierarchical cluster analysis. Anderberg (1973, p. 162–163) describes the *k*-means method more fully.

More information on the BMDP programs is available by writing to: BMDP Program Librarian, Health Services Computing Facility, AV-11, CHS, University of California, Los Angeles, California 90024, U.S.A.

A2.3 CLUSTAN PROGRAMS

CLUSTAN is primarily a statistical package for both hierarchical and nonhierarchical cluster analysis, although it also includes principal component analysis and discriminant function analysis. It is supported by the CLUSTAN Project, located in Edinburgh, Scotland, under the direction of Dr. David Wishart, an early contributor to the theory of cluster analysis. A manual by Wishart (1982) describes the subprograms and gives their input specifications.

For hierarchical cluster analysis, the package makes available a large set of resemblance coefficients: 13 for quantitative and 27 for qualitative data. Moreover, in the event you want to use a resemblance coefficient that is not included in the CLUSTAN menu—one that is obscure or one of your invention—you can call a function that allows you to enter the equation of any resemblance coefficient.

CLUSTAN optionally allows the following eight clustering methods for hierarchical clustering: SLINK, CLINK, UPGMA, centroid method, median method, Ward's method, flexible method, and McQuitty's similarity analysis method. The package has options for drawing trees and a variety of scatter plots and interpretative diagrams.

For nonhierarchical cluster analysis, CLUSTAN will execute the *k*-means method and several other methods. You have the option of specifying the *k* cluster seeds to initiate the cluster-finding process, or CLUSTAN will choose a good starting set. In addition, you can specify any of more than a dozen measures of between-cluster similarity to be used.

CLUSTAN is set up to receive data that have already been entered into SPSS, the Statistical Package for the Social Sciences (*SPSS X User's Guide*, 1983). Because SPSS contains several multivariate analysis procedures—such as factor analysis, discriminant function analysis, and canonical correlation analysis—the cluster analysis programs of CLUSTAN make a good adjunct.

To run CLUSTAN, a simple command language is used. Once you have accessed the program, you enter procedure keywords to initiate the desired sequence of data processing steps, and for each keyword you enter the option codes that specify the task you want to perform. Apart from the data, a program of key words and option codes usually amounts to not much more than a dozen lines. Certain graph-plotting procedures use the CalComp 1620 graphics library.

CLUSTAN is installed worldwide at several hundred universities and research institutes. It is available by lease from: CLUSTAN Project, 16 Kingsburgh Road, Murrayfield, Edinburgh EH12 6DZ, Scotland.

A2.4 NTSYS PROGRAMS

NTSYS stands for Numerical Taxonomy System of Multivariate Programs. As this suggests, the programs are designed primarily for research in numerical taxonomy. This specialization shows itself in two ways. First, some of its features—for example, the subprogram for principal coordinate analysis—were designed for numerical taxonomic research and are not used much in other fields. Second, the manual that tells how to run the programs is couched in the language of numerical taxonomy. Despite this, the programs are useful for any subject matter if the user makes a little effort to decode special terms. This is not hard to do because the programs follow the presentation of Sneath and Sokal's (1973) book, *Numerical Taxonomy*. The package is supported by Dr. F. James Rohlf of The State University of New York.

NTSYS is a collection of subroutines that the user can arrange sequentially to operate on data matrices, resemblance matrices, cophenetic matrices, and so on. For example, arranging the subroutines and setting the parameters in one way will give a cluster analysis, while arranging in another way will give a principal component analysis. In addition to performing hierarchical and nonhierarchical cluster analysis, the system will also perform principal component analysis, principal coordinate analysis, factor analysis, and multidimensional scaling. NTSYS also has a set of subroutines for displaying the results of these analyses in tabular and graphical forms.

Almost all of the calculations for hierarchical cluster analysis described in this book can be executed by using NTSYS. The program gives a choice of nine quantitative resemblance coefficients, 16 qualitative resemblance coefficients, and eight clustering methods (these include UPGMA, SLINK, CLINK, WPGMA, centroid method, Ward's method and the flexible clustering method). It draws a tree in the style of those presented in this book, one that is suitable for direct publication.

Other subroutines allow either the data matrix or the resemblance matrix to be entered as input, the data matrix to be standardized by using a variety of standardizing functions, and the cophenetic matrix and cophenetic correlation coefficient to be calculated. The program also allows several cluster analyses to be carried out in one run, either based on different data matrices, or based on the same data matrix but different resemblance coefficients and/or clustering methods. It will compute the various matrix correlations that are described in Chapter 14.

The program for nonhierarchical cluster analysis uses a variant of the k-means method called **function point clustering**. Instead of specifying the value of k for how many clusters you want, you specify a parameter W. The method uses this parameter in fitting a surface in the attribute space, such that peaks form over the regions in which the sample objects are the densest. All objects under the same peaks of the surface are considered to belong to the same cluster. The method uses nonparametric density estimation to find the peaks and hence to find the clusters.

NTSYS is written in Fortran for IBM computers, but with slight adaptation it can be run on virtually any make of mainframe computer that supports a Fortran compiler. The program can be purchased from: Dr. F. James Rohlf, Department of Ecology and Evolution, The State University of New York, Stony Brook, New York 11794, U.S.A.

A2.5 CLUSTAR AND CLUSTID PROGRAMS

CLUSTAR and CLUSTID are Fortran IV programs that will execute all of the procedures presented in this book. CLUSTAR performs hierarchical cluster analysis, and CLUSTID identifies unknown objects into classifications that have been created by using CLUSTAR. Because making classifications is only one of the uses of cluster analysis, it follows that CLUSTAR has more uses than CLUSTID. Both are stand-alone programs designed specifically for hierarchical cluster analysis.

The programs optionally allow you to choose from 10 quantitative and 14 qualitative resemblance coefficients. Four clustering methods are provided: UPGMA, SLINK, CLINK, and Ward's method. You can input a data matrix or a resemblance matrix, standardize the data matrix according to any of the standardizing functions described in Chapter 7, and when the tree is obtained you can optionally calculate the cophenetic correlation coefficient and rearrange the data and resemblance matrices. Mixtures of quantitative and qualitative attributes can be handled by using the combined resemblance matrix approach presented in Chapter 12. Multiple cluster analyses can be done sequentially within a single run, and matrix correlations between resemblance matrices and between cophenetic matrices can be computed as described in Chapter 14.

Figure A2.1,a shows the input to a CLUSTAR run. The data are stream quality data taken from Darby et al. (1976). The objects are 24 streams in Allegheny County, Pennsylvania (identifiable in the figure as STR). The 12 attributes are measures of water quality, such as pH, dissolved oxygen, total coliform, and total

```
- - - - - - TITLE OF THIS RUN - - - - -
WATER QUALITY DATA FOR 24 STREAMS IN ALLEGHENY COUNTY, PENNSYLVANIA
*NAME
        OBJS   24
  STR 01  STR 02  STR 03  STR 04  STR 05  STR 06  STR 07  STR 08  STR 09  STR 10
  STR 11  STR 12  STR 13  STR 14  STR 15  STR 16  STR 17  STR 18  STR 19  STR 20
  STR 21  STR 22  STR 23  STR 24
*NAME
        ATRS   12
  ATR 01  ATR 02  ATR 03  ATR 04  ATR 05  ATR 06  ATR 07  ATR 08  ATR 09  ATR 10
  ATR 11  ATR 12
*INPT
        DATA     0    24    12     1     1     9     5 OBJS ATRS
(48F2.0)
 1 1 0 1 1 1 1 1 1 1 1 0 1 0 1 1 0 0 1 0 1 1 1 0
 1 1 1 1 1 1 1 1 1 1 1 0 1 1 1 1 1 1 0 1 1 1 1 1
 0 1 1 1 1 1 1 0 1 1 0 1 0 1 1 1 1 1 0 1 0 1 1 0
 1 1 0 0 0 0 1 1 1 1 1 0 0 0 0 1 0 0 1 0 1 1 0 1
 1 1 1 0 1 0 1 1 1 0 1 0 1 0 0 1 0 1 1 1 1 1 1 0
 1 1 1 1 1 1 1 1 1 1 1 1 1 1 1 1 1 1 1 1 1 1 1 1
 9 1 9 1 1 9 9 1 9 9 1 1 1 9 1 1 0 1 9 0 1 1 1 0
 9 1 9 1 1 9 9 1 9 9 1 1 1 9 1 1 1 1 9 1 1 1 1 1
 9 1 9 0 1 9 9 0 9 9 0 0 1 9 0 0 0 0 9 0 0 1 0 1
 9 1 9 0 1 9 9 1 9 9 1 1 1 9 1 1 0 1 9 0 0 1 0 1
 9 1 9 0 1 9 9 0 9 9 0 0 0 9 0 0 0 0 9 1 0 1 0 0
 9 1 9 1 1 9 9 1 9 9 1 1 1 9 1 1 1 1 9 1 1 1 1 1
*SIMQ
        DATA RESM     3     1                   OBJS
*CLST
        RESM TREE     3     1     1                  1 OBJS
*PNCH
        TREE
*END
```

(*a*)

Figure A2.1. (*a*) Example of the input instructions of a CLUSTAR run; (*b–f*) output of the same CLUSTAR run. (reduced in size from the computer printout). Output: (*b*) data matrix; (*c*) resemblance matrix; (*d*) tree drawn on a line printer; (*e*) cophenetic matrix; (*f*) tree drawn on a CalComp plotter.

```
- - - - - - DATA MATRIX - - - - - -

INPUT FORMAT : (48F2.0)
MATRIX NAME : DATA
TYPE OF MATRIX : DATA
NUMBER OF OBJECTS :   24
NUMBER OF ATTRIBUTES :   12
MISSING VALUE CODE :        9.00
OUTPUT OPTION  :    1    OBJS    ATRS

- - - - - - DATA MATRIX - - - - - -
```

	STR 01	STR 02	STR 03	STR 04	STR 05	STR 06	STR 07	STR 08	STR 09	STR 10	STR 11	STR 12
ATR 01	1.0000	1.0000	0.0000	1.0000	1.0000	1.0000	1.0000	1.0000	1.0000	1.0000	1.0000	0.0000
ATR 02	1.0000	1.0000	1.0000	1.0000	1.0000	1.0000	1.0000	1.0000	1.0000	1.0000	1.0000	0.0000
ATR 03	0.0000	1.0000	1.0000	1.0000	1.0000	1.0000	1.0000	0.0000	1.0000	1.0000	0.0000	1.0000
ATR 04	1.0000	1.0000	0.0000	0.0000	0.0000	0.0000	1.0000	1.0000	1.0000	1.0000	1.0000	0.0000
ATR 05	1.0000	1.0000	1.0000	0.0000	1.0000	0.0000	1.0000	1.0000	1.0000	0.0000	1.0000	0.0000
ATR 06	1.0000	1.0000	1.0000	1.0000	1.0000	1.0000	1.0000	1.0000	1.0000	1.0000	1.0000	1.0000
ATR 07	9.0000	1.0000	9.0000	1.0000	1.0000	9.0000	9.0000	1.0000	9.0000	9.0000	1.0000	1.0000
ATR 08	9.0000	1.0000	9.0000	1.0000	1.0000	9.0000	9.0000	1.0000	9.0000	9.0000	1.0000	1.0000
ATR 09	9.0000	1.0000	9.0000	0.0000	1.0000	9.0000	9.0000	0.0000	9.0000	9.0000	0.0000	0.0000
ATR 10	9.0000	1.0000	9.0000	0.0000	1.0000	9.0000	9.0000	1.0000	9.0000	9.0000	1.0000	1.0000
ATR 11	9.0000	1.0000	9.0000	0.0000	1.0000	9.0000	9.0000	0.0000	9.0000	9.0000	0.0000	0.0000
ATR 12	9.0000	1.0000	9.0000	1.0000	1.0000	9.0000	9.0000	1.0000	9.0000	9.0000	1.0000	1.0000

	STR 13	STR 14	STR 15	STR 16	STR 17	STR 18	STR 19	STR 20	STR 21	STR 22	STR 23	STR 24
ATR 01	1.0000	0.0000	1.0000	1.0000	0.0000	0.0000	1.0000	0.0000	1.0000	1.0000	1.0000	0.0000
ATR 02	1.0000	1.0000	1.0000	1.0000	1.0000	1.0000	0.0000	1.0000	1.0000	1.0000	1.0000	1.0000
ATR 03	0.0000	1.0000	1.0000	1.0000	1.0000	1.0000	0.0000	1.0000	0.0000	1.0000	1.0000	0.0000
ATR 04	0.0000	0.0000	0.0000	1.0000	0.0000	0.0000	1.0000	0.0000	1.0000	1.0000	0.0000	1.0000
ATR 05	1.0000	0.0000	0.0000	1.0000	0.0000	1.0000	1.0000	1.0000	1.0000	1.0000	1.0000	0.0000
ATR 06	1.0000	1.0000	1.0000	1.0000	1.0000	1.0000	1.0000	1.0000	1.0000	1.0000	1.0000	1.0000
ATR 07	1.0000	9.0000	1.0000	1.0000	0.0000	1.0000	9.0000	0.0000	1.0000	1.0000	1.0000	0.0000
ATR 08	1.0000	9.0000	1.0000	1.0000	1.0000	1.0000	9.0000	1.0000	1.0000	1.0000	1.0000	1.0000
ATR 09	1.0000	9.0000	0.0000	0.0000	0.0000	0.0000	9.0000	0.0000	0.0000	1.0000	0.0000	1.0000
ATR 10	1.0000	9.0000	1.0000	1.0000	0.0000	1.0000	9.0000	0.0000	0.0000	1.0000	0.0000	1.0000
ATR 11	0.0000	9.0000	0.0000	0.0000	0.0000	0.0000	9.0000	1.0000	0.0000	1.0000	0.0000	0.0000
ATR 12	1.0000	9.0000	1.0000	1.0000	1.0000	1.0000	9.0000	1.0000	1.0000	1.0000	1.0000	1.0000

(*b*)

```
- - - - RESEMBLANCE MATRIX FOR BINARY DATA - - - -

DATA MATRIX NAME : DATA
RESEMBLANCE MATRIX NAME : RESM
RESEMBLANCE COEFFICIENT : SIMPLE MATCHING
VALUE FOR CONVERTING
METRIC TO BINARY DATA :     0.
USE ATTRIBUTES  1  THRU 12
OUTPUT OPTION :  1    OBJS

- - - - RESEMBLANCE MATRIX FOR BINARY DATA - - - -
```

	STR 01	STR 02	STR 03	STR 04	STR 05	STR 06	STR 07	STR 08	STR 09	STR 10	STR 11	STR 12
STR 02	0.8333											
STR 03	0.5000	0.6667										
STR 04	0.5000	0.5833	0.6667									
STR 05	0.6667	0.9167	0.8333	0.6667								
STR 06	0.5000	0.6667	0.6667	1.0000	0.8333							
STR 07	0.8333	1.0000	0.6667	0.6667	0.8333	0.6667						
STR 08	1.0000	0.7500	0.5000	0.6667	0.6667	0.5000	0.8333					
STR 09	0.8333	1.0000	0.6667	0.6667	0.8333	0.6667	1.0000	0.8333				
STR 10	0.6667	0.8333	0.5000	0.8333	0.6667	0.8333	0.8333	0.6667	0.8333			
STR 11	1.0000	0.7500	0.5000	0.6667	0.6667	0.5000	0.8333	1.0000	0.8333	0.6667		
STR 12	0.1667	0.5000	0.6667	0.7500	0.5833	0.6667	0.3333	0.5833	0.3333	0.5000	0.5833	
STR 13	0.8333	0.7500	0.6667	0.6667	0.8333	0.6667	0.6667	0.8333	0.6667	0.5000	0.8333	0.5833
STR 14	0.3333	0.5000	0.8333	0.8333	0.6667	0.8333	0.5000	0.3333	0.5000	0.6667	0.3333	0.8333
STR 15	0.5000	0.6667	0.6667	0.9167	0.7500	1.0000	0.6667	0.7500	0.6667	0.8333	0.7500	0.8333
STR 16	0.8333	0.8333	0.6667	0.7500	0.7500	0.6667	1.0000	0.9167	1.0000	0.8333	0.9167	0.6667
STR 17	0.3333	0.4167	0.8333	0.8333	0.5000	0.8333	0.5000	0.5000	0.5000	0.6667	0.5000	0.7500
STR 18	0.5000	0.6667	1.0000	0.7500	0.7500	0.6667	0.6667	0.7500	0.6667	0.5000	0.7500	0.8333
STR 19	0.8333	0.6667	0.3333	0.3333	0.5000	0.3333	0.6667	0.8333	0.6667	0.5000	0.8333	0.3333
STR 20	0.5000	0.5833	1.0000	0.6667	0.6667	0.6667	0.6667	0.5000	0.6667	0.5000	0.5000	0.5833
STR 21	1.0000	0.6667	0.5000	0.7500	0.5833	0.5000	0.8333	0.9167	0.8333	0.6667	0.9167	0.5000
STR 22	0.8333	1.0000	0.6667	0.5833	0.9167	0.6667	1.0000	0.7500	1.0000	0.8333	0.7500	0.5000
STR 23	0.6667	0.6667	0.8333	0.9167	0.7500	0.8333	0.8333	0.7500	0.8333	0.6667	0.7500	0.6667
STR 24	0.6667	0.5833	0.5000	0.5000	0.5000	0.5000	0.5000	0.6667	0.5000	0.6667	0.6667	0.5833

	STR 13	STR 14	STR 15	STR 16	STR 17	STR 18	STR 19	STR 20	STR 21	STR 22	STR 23
STR 14	0.5000										
STR 15	0.7500	0.8333									
STR 16	0.7500	0.5000	0.8333								
STR 17	0.5000	1.0000	0.7500	0.5833							
STR 18	0.7500	0.8333	0.8333	0.8333	0.7500						
STR 19	0.6667	0.1667	0.3333	0.6667	0.1667	0.3333					
STR 20	0.5000	0.8333	0.5833	0.5833	0.8333	0.7500	0.3333				
STR 21	0.7500	0.3333	0.6667	0.8333	0.5833	0.6667	0.8333	0.5833			
STR 22	0.7500	0.5000	0.6667	0.8333	0.4167	0.6667	0.6667	0.5833	0.6667		
STR 23	0.7500	0.6667	0.8333	0.8333	0.7500	0.8333	0.5000	0.7500	0.8333	0.6667	
STR 24	0.6667	0.6667	0.5833	0.5833	0.6667	0.5833	0.5000	0.5000	0.5833	0.5833	0.4167

(*c*)

iron (identifiable as ATR). A value of 1 indicates that a stream meets the applicable water quality standard; 0 indicates a violation. A value of 9 indicates no data were available, and is treated as a missing value. A cluster analysis will show streams that are of similar quality, and this could be important information for environment management. For example, perhaps streams that fail to meet the standard and have similar quality profiles also have similar causes of pollution, or perhaps they will require similar controls to be exercised.

Figure A.2.1,b–e shows the output for this run. It is clearly labeled. We could have asked for the rearranged data and resemblance matrices, for they would have been informative, but to conserve space we did not. CLUSTAR can also be used to label the names of the objects and attributes, which then appear on the data matrix, and with the object names labeling the tree's branch tips. Then we would know at a glance, for example, that object 1 above is Big Sewickly Creek and object 24 is

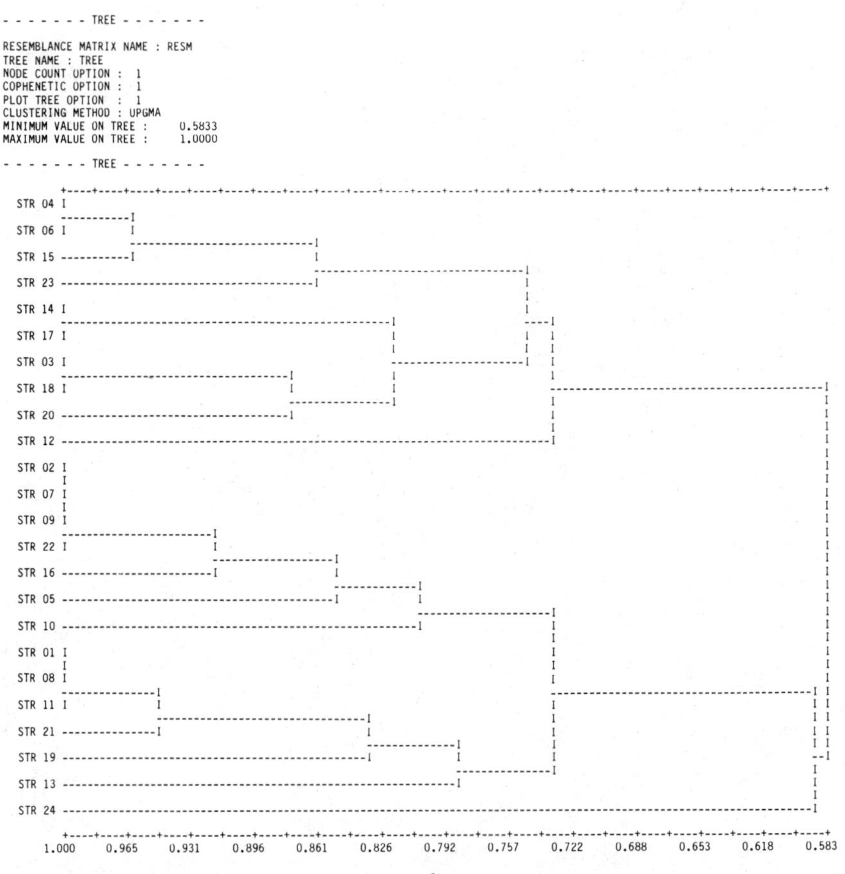

(*d*)

COPHENETIC CORRELATION MATRIX

	STR 01	STR 02	STR 03	STR 04	STR 05	STR 06	STR 07	STR 08	STR 09	STR 10	STR 11	STR 12
STR 01	0.7302											
STR 02	0.5833	0.5833										
STR 03	0.5833	0.5833	0.7458									
STR 04	0.7302	0.8500	0.5833	0.5833								
STR 05	0.5833	0.5833	0.7458	1.0000	0.5833							
STR 06	0.7302	1.0000	0.5833	0.5833	0.8500	0.5833						
STR 07	1.0000	0.7302	0.5833	0.5833	0.7302	0.5833	0.7302					
STR 08	0.7302	1.0000	0.5833	0.5833	0.8500	0.5833	1.0000	0.7302				
STR 09	0.7302	0.8056	0.5833	0.5833	0.8056	0.5833	0.8056	0.7302	0.8056			
STR 10	1.0000	0.7302	0.5833	0.5833	0.7302	0.5833	0.7302	1.0000	0.7302	0.7302		
STR 11	0.5833	0.5833	0.7315	0.7315	0.5833	0.7315	0.5833	0.5833	0.5833	0.5833	0.5833	
STR 12	0.7833	0.7302	0.5833	0.5833	0.7302	0.5833	0.7302	0.7833	0.7302	0.7302	0.7833	0.5833
STR 13	0.5833	0.5833	0.8194	0.7458	0.5833	0.7458	0.5833	0.5833	0.5833	0.5833	0.5833	0.7315
STR 14	0.5833	0.5833	0.7458	0.9583	0.5833	0.9583	0.5833	0.5833	0.5833	0.5833	0.5833	0.7315
STR 15	0.7302	0.9167	0.5833	0.5833	0.8500	0.5833	0.9167	0.7302	0.9167	0.8056	0.7302	0.5833
STR 16	0.5833	0.5833	0.8194	0.7458	0.5833	0.7458	0.5833	0.5833	0.5833	0.5833	0.5833	0.7315
STR 17	0.5833	0.5833	1.0000	0.7458	0.5833	0.7458	0.5833	0.5833	0.5833	0.5833	0.5833	0.7315
STR 18	0.8333	0.7302	0.5833	0.5833	0.7302	0.5833	0.7302	0.8333	0.7302	0.7302	0.8333	0.5833
STR 19	0.5833	0.5833	0.8750	0.7458	0.5833	0.7458	0.5833	0.5833	0.5833	0.5833	0.5833	0.7315
STR 20	0.9444	0.7302	0.5833	0.5833	0.7302	0.5833	0.7302	0.9444	0.7302	0.7302	0.9444	0.5833
STR 21	0.7302	1.0000	0.5833	0.5833	0.8500	0.5833	1.0000	0.7302	1.0000	0.8056	0.7302	0.5833
STR 22	0.5833	0.5833	0.7458	0.8611	0.5833	0.8611	0.5833	0.5833	0.5833	0.5833	0.5833	0.7315
STR 23	0.5897	0.5897	0.5833	0.5833	0.5897	0.5833	0.5897	0.5897	0.5897	0.5897	0.5897	0.5833

COPHENETIC CORRELATION MATRIX

	STR 13	STR 14	STR 15	STR 16	STR 17	STR 18	STR 19	STR 20	STR 21	STR 22	STR 23
STR 01	0.5833										
STR 02	0.5833	0.7458									
STR 03	0.7302	0.5833	0.5833								
STR 04	0.5833	1.0000	0.7458	0.5833							
STR 05	0.5833	0.8194	0.7458	0.5833	0.8194						
STR 06	0.7833	0.5833	0.5833	0.7302	0.5833	0.5833					
STR 07	0.5833	0.8194	0.7458	0.5833	0.8194	0.8750	0.5833				
STR 08	0.7833	0.5833	0.5833	0.7302	0.5833	0.5833	0.8333	0.5833			
STR 09	0.7302	0.5833	0.5833	0.9167	0.5833	0.5833	0.7302	0.5833	0.7302		
STR 10	0.5833	0.7458	0.8611	0.5833	0.7458	0.7458	0.5833	0.7458	0.5833	0.5833	
STR 11	0.5897	0.5833	0.5833	0.5897	0.5833	0.5833	0.5897	0.5833	0.5897	0.5897	0.5833

COPHENETIC CORRELATION = 0.7139

(*e*)

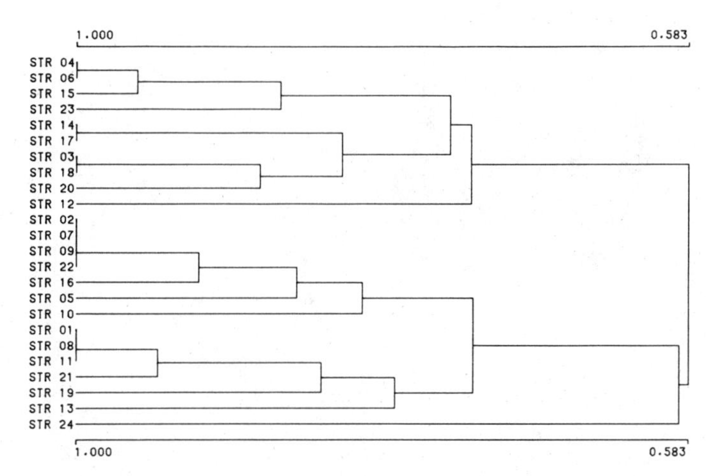

(*f*)

Turtle Creek. CLUSTAR also has a subprogram which draws the tree on a CalComp plotter. (Figure A2.1,f.) Both programs can be run from cards, or interactively from a terminal.

The largest data matrix that CLUSTAR can handle depends on the computer system it is used on. The resemblance matrix is stored in the computer's memory, and as larger computers become available CLUSTAR will be able to handle more objects. Currently it is being used on medium sized computers, such as Digital Equipment Corporation's VAX 11/780, for data matrices having upwards to 400 objects and as many attributes. The cluster analysis requires only a few minutes of "central processor time."

The programs and users manual are available from: Lifetime Learning Publications, Ten Davis Drive, Belmont, California 94002.

A2.6 OTHER CLUSTER ANALYSIS PROGRAMS

The following books contain listings of Fortran programs for both hierarchical and nonhierarchical cluster analysis: *Cluster Analysis for Applications*, Anderberg (1973); *Clustering Algorithms*, Hartigan (1975); *Computational Methods of Multivariate Analysis*, Mather (1976); *Cluster Analysis Algorithms*, Späth (1980).

The following books contain Fortran programs for hierarchical cluster analysis: *Statistics and Data Analysis in Geology*, Davis (1973); *Applied Multivariate Analysis*, Overall and Klett (1972). In a journal, Froidevaux et al. (1977) have published the listing of a Fortran program for nonhierarchical cluster analysis that can handle up to 1,000 objects and ten attributes in a Q-analysis.

Answers to Exercises

2.1 The initial resemblance matrix values are

$$e_{12} = ((2-9)^2 + (2-5)^2)^{1/2} = 7.6;$$
$$e_{13} = ((2-4)^2 + (2-3)^2)^{1/2} = 2.2;$$
$$e_{14} = ((2-10)^2 + (2-2)^2)^{1/2} = 8.0;$$
$$e_{15} = ((2-3)^2 + (2-10)^2)^{1/2} = 8.1;$$
$$e_{23} = ((9-4)^2 + (5-3)^2)^{1/2} = 5.4;$$
$$e_{24} = ((9-10)^2 + (5-2)^2)^{1/2} = 3.2;$$
$$e_{25} = ((9-3)^2 + (5-10)^2)^{1/2} = 7.8;$$
$$e_{34} = ((4-10)^2 + (3-2)^2)^{1/2} = 6.1;$$
$$e_{35} = ((4-3)^2 + (3-10)^2)^{1/2} = 7.1;$$
$$e_{45} = ((10-3)^2 + (2-10)^2)^{1/2} = 10.6.$$

These ten values form the initial resemblance matrix:

	Cluster				
Cluster	1	2	3	4	5
1	—	—	—	—	—
2	7.6	—	—	—	—
3	2.2	5.4	—	—	—
4	8.0	3.2	6.1	—	—
5	8.1	7.8	7.1	10.6	—

Step 1. Merge clusters 1 and 3, giving clusters 2, 4, 5, and (13) at the value of $e_{13} = 2.2$.

Revise the resemblance matrix:

$$e_{2(13)} = \tfrac{1}{2}(e_{12} + e_{23}) = 1/2(7.6 + 5.4) = 6.5;$$
$$e_{4(13)} = \tfrac{1}{2}(e_{14} + e_{34}) = 1/2(8.0 + 6.1) = 7.0;$$
$$e_{5(13)} = \tfrac{1}{2}(e_{15} + e_{35}) = 1/2(8.1 + 7.1) = 7.6.$$

Revised resemblance matrix:

	Cluster			
Cluster	2	4	5	(13)
2	—	—	—	—
4	3.2	—	—	—
5	7.8	10.6	—	—
(13)	6.5	7.0	7.6	—

Step 2. Merge clusters 2 and 4, giving clusters 5, (13), (24) at the value of $e_{24} = 3.2$. Revise the resemblance matrix:

$$e_{5(24)} = \tfrac{1}{2}(e_{25} + e_{45}) = \tfrac{1}{2}(7.8 + 10.6) = 9.2;$$
$$e_{(13)(24)} = \tfrac{1}{4}(e_{12} + e_{14} + e_{23} + e_{34}) = \tfrac{1}{4}(7.6 + 8.0 + 5.4 + 6.1) = 6.8.$$

Revised resemblance matrix:

	Cluster		
Cluster	5	(13)	(24)
5	—	—	—
(13)	7.6	—	—
(24)	9.2	6.8	—

Step 3. Merge clusters (13) and (24), giving 5, (1234) at the value of $e_{(13)(24)} = 6.8$. Revise the resemblance matrix:

$$e_{5(1234)} = \tfrac{1}{4}(e_{15} + e_{25} + e_{35} + e_{45}) = \tfrac{1}{4}(8.1 + 7.8 + 7.1 + 10.6) = 8.4.$$

Revised resemblance matrix:

	Cluster	
Cluster	5	(1234)
5	—	—
(1234)	8.4	—

Step 4. Merge clusters 5 and (1234), giving (12345) at the value of $e_{5(1234)} = 8.4$. The tree appears in Figure E.1.

2.2 $e_{(256)(134)} = 1/9(e_{12} + e_{23} + e_{24} + e_{15} + e_{35} + e_{45} + e_{16} + e_{36} + e_{46})$

$= 1/9(6.1 + 7.3 + 10.0 + 7.1 + 8.9 + 12.0 + 7.8 + 7.3 + 8.2)$

$= 8.3.$

The number of e_{jk}-values that need to be averaged is always the product of the number of object in the jth and kth clusters. In this case, (3)(3) = 9.

7.1 The data matrix and row summary statistics are:

	Object							
Attribute	1	2	3	4	$\bar{X}_i$	S_i	$RMAX_i$	$RMIN_i$
1	10	14	10	10	11	2.0	14	10
2	66	20	10	80	44	34.2	80	10
3	−30	0	−10	−20	−15	12.9	0	−30

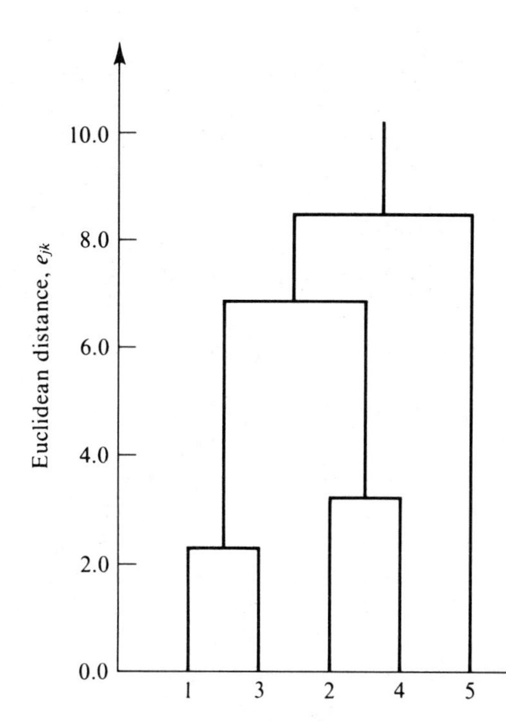

Figure E.1. Tree for Exercise 2.1.

(a) By equation 7.1, the standardized data matrix is,

	Object			
Attribute	1	2	3	4
1	−0.50	1.5	−0.50	−0.50
2	0.64	−0.70	−0.99	1.05
3	−1.16	1.16	0.39	−0.39

Here's where some of the values come from: $Z_{11} = (10 - 11)/2 = -0.50$; $Z_{21} = (66 - 44)/34.2 = 0.64$; $Z_{31} = [-30 - (-15)]/12.9 = -1.16$.

(b) By equation 7.3, the standardized data matrix is,

	Object			
Attribute	1	2	3	4
1	0.0	1.0	0.0	0.0
2	0.8	0.14	0.0	1.0
3	0.0	1.0	0.67	0.33

Here's where some of the values come from: $Z_{11} = (10 - 10)/(14 - 10) = 0$; $Z_{21} = (66 - 10)/(80 - 10) = 0.8$; $Z_{31} = [-30 - (-30)]/(0 - (-30)) = 0$.

7.2 The data matrix and column summary statistics are:

	Object			
Attribute	1	2	3	4
1	10	14	10	10
2	66	20	10	80
3	−30	0	−10	−20
$\overline{X}_j$:	15.3	11.3	3.33	23.3
S_j:	48.2	10.3	11.5	51.3

By equation 7.5, the standardized data matrix is

	Object			
Attribute	1	2	3	4
1	−0.11	0.26	0.58	−0.26
2	1.05	0.84	0.58	1.11
3	−0.94	−1.10	−1.16	−0.84

Here's where some of the values come from: $Z_{11} = (10 - 15.3)/48.2 = -0.11$; $Z_{23} = (10 - 3.33)/11.5 = 0.58$; $Z_{34} = (-20 - 23.3)/51.3 = -0.84$.

7.3 Note that $S_2 = 0.0$, and $\text{RMIN}_2 = \text{RMAX}_2$. Thus the denominators of equations 7.1 and 7.3 are zero, and Z_{ij} is incalculable. In general, when the values in a row are all the same, then equations 7.1 and 7.3 cannot be used; and when the values in a column are the same then equations 7.5 and 7.7 cannot be used.

8.1
$$\begin{aligned}
d_{16} &= ((124 - 107)^2 + (167 - 195)^2 + (83 - 115)^2 \\
&\quad + (63 - 38)^2 + (91 - 103)^2 + \\
&\quad + (106 - 127)^2)/(6))^{1/2} = 23.5; \\
q_{16}^2 &= [1/(6)^2][(124 + 167 + 83 + 63 + 91 + 106) - \\
&\quad (107 + 195 + 115 + 38 + 103 + 127)]^2 \\
&= 1/36[(634) - (685)]^2 = 72.25; \\
z_{16} &= \{6/5[(23.5)^2 - (72.25)]\}^{1/2} = 24.0; \\
c_{16} &= 80{,}607/[(73{,}640)^{1/2}(90{,}881)^{1/2}] = 0.985; \\
r_{16} &= \frac{80{,}607 - (1/6)(634)(685)}{\{[73{,}640 - (1/6)(634)^2][90{,}881 - (1/6)(685)^2]\}^{1/2}} = 0.896; \\
a_{16} &= (1/6)[|124 - 107|/(124 + 107) + \cdots + |106 - 127|/(106 + 127)] \\
&= 0.119; \\
b_{16} &= (|124 - 107| + \cdots + |106 - 127|)/[(124 + 107) \\
&\quad + \cdots + (106 + 127)] = 0.102.
\end{aligned}$$

8.2 When objects j and k are identical, then $X_{ij} - X_{ik} = 0$ for all i, and $m_{jk} = 0.0$. Thus, m_{jk} is a dissimilarity coefficient; m_{jk} can be no larger than 1.0.

8.3 When $n = 1$, d_{jk}, c_{jk}, a_{jk}, and b_{jk} can be calculated. However, c_{jk} is not meaningful because, always, $c_{jk} = 1.0$ when $n = 1$. Because of the factor $(n - 1)$ in the denominator of z_{jk}, it cannot be calculated except when $n \geq 2$. When $n = 1$, r_{jk} contains a value of 0.0 in its denominator irrespective of the data used and cannot, therefore, be calculated. When $n = 2$, r_{jk} can be

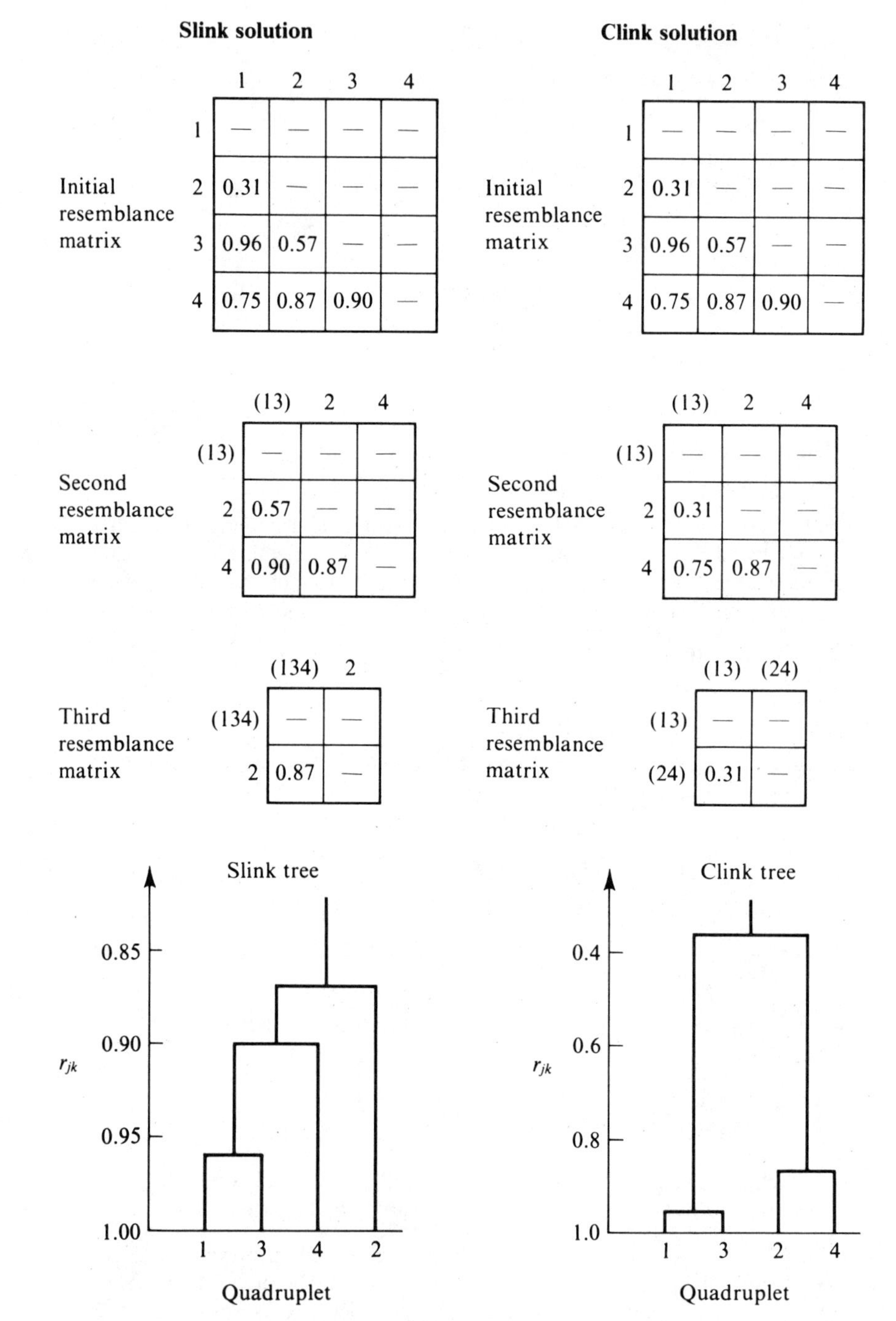

Figure E.2. Initial resemblance matrices, revised resemblance matrices, and trees for SLINK and CLINK clustering methods for Exercise 9.1.

calculated but is always equal to 1.0; thus, r_{jk} is meaningful only when $n \geq 3$. Coefficients a_{jk}, b_{jk}, and d_{jk} can be calculated when $n = 1$, and their values will be meaningful.

8.4 As can be seen from the structures of their equations, d_{jk}, a_{jk}, and b_{jk} will be invariant. But calculation of z_{12}, c_{12}, and r_{12} for the two data matrices containing a variable coded oppositely shows that z_{jk}, c_{jk}, and r_{jk} depend on the direction of coding.

Cohen (1969), aware of the sensitivity of r_{jk} to the direction of coding and aware that it is probably the most-often-used resemblance coefficient in psychological research, proposed a **profile similarity coefficient** that is invariant to the direction of coding. Cohen's coefficient has the same mathematical formulation as r_{jk} except that n, the number of attributes, applies to an augmented data matrix. The augmented data matrix is obtained by adding to the original data matrix dummy attributes, one for each real attribute in the original data matrix. The dummy attributes have scales that run in directions opposite to the scales of their respective attributes in the original data matrix.

9.1 Figure E.2 shows the initial resemblance matrix of all r_{jk} and the revised resemblance matrices, and the trees for the SLINK and CLINK clustering methods.

SLINK solution:

1. Merge clusters 1 and 3, giving (13), 2, and 4 at the value of $r_{13} = 0.96$. Revise values in the resemblance matrix:

$$r_{(13)2} = \max(r_{12}, r_{23}) = \max(0.31, 0.57) = 0.57;$$
$$r_{(13)4} = \max(r_{14}, r_{34}) = \max(0.75, 0.90) = 0.90.$$

2. Merge clusters (13) and 4, giving (134) and 2 at the value of $r_{(13)4} = 0.90$. Revise values in the resemblance matrix:

$$r_{(134)2} = \max(r_{12}, r_{23}, r_{24}) = \max(0.31, 0.57, 0.87) = 0.87.$$

3. Merge clusters (134) and 2, giving (1234) at the value of $r_{(134)2} = 0.87$.

CLINK solution:

1. Merge clusters 1 and 3, giving (13), 2, and 4 at the value of $r_{13} = 0.96$. Revise values in the resemblance matrix:

$$r_{(13)2} = \min(r_{12}, r_{23}) = \min(0.31, 0.57) = 0.31;$$
$$r_{(13)4} = \min(r_{14}, r_{34}) = \min(0.75, 0.90) = 0.75.$$

2. Merge clusters 2 and 4, giving (13) and (24) at the value of $r_{24} = 0.87$. Revise values in the resemblance matrix:

$$r_{(13)(24)} = \min(r_{12}, r_{14}, r_{23}, r_{34}) = \min(0.31, 0.75, 0.57, 0.90) = 0.31.$$

3. Merge clusters (13) and (24), giving (1234) at the value of $r_{(13)(24)} = 0.31$.

10.1 The initial resemblance matrix and the tree that results from the clustering steps is shown in Figure E.3,a, b. The Yoeman Yodeler is unique. It is reasonable to include d (0–0 matches) in the resemblance coefficient because two knives lacking the same feature—a can opener, let us say—should have this count (by most opinions) toward their mutual similarity.

10.2 The Jaccard coefficient,

$$\frac{a}{a+b+c},$$

can be rewritten as

$$\frac{1}{1+\left(\frac{b+c}{a}\right)}.$$

The Sorenson coefficient,

$$\frac{2a}{2a+b+c},$$

can be rewritten as

$$\frac{1}{1+\left(\frac{b+c}{2a}\right)}.$$

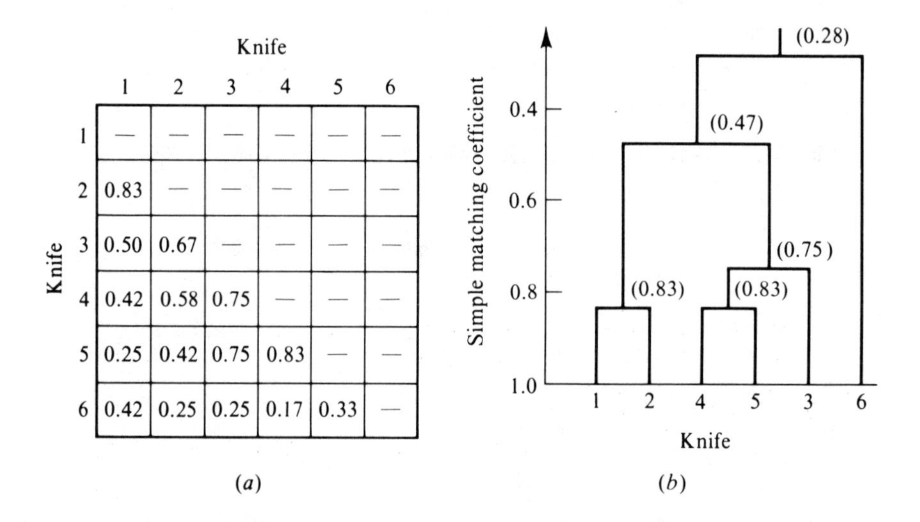

Knife	1	2	3	4	5	6
1	—	—	—	—	—	—
2	0.83	—	—	—	—	—
3	0.50	0.67	—	—	—	—
4	0.42	0.58	0.75	—	—	—
5	0.25	0.42	0.75	0.83	—	—
6	0.42	0.25	0.25	0.17	0.33	—

Figure E.3. Resemblance matrix and tree for Exercise 10.1.

Since for any values of a, b, and c,

$$\frac{b + c}{a} \geq \frac{b + c}{2a},$$

it follows that the denominator of the rewritten Jaccard coefficient can be larger, but not smaller, than that of the rewritten Sorenson coefficient. And these properties of the denominators imply inverse properties for the coefficients as a whole; namely, the rewritten Jaccard coefficient can be no larger than the rewritten Sorenson coefficient, regardless of the values of a, b, and c.

Glossary of Terms

Additive translation of data profiles. Two data profiles that have the same shape but are displaced from each other by the addition of a constant. Compare with **proportional translation**.

Asymmetric resemblance matrix. One in which the similarity of object j to object k is not the same as the similarity of object k to object j.

Binary attributes. Qualitative attributes that have two states, such as presence/absence, like/dislike, yes/no, and male/female.

Block clustering. A method of cluster analysis that simultaneously performs an R-analysis and a Q-analysis on a data matrix.

Classification. A set of classes, each class containing similar objects. "Classify" is used as a verb meaning "to make a classification." Compare with **identification**.

CLINK clustering method. CLINK is short for "complete linkage clustering method." When two clusters are joined by this method, their similarity is that of their most dissimilar pair of objects, one in each cluster. Also called "furthest neighbor clustering method."

Cluster. A set of objects that are similar to each other.

Cluster analysis. A generic term for mathematical methods that show which objects in a set are similar to each other.

Clustering method. The method that uses a resemblance matrix as its input and creates a tree showing the similarities among objects as its output. Such a method groups similar objects into clusters.

Confusion data. Data that record the frequency with which different stimuli are confused with each other. The confusion between stimuli is taken as a measure of their similarity. The more confusion between two stimuli when they are identified by people or machines, the more similar the stimuli are. Confusion data can be directly entered into the resemblance matrix and cluster-analyzed, bypassing the data matrix.

Cophenetic correlation. The value of the Pearson product-moment correlation coefficient computed between the values in the resemblance matrix (X) and the values in the cophenetic matrix (Y). Denoted by the symbol $r_{X,Y}$ or r_{coph}, it measures the amount of distortion that the clustering method causes in representing the similarities among pairs of objects as a tree.

Cophenetic matrix. Represents the tree in the form of a matrix. Shows the level of similarity that relates each pair of objects in the tree.

Data matrix. The input data to a cluster analysis. The columns of the matrix stand for the objects whose similarities to other objects are to be estimated. The rows stand for the attributes describing the objects.

Data profile. A graphical plot of the values of data for an object or an attribute. To obtain the data profile of an object, its attribute values are plotted and adjacent values are connected by line segments. For the data profile of an attribute, its values for each object are plotted and connected.

Descriptive attributes. Attributes that describe the size, shape, chemical properties, and so on, of the objects in a cluster analysis. Compare with "functional attributes."

Dissimilarity coefficient. A resemblance coefficient: the smaller its value, the more similar the two objects are. Compare with **similarity coefficient**.

Framing alternatives. The options available in a cluster analysis for framing the data. Includes the options for choosing objects, attributes, and scales of measurement, as well as methods of standardization, resemblance coefficients, clustering methods, and so on.

Framing decisions. The set of decisions that the researcher makes to frame the analysis. By analyzing the framing alternatives, the researcher selects the best options—those felt to have the best chance of achieving the research goal. These are the framing decisions.

Functional attributes. Attributes that describe an object's effect on something else or an object's reaction to an external agent. Compare with **descriptive attributes**.

General-purpose classification. A classification, usually made by a taxonomist, to serve generally in science however other scientists see fit to use it in their theories. Compare with **specific-purpose classification**.

Hierarchical cluster analysis. A term for those methods of cluster analysis that show the similarities among objects as a hierarchy. Compare with **nonhierarchical cluster analysis**.

Hypothetico-deduction. A scientific method used to test research hypotheses about natural processes. From the assumed truth of the hypothesis, a test prediction is deduced and compared to experimental fact. The outcome either confirms or disproves the hypothesis.

Identification. The act of establishing what class in a classification an object belongs to. Compare with **classification**.

Information. Any difference among data that makes a difference in what a person believes.

Multistate attributes. Qualitative attributes that have more than two states, such as yes/no/uncertain, or north-/south-/east-/west-facing.

Nonhierarchical cluster analysis. A term for those methods of cluster analysis that show the similarities among objects at a specified level of similarity. Compare with **hierarchical cluster analysis**.

Numerical taxonomy. The profession that uses multivariate methods—such as cluster analysis, principal component analysis, multidimensional scaling—with multivariate data, to help make classifications. Initially it referred to biological taxonomy but now it refers to all fields (geological taxonomy, ecological taxonomy, and so on).

OTU. Short for "operational taxonomic unit," this is a specialized term used in biological numerical taxonomy for the biological specimens that are the objects to be clustered.

Pattern of data. A design made by the relations among a set of objects. The most basic pattern is the **first-order pattern**. A **second-order pattern** is the pattern of relations among two or more first-order patterns. A **third-order pattern** is the pattern of relations among two or more second-order patterns. The relations can be relations of similarities or of associations.

Planning. Decision-making that aims to change some state of the world from what it is to some state that people think it ought to be.

Practical significance. When a value or pattern of data takes on meaning that, within the context of the field of study, is felt to be nontrivial. In such an instance, the data becomes information for science or for planning and management—a basis for making decisions.

Proportional translation of data profiles. Two data profiles that are displaced from each other by the multiplication of a constant. Compare with **additive translation**.

Q-analysis and R-analysis. A Q-analysis is a cluster analysis of objects, whereas an R-analysis is a cluster analysis of attributes. The terms are also used in describing other multivariate methods—for example, factor analysis.

Qualitative attributes. Attributes measured on a nominal scale of measurement, as, for example, those that are binary or multistate.

Quantitative attributes. Attributes measured on ordinal, interval, or ratio scales of measurement.

Rearranged data matrix. The data matrix with its columns arranged in the order the objects appear in the tree.

Rearranged resemblance matrix. The resemblance matrix with its rows and columns arranged in the order the objects appear in the tree.

Research goal. The purpose of the cluster analysis—that is, what the researcher intends it to achieve. This book breaks the research goals addressed with cluster analysis into the following six categories, the first five of which are for science and the last for planning and management:

1. Create a scientific question that stimulates further research.
2. Create a research hypothesis in answer to a scientific question.
3. Test a research hypothesis with an experiment designed either to confirm or disprove the hypothesis.
4. Make a general-purpose classification (one meant to serve science broadly and generally in unspecified scientific theories and laws).
5. Make a specific-purpose classification (one that is related to other specific variables of interest). This may be an ultimate research goal in itself, or it may be a subgoal of any of the first three research goals above.
6. Facilitate planning and management. Rather than assessing similarities among objects to create scientific knowledge, this purpose requires they be assessed in order to plan and manage better. The need to find out which objects in a set are similar and dissimilar is a recurring need in planning and management.

Resemblance coefficient and matrix. A **resemblance coefficient** is either a similarity or a dissimilarity coefficient, and it measures the similarity between a pair of objects. A **resemblance matrix** stores the values of the resemblance coefficient for all pairs of objects.

Retroduction. A scientific method used to create research hypotheses about natural processes. Argues from observed facts to a hypothetical reason that can explain the facts.

Science. As a professional activity, science is the activity that seeks to create knowledge. As a process, science is the planning, the methods, the standards, and the creativity that are integrated to validate or disprove ideas with respect to experimental fact. Validated or confirmed ideas become knowledge.

Similarity coefficient. A resemblance coefficient for which the larger its value, the more similar the two objects being compared are. Compare with **dissimilarity coefficient**.

Size displacement of data profiles. When one object's data profile is, attribute for attribute, larger or smaller than another's, the two profiles are separated by a size displacement. For example, two data profiles may have nearly the same shapes but be displaced in size.

SLINK clustering method. SLINK is short for "single linkage clustering method." When two clusters are joined by this method their similarity is that of their most similar pair of objects, one in each cluster. Also called "nearest neighbor clustering method."

Specific-purpose classification. A classification, generally made by a scientist, to serve a specific purpose (to be related to a few specific variables chosen for their interest). Compare with **general-purpose classification**.

Standardized data matrix. The matrix, having the same number of columns and rows as the data matrix, that results from standardizing the data by use of one of the standardizing functions of Chapter 7.

Standardizing function. An equation used to standardize the attributes of the data matrix, thereby producing the standardized data matrix.

Tree. The main output of a hierarchical cluster analysis. The tree shows the hierarchy of similarities between all pairs of objects. Also called a "phenogram" or a "dendrogram."

UPGMA clustering method. A clustering method that forms clusters based on the average value of similarity between the two clusters being merged. Also called "average linkage clustering method."

Validating criteria. The standards that researchers use when they judge the validity and usefulness of their research.

Ward's clustering method. A clustering method that assigns objects to clusters in such a way that a sums-of-squares index E is minimized.

References

Adam, D. P. 1974. Palynological applications of principal component and cluster analysis. *Journal of Research of the U.S. Geological Survey*. *2*:727–741.

Adams, J. M., Attinger, E. O., and Attinger, F. M. 1978. Cluster analysis of a respiratory time series. *Biological Cybernetics*. *28*: 183–190.

Agnew, N. M., and Pyke, S. W. 1978. *The Science Game*. Englewood Cliffs, N.J.: Prentice-Hall, Inc.

Alexander, R. W. 1976. Hierarchical grouping analysis and skeletal materials. *American Journal of Physical Anthropology*. *45*: 39–44.

Ali, S. A., Gross, M. G., and Kishpaugh, J. R. L. 1975. Cluster analysis of marine sediments and waste deposits in New York Bight. *Environmental Geology*. *1*: 143–148.

Anderberg, M. R. 1973. *Cluster Analysis for Applications*. New York: Academic Press.

Arkley, R. J. 1976. Statistical methods in soil classification research. In N. C. Brady (Ed.), *Advances in Agronomy* (vol. 28). New York: Academic Press.

Armstrong, D. M. 1977. Ecological distribution of small mammals in the Upper Williams Fork Basin, Grand County, Colorado. *The Southwestern Naturalist*. *22*: 289–304.

Arno, S. F., and Pfister, R. D. 1977. Habitat types: an improved system for classifying Montana's forests. *Western Wildlands*. Spring: 6–11.

Ashley, R. P., and Lloyd, J. W. 1978. An example of the use of factor analysis and cluster analysis in groundwater chemistry interpretations. *Journal of Hydrology*. *39*: 355–364.

Aspey, W. P., and Blankenship, J. E. 1977. Spiders & snails & statistical tales: application of multivariate analyses to diverse ethological data. In B. A. Hazlett (Ed.), *Quantitative Methods in the Study of Animal Behavior*. New York: Academic Press.

Aspey, W. P., and Blankenship, J. E. 1978. Comparative ethometrics: congruence of different multivariate analyses applied to the same ethological data. *Behavioural Processes*. *3*: 173–195.

Bailey, K. D. 1975. Cluster analysis. In D. R. Heise (Ed.), *Sociological Methodology*. San Francisco: Jossey-Bass.

Balakrishnan, V., Sanghvi, L. D., and Kirk, R. L. 1975. *Genetic Diversity among Australian Aborigines*. Canberra, Australia: Australian Institute of Aboriginal Studies.

Baroni-Urbani, C., and Buser, M. W. 1976. Similarity of binary data. *Systematic Zoology*. *25*: 251–259.
Bartko, J. J., and Gulbinat, W. 1980. *Multivariate Statistical Methodologies Used in the International Pilot Study of Schizophrenia*. Publication No. ADM 80-630. Rockville, Maryland: National Institute of Mental Health.
Bateson, G. 1980. *Mind and Nature*. New York: Bantam Books.
Bättig, K., Zeier, H., Muller, R., and Buzzi, R. 1980. A field study on vegetative effects of aircraft noise. *Archives of Environmental Health*. *35*: 228–235.
Beal, G. M., Power, R. C., and Coward, E. W. (Eds.). 1971. *Sociological Perspective of Domestic Development*. Ames, Iowa: Iowa State University Press.
Bell, H. 1976. *Geochemical Reconnaissance Using Heavy Minerals from Small Streams in Central South Carolina*. Geological Survey Bulletin 1404. Washington, D.C.: U.S. Government Printing Office. 1–22.
Bentler, P. M., Lettieri, D. J., and Austin, G. A. 1976. *Data Analysis Strategies and Designs for Substance Abuse Research*. Research Issues 13. Rockville, Maryland: National Institute on Drug Abuse.
Berger, P. L., and Luckman, T. 1967. *The Social Construction of Reality*. Garden City, New York: Anchor Books.
Betters, D. R., and Rubingh, J. L. 1978. Suitability analysis and wildland classification: an approach. *Journal of Environmental Management*. *7*: 59–72.
Bijnen, E. J. 1973. *Cluster Analysis*. The Netherlands: Tilburg University Press.
Bishop, Y. M. M., Fienberg, S. E., and Holland, P. W. 1975. *Discrete Multivariate Analysis: Theory and Practice*. Cambridge, Mass.: MIT Press.
Blashfield, R. K. 1976. Questionnaire on cluster analysis software. *Classification Society Bulletin*. *3*(4): 25–42.
Blashfield, R. K., and Morey, L. C. 1979. The classification of depression through cluster analysis. *Comprehensive Psychiatry*. *20*: 516–527.
Blin, J., and Cohen, C. 1977. Technological similarity and aggregation in input-output systems: a cluster-analytic approach. *The Review of Economics and Statistics*. *59*: 82–91.
Blong, R. J. 1973. A numerical classification of selected landslides of the debris slide–avalanche–flow type. *Engineering Geology*. *7*: 99–114.
Bock, C. E., Mitton, J. B., and Lepthien, L. W. 1978. Winter biogeography of North American Fringillidae (Aves): a numerical analysis. *Systematic Zoology*. *27*: 411–420.
Boesch, D. F. 1977. *Application of Numerical Classification in Ecological Investigations of Water Pollution*. Report No. EPA-600/3-77-033. Corvallis, Oregon: U.S. Environmental Protection Agency.
Bonde, G. J. 1975. The genus *Bacillus*. *Danish Medical Bulletin*. *22*(2): 41–61.
Borchardt, G. A., Aruscavage, P. J., and Miller, H. T. 1972. Correlation of the Bishop Ash, a Pleistocene marker bed, using instrumental neutron activation analyses. *Journal of Sedimentary Petrology*. *42*: 301–306.
Boyce, A. J. 1969. Mapping diversity: a comparative study of some numerical methods. In A. J. Cole (Ed.), *Numerical Taxonomy*. New York: Academic Press.
Bradfield, G. E., and Orlóci, L. 1975. Classification of vegetation data from an open beach environment in southwestern Ontario: cluster analysis followed by generalized distance assignment. *Canadian Journal of Botany*. *53*: 495–502.
Cain, A. J., and Harrison, G. A. 1958. An analysis of the taxonomists' judgement of

affinity. *Proceedings of the Zoological Society of London. 131*: 85–98.
Cairns, J., Ruthven, J. A., and Kaesler, R. L. 1974. Distribution of protozoa in a small stream. *The American Midland Naturalist. 92*: 406–414.
Chadwick, G. 1971. *A Systems View of Planning.* Oxford: Pergamon Press.
Chalmers, A. F. 1976. *What Is This Thing Called Science?* St. Lucia, Queensland: University of Queensland Press.
Charlton, T. H., Grove, D. C., and Hopke, P. K. 1978. The Paredón, Mexico, obsidian source and early formative exchange. *Science. 201*: 807–809.
Chatfield, C., and Collins, A. J. 1980. *Introduction to Multivariate Analysis.* London: Chapman and Hall.
Cheetham, A. H., and Hazel, J. E. 1969. Binary (presence-absence) similarity coefficients. *Journal of Paleontology. 43*: 1130–1136.
Chu, K. C. 1974. Applications of artificial intelligence to chemistry. *Analytical Chemistry. 46*: 1181–1187.
Clifford, H. T., and Stephenson, W. 1975. *An Introduction to Numerical Classification.* New York: Academic Press.
Clover, R. C. 1979. Phenetic relationships among populations of *Podarcis sicula* and *P. melisellensis* (Sauria: Lacertidae) from islands in the Adriatic Sea. *Systematic Zoology. 28*: 284–298.
Cohen, J. 1969. r_c: A profile similarity coefficient invariant over variable reflection. *Psychological Bulletin. 71*: 281–284.
Colgan, P. W. 1978. *Quantitative Ethology.* New York: John Wiley & Sons.
Computerworld (Eds.). 1980. *14(52)*: 1, 10–20.
Crecellus, E. A., Lepel, E. A., Laul, J. C., Rancitelli, L. A., and McKeever, R. L. 1980. Background air particulate chemistry near Colstrip, Montana. *Environmental Science and Technology. 14*: 422–428.
Crossman, J. S., Kaesler, R. L., and Cairns, J. 1974. The use of cluster analysis in the assessment of spills of hazardous materials. *The American Midland Naturalist. 92*: 94–114.
Dale, M. B. 1978. Pattern seeking methods in vegetation studies. In L. 't Mannetje (Ed.), *Measurement of Grassland Vegetation and Animal Production.* Farnham Royal, Bucks, England: Commonwealth Agricultural Bureaux.
Darby, W. P., McMichael, F. C., and Dunlap, R. W. 1976. Urban watershed management using activity indicators to predict water quality. *Water Resources Research. 12*: 245–252.
Davis, J. C. 1973. *Statistics and Data Analysis in Geology.* New York: John Wiley & Sons.
Davis, J. H. C. 1975. Cluster analysis as an aid to selection for diversity in dry beans. *Annual Report of the Bean Improvement Cooperative. 18*: 21–23.
Derr, D. A., and Nagle, G. 1977. Land transfers at the rural/urban fringe: the New Jersey experience. *American Society of Farm Managers and Rural Appraisers Journal. 41*: 30–35.
Dixon, W. J. (Ed.). 1983. *BMDP Statistical Software, 1983 Printing with Additions.* Berkeley, Calif.: University of California Press.
Doran, J. E., and Hodson, F. R. 1975. *Mathematics and Computers in Archaeology.* Cambridge, Mass.: Harvard University Press.
Doyle, L. J., and Feldhausen, P. H. 1981. Bottom sediments of the Eastern Gulf of Mexico examined with traditional and multivariate statistical methods. *Mathematical Geology. 13*: 93–117.

Duran, B. S., and Odell, P. L. 1974. *Cluster Analysis*. Berlin: Springer-Verlag.
Ekehammar, B., and Magnusson, D. 1978. Co-frequency ratings of anxiety reactions. *Perceptual and Motor Skills*. *46*: 1215–1224.
Everitt, B. S. 1977*a*. *The Analysis of Contingency Tables*. New York: Halsted Press.
Everitt, B. S. 1977*b*. Cluster analysis. In O'Muircheartaigh, C. A., and Payne, C. (Eds.), *Exploring Data Structures*. Vol. 1. New York: John Wiley & Sons.
Everitt, B. S. 1979. Unresolved problems in cluster analysis. *Biometrics*. *35*: 169–181.
Everitt, B. S. 1980. *Cluster Analysis*. New York: Halsted Press.
Faludi, A. 1973. *Planning Theory*. Oxford: Pergamon Press.
Fanizza, G., and Bogyo, T. P. 1976. A cluster analysis of almond varieties in Apulia. *Riv. Ortoflorofrutt. It.* *60*: 277–281.
Farris, J. S. 1969. On the cophenetic correlation coefficient. *Systematic Zoology*. *18*: 279–285.
Farris, J. S. 1973. On comparing the shapes of taxonomic trees. Systematic Zoology. *22*: 50–54.
Fesenmaier, D. R., Goodchild, M. F., and Morrison, S. 1979. The spatial structure of the rural-urban fringe: a multivariate approach. *Canadian Geographer*. *23*: 255–265.
Fienberg, S. E. 1977. *The Analysis of Cross-classified Categorical Data*. Cambridge, Mass.: MIT Press.
Fillenbaum, S., and Rapoport, A. 1971. *Structures in the Subjective Lexicon*. New York: Academic Press.
Financial Times of Canada (Eds.). 1977. What Canadians think is important. *Financial Times of Canada*. September 26, p. 9.
Fox, L., Mayer, K. E., and Forbes, A. R. 1983. Classification of forest resources with Landsat data. *Journal of Forestry*. *81*: 283–287.
Frankfurter, G. M., and Phillips, H. E. 1980. Portfolio selection: an analytic approach for selecting securities from a large universe. *Journal of Financial and Quantitative Analysis*. *15*: 357–377.
Froidevaux, R., Jaquet, J. M., and Thomas, R. L. 1977. AGCL, a Fortran IV program for agglomerative, nonhierarchical Q-mode classification of large data sets. *Computer and Geosciences*. *3*: 31–48.
Fowlkes, E. B., and Mallows, C. L. 1983. A method for comparing two hierarchical clusterings. *Journal of the American Statistical Association*. *78*: 553–569.
Fujii, K. 1969. Numerical taxonomy of ecological characteristics and the niche concept. *Systematic Zoology*. *18*: 151–153.
Gaikawad, S. T., Goel, S. K., and Raghumohan, N. G. 1977. Numerical classification of some Indian soils. *Journal of the Indian Society of Soil Science*. *25*(3): 288–297.
Galtung, J. 1971. A structural theory of imperialism. *Journal of Peace Research*. *8*: 81–118.
Gardner, M. 1975. Mathematical games. *Scientific American*. *233*(6): 116–119.
Ghent, A. W. 1963. Kendall's "tau" coefficient as an index of similarity in comparisons of plant or animal communities. *The Canadian Entomologist*. *95*: 568–575.
Gidengil, E. L. 1978. Centres and peripheries: an empirical test of Galtung's theory of imperialism. *Journal of Peace Research*. *15*: 51–66.
Gilmour, J. S. L. 1951. The development of taxonomic theory since 1851. *Nature*. *168*: 400–402.

Gnanadesikan, R. 1977. *Methods of Statistical Data Analysis of Multivariate Observations*. New York: John Wiley & Sons.

Golovnja, R. V., Jakovleva, V. N., Cesnokova, A. E., Matveeva, L. V., and Borisov, Ju. A. 1981. A method for selecting panel for hedonic assessment of new food products. *Die Nahrung*. *25*: 53–58.

Goodall, D. W. 1973. Numerical classification. In R. H. Whittaker (Ed.), *Handbook of Vegetation Science, Part V: Ordination and Classification of Vegetation*. The Hague: Dr. W. Junk B. V.

Gower, J. C. 1971. A general coefficient of similarity and some of its properties. *Biometrics*. *27*: 857–874.

Grant, P. G., and Hoff, C. J. 1975. The skin of primates. *American Journal of Physical Anthropology*. *42*: 151–166.

Grant, W. S., Milner, G. B., Krasnowski, P., and Utter, F. M. 1980. Use of biochemical genetic variants for identification of sockeye salmon (*Oncorhynchus nerka*) stocks in Cook Inlet, Alaska. *Canadian Journal of Fisheries and Aquatic Science*. *37*: 1236–1247.

Grant, W. S., and Utter, F. M. 1980. Biochemical genetic variation in walleye pollock, *Theragra chalcogramma*: population structure in the southeastern Bering Sea and the Gulf of Alaska. *Canadian Journal of Fisheries and Aquatic Science*. *37*: 1093–1100.

Green, R. H. 1979. *Sampling Design and Statistical Methods for Environmental Biologists*. New York: John Wiley & Sons.

Green, R. H. 1980. Multivariate approaches in ecology: the assessment of ecological similarity. In R. F. Johnston, P. W. Frank, and C. D. Michener (Eds.), *Annual Review of Ecology and Systematics. Vol. 11*. Palo Alto, California: Annual Reviews, Inc.

Green, R. H., and Vascotto, G. L. 1978. A method for the analysis of environmental factors controlling patterns of species composition in aquatic communities. *Water Research*. *12*: 583–590.

Gregor, T. 1979. Short people. *Natural History*. *88*: 14.

Gupta, M. C., and Huefner, R. J. 1972. A cluster analysis study of financial ratios and industry characteristics. *Journal of Accounting Research*. *10*: 77–95.

Guralnick, E. 1976. The proportions of some archaic Greek sculptured figures: a computer analysis. *Computers and the Humanities*. *10*: 153–169.

Hansch, C., and Unger, S. H. 1973. Strategy in drug design. Cluster analysis as an aid in the selection of substituents. *Journal of Medicinal Chemistry*. *16*: 1217–1222.

Hanson, N. R. 1958. *Patterns of Discovery*. Cambridge: Cambridge University Press.

Harbaugh, J. W., and Merriam, D. F. 1968. *Computer Applications in Stratigraphic Analysis*. New York: John Wiley & Sons.

Harris, A. H., and Porter, L. S. W. 1980. Late Pleistocene horses of Dry Cave, Eddy County, New Mexico. *Journal of Mammalogy*. *61*: 46–65.

Harris, J. E., Wente, E. F., Cox, C. F., Nawaway, I. E., Kowalski, C. J., Storey, A. T., Russell, W. R., Ponitz, P. V., and Walker, G. F. 1978. Mummy of the "Elder Lady" in the tomb of Amenhotep II: Egyptian Museum catalog number 61070. *Science*. *200*: 1149–1151.

Hartigan, J. A. 1975. *Clustering Algorithms*. New York: John Wiley & Sons.

Harvey, D. 1969. *Explanation in Geography*. London: Edward Arnold.

Hays, W. L. 1963. *Statistics*. New York: Holt, Rinehart and Winston.

Hazel, J. E. 1970. Binary coefficients and clustering in biostratigraphy. *Geological Society of America Bulletin. 81*: 3237–3252.

Hecht, A. D., Bé, A. W. H., and Lott, L. 1976. Ecologic and paleoclimatic implications of morphologic variation of *Orbulina universa* in the Indian Ocean. *Science. 194*: 422–424.

Herrin, C. S., and Oliver, J. H. 1974. Numerical taxonomic studies of parthenogenetic and bisexual populations of *Haemaphysalis longicornis* and related species (Acari: Ixodidae). *The Journal of Parasitology. 60*: 1025–1035.

Hesse, M. B. 1966. *Models and Analogies in Science*. Notre Dame, Indiana: University of Notre Dame Press.

Hicks, R. R., and Burch, J. B. 1977. Numerical taxonomy of southern red oak in the vicinity of Nacogdoches, Texas. *Forest Science. 23*: 290–298.

Hodson, F. R., Sneath, P. H. A., and Doran, J. E. 1966. Some experiments in the numerical analysis of archaeological data. *Biometrika. 53*: 311–324.

Holland, R. E., and Claflin, L. W. 1975. Horizontal distribution of planktonic diatoms in Green Bay, mid-July 1970. *Limnology and Oceanography. 20*: 365–378.

Holley, J. W., and Guilford, J. P. 1964. A note on the G index of agreement. *Educational and Psychological Measurement. 24*: 749–753.

Hughes, M. M. 1979. Exploration and play re-visited: a hierarchical analysis. *International Journal of Behavioral Development. 2*: 225–233.

Imbrie, J., and Purdy, E. G. 1962. Classification of modern Bahamian carbonate sediments. In W. E. Ham (Ed.), *Classification of Carbonate Rocks*. Tulsa, Oklahoma: The American Association of Petroleum Geologists.

Imhof, M., Leary, R., and Booke, H. E. 1980. Population or stock structure of lake whitefish, *Coregonus clupeaformis*, in northern Lake Michigan as assessed by isozyme electrophoresis. *Canadian Journal of Fisheries and Aquatic Science. 37*: 783–793.

Jacobsen, T. D. 1979. Numerical analysis of variation between *Allium cernuum* and *Allium stellatum* (Liliaceae). *Taxon. 28*: 517–523.

Jeffers, N. J. R. 1978. *An Introduction to Systems Analysis: With Ecological Applications*. Baltimore: University Park Press.

Jensen, R. E. 1971. A cluster analysis study of financial performance of selected business firms. *The Accounting Review. 46*: 36–56.

Johnson, K. R., and Albani, A. D. 1973. Biotopes of recent benthonic foraminifera in Pitt Water, Broken Bay, N.S.W. (Australia). *Palaeogeography, Palaeoclimatology, Palaeoecology. 14*: 265–276.

Juffs, H. S. 1973. Identification of *Pseudomonas* spp. isolated from milk produced in south eastern Queensland. *Journal of Applied Bacteriology. 36*: 585–598.

Kaesler, R. L., Mulvany, P. S., and Kornicker, L. S. 1977. Delimitation of the Antarctic convergence by cluster analysis and ordination of benthic myodocopid Ostracoda. In H. Loffler and D. Danielopol (Eds.), *Aspects of Ecology and Zoogeography of Recent Fossil Ostracoda*. The Hague: Dr. W. Junk B. V.

Kendall, M. G. 1970. *Rank Correlation Methods* (4th ed.). London: Charles Griffin and Company Limited.

Kendig, H. 1976. Cluster analysis to classify residential areas: a Los Angeles application. *Journal of the American Institute of Planners. 42*: 286–294.

King, J. R. 1980. Machine-component group formation in group technology. *Omega. 8*: 193–199.

King, L. J. 1969. *Statistical Analysis in Geography*. Englewood Cliffs, N.J.: Prentice-Hall.

Kirkpatrick, M. 1981. Spatial and age dependent patterns of growth in New England black birch. *American Journal of Botany*. *68*: 535–543.

Lance, G. N., and Williams, W. T. 1967. A general theory of classificatory sorting strategies. I. Hierarchical systems. *Computer Journal*. *9*: 373–380.

Langohr, R., Scoppa, C. O., and Van Wambeke, A. 1976. The use of a comparative particle size distribution index for the numerical classification of soil parent materials: application to mollisols of the Argentinian pampa. *Geoderma*. *15*: 305–312.

Larson, K. R., and Tanner, W. W. 1974. Numeric analysis of the lizard genus *Sceloporus* with special reference to the cranial osteology. *The Great Basin Naturalist*. *34*: 1–41.

Lasker, G. W. 1977. A coefficient of relationship by isonymy: a method for estimating the genetic relationship between populations. *Human Biology*. *49*: 489–493.

Lasker, G. W. 1978. Relationships among the Otmoor villages and surrounding communities as inferred from surnames contained in the current register of electors. *Annals of Human Biology*. *5*: 105–111.

Laumann, E. O. 1973. *Bonds of Pluralism: The Form and Substance of Urban Social Networks*. New York: John Wiley & Sons.

Lawson, B. 1980. *How Designers Think*. Westfield, N.J.: Eastfield Editions, Inc.

Legendre, L., and Legendre, P. 1983. *Numerical Ecology*. Amsterdam: Elsevier Scientific Publishing Company.

Leopold, L. B., and Marchand, M. O. 1968. On the quantitative inventory of the riverscape. *Water Resources Research*. *4*: 709–717.

Lessig, V. P., and Tollefson, J. O. 1971. Market segmentation through numerical taxonomy. *Journal of Marketing Research*. *8*: 480–487.

Levin, J., and Brown, M. 1979. Scaling a conditional proximity matrix to symmetry. *Psychometrika*. *44*: 239–243.

Levine, D. M. 1977. Multivariate analysis of the visual information processing of numbers. *The Journal of Multivariate Behavioral Research*. *12*: 347–355.

Levy, P. S., and Lemeshow, S. 1980. *Sampling for the Health Professions*. Belmont, California: Lifetime Learning Publications.

Lighthart, B. 1975. A cluster analysis of some bacteria in the water column of Green Lake, Washington. *Canadian Journal of Microbiology*. *21*: 392–394.

Lindeman, R. H., Merenda, P. F., and Golf, R. Z. 1980. *Introduction to Bivariate and Multivariate Analysis*. Glenview, Illinois: Scott, Foresman and Company.

Lissitz, R. W., Mendoza, J. L., Huberty, C. J., and Markos, H. V. 1979. Some further ideas on a methodology for determining job similarities/differences. *Personnel Psychology*. *32*: 517–528.

Lorr, M. 1976. Cluster and typological analysis. In P. M. Bentler, D. J. Lettieri, and G. A. Austin (Eds.), *Data Analysis Strategies and Designs for Substance Abuse Research*. Research Issues 13. Rockville, Maryland: National Institute on Drug Abuse.

Lytle, T. F., and Lytle, J. S. 1979. Sediment hydrocarbons near an oil rig. *Estuarine and Coastal Marine Science*. *9*: 319–330.

Martin, D. J. 1979. Songs of the fox sparrow. II. Intra- and interpopulation variation. *Condor*. *81*: 173–184.

Mather, P. M. 1976. *Computational Methods of Multivariate Analysis in Physical Geography*. New York: John Wiley & Sons.

McCormick, W. T., Schweitzer, P. J., and White, T. W. 1972. Problem decomposition and data reorganization by a clustering technique. *Operations Research*. *20*: 993–1009.

McCutchen, M. H. 1980. Use of cluster analysis and discriminant analysis to classify synoptic weather types in southern California. In *Eighth Conference on Weather Forecasting and Analysis* (Denver, Colo., June 10–13, 1980). Boston, Mass.: American Meteorological Society.

McHenry, H. M. 1973. Early hominid humerus from East Rudolf, Kenya. *Science*. *180*: 739–741.

Mendenhall, W., Ott, L., and Scheaffer, R. L. 1971. *Elementary Survey Sampling*. Belmont, California: Wadsworth Publishing Co.

Miller, G. A. 1969. A psychological method to investigate verbal concepts. *Journal of Mathematical Psychology*. *6*: 169–191.

Miller, G. A., and Johnson-Laird, P. N. 1976. *Language and Perception*. Cambridge, Massachusetts: Harvard University Press.

Mobley, W. H., and Ramsay, R. S. 1973. Hierarchical clustering on the basis of inter-job similarity as a tool in validity generalization. *Personnel Psychology*. *26*: 213–225.

Monge, P. R., and Cappella, J. N. (Eds.). 1980. *Multivariate Techniques in Human Communication Research*. New York: Academic Press.

Morgan, B. J. T. 1973. Cluster analyses of two acoustic confusion matrices. *Perception & Psychophysics*. *13*: 13–24.

Morgan, B. J. T., Simpson, M. J. A., Hanby, J. P., & Hall-Craggs, J. 1976. Visualizing interaction and sequential data in animal behaviour: theory and application of cluster-analysis methods. *Behaviour*. *56*: 1–43.

Moskowitz, H. R., and Klarman, L. 1977. Food compatibilities and menu planning. *Canadian Institute of Food Science and Technology*. *10*: 257–264.

Moss, W. W., Posey, F. T., and Peterson, P. C. 1980. A multivariate analysis of the infrared spectra of drugs of abuse. *Journal of Forensic Sciences*. *25*: 304–313.

Mueller-Dombois, D., and Ellenberg, H. 1974. *Aims and Methods of Vegetation Ecology*. New York: John Wiley & Sons.

Mungomery, V. E., Shorter, R., and Byth, D. E. 1974. Genotype X environment interactions and environmental adaptation. I. Pattern analysis—application to soya bean populations. *Australian Journal of Agriculture Research*. *25*: 59–72.

Neogy, S., and Rao, S. V. L. N. 1981. Utilisation of cluster analyses in iron ore mine planning studies. *Indian Journal of Earth Sciences*. *8*(1): 10–15.

Olson, A. C., Larson, N. M., and Heckman, C. A. 1980. Classification of cultured mammalian cells by shape analysis and pattern recognition. *Proceedings of the National Academy of Sciences, USA*. *77*: 1516–1520.

Orlóci, L. 1975. *Multivariate Analysis in Vegetation Research*. The Hague: Dr. W. Junk B. V.

Otter, C. J. D., Behan, M., and Maes, F. W. 1980. Single cell responses in female *Pieris brassicae* (Lepidoptera: Pieridae) to plant volatiles and conspecific egg odours. *Journal of Insect Physiology*. *26*: 465–472.

Overall, J. E., and Klett, C. J. 1972. *Applied Multivariate Analysis*. New York: McGraw Hill Book Company.

Penrose, L. S. 1953. Distance, size and shape. *Annals of Eugenics*. *18*: 337–343.

Phipps, J. B. 1971. Dendrogram topology. *Systematic Zoology. 20*: 306–308.
Platts, W. S. 1979. Relationships among stream order, fish populations, and aquatic geomorphology in an Idaho river drainge. *Fisheries. 4*: 5–9.
Polya, G. 1954. *Induction and Analogy in Mathematics. Vol. 1.* Princeton: Princeton University Press.
Polya, G. 1957. *How to Solve It.* Garden City, New York: Doubleday Anchor Books.
Popper, K. R. 1968. *Conjectures and Refutations.* New York: Harper Torchbooks.
Pratt, V. 1972. Biological classification. *British Journal of the Philosophy of Science. 23*: 305–327.
Radloff, D. L., and Betters, D. R. 1978. Multivariate analysis of physical site data for wildland classification. *Forest Science. 24*: 2–10.
Reid, A. H., Ballinger, B. R., and Heather, B. B. 1978. Behavioural syndromes identified by cluster analysis in a sample of 100 severely and profoundly retarded adults. *Psychological Medicine. 8*: 399–412.
Reynolds, C. M. 1979. The heronries census: 1972–1977 population changes and a review. *Bird Study. 26*: 7–12.
Robins, J. D., and Schnell, G. D. 1971. Skeletal analysis of the *Ammodramus-Ammospiza* grassland sparrow complex: a numerical taxonomic study. *Auk. 88*: 567–590.
Rohlf, F. J. 1974. Methods of comparing classifications. In R. F. Johnston, P. W. Frank, and C. D. Michener (Eds.), *Annual Review of Ecology and Systematics, Vol. 5.* Palo Alto, California: Annual Reviews Inc.
Romesburg, H. C. 1981. Wildlife science: gaining reliable knowledge. *Journal of Wildlife Management. 45*: 293–313.
Rompelman, O., van Kampen, W. H. A., Backer, E., and Offerhaus, R. E. 1980. Heart rate variability in relation to psychological factors. *Ergonomics. 23*: 1101–1115.
Rothkopf, E. Z. 1957. A measure of stimulus similarity and errors in some paired-associate learning tasks. *Journal of Experimental Psychology. 53*: 94–101.
Russell, B. 1974. *The art of Philosophizing and Other Essays.* Totowa, N.J.: Littlefield, Adams and Company.
Sale, P. F., McWilliam, P. S., and Anderson, D. T. 1978. Faunal relationships among the near-reef zooplankton at three locations on Heron Reef, Great Barrier Reef, and seasonal changes in this fauna. *Marine Biology. 49*: 133–145.
Salter, R. E., and Hudson, R. J. 1978. Habitat utilization by feral horses in western Alberta. *Le Nautraliste Canadien. 105*: 309–321.
Salzano, F. M., Pages, F., Neel, J. V., Gershowitz, H., Tanis, R. J., Moreno, R., and Franco, M. H. L. P. 1978. Unusual blood genetic characteristics among the Ayoreo Indians of Bolivia and Paraguay. *Human Biology. 50*: 121–136.
SAS Supplemental Library User's Guide, 1980 Edition. 1980. Cary, North Carolina: SAS Institute Inc.
SAS User's Guide: Statistics, 1982 Edition. 1982. Cary, North Carolina: SAS Institute Inc.
Schmitt, D. P., and Norton, D. C. 1972. Relationships of plant parasitic nematodes to sites in native Iowa praries. *Journal of Nematology. 4*: 200–206.
Scott, D. B., Medioli, F. S., and Schafer, C. T. 1977. Temporal changes in foraminiferal distributions in Miramichi River estuary, New Brunswick. *Canadian*

Journal of the Earth Sciences. *14*: 1566–1587.
Searles, H. L. 1968. *Logic and Scientific Method*. New York: Ronald Press Co.
Sethi, S. P., and Holton, R. H. 1973. Country typologies for the multinational corporation: a new basic approach. *California Management Review*. *15*(3): 105–118.
Shepard, R. N. 1963. Analysis of proximities as a technique for the study of information processing in man. *Human Factors*. *5*: 33–48.
Shepard, R. N. 1980. Multidimensional scaling, tree-fitting, and clustering. *Science*. *210*: 390–398.
Siegel, S. 1956. *Nonparametric Statistics for the Behavioral Sciences*. New York: McGraw-Hill Book Company.
Sigleo, A. C. 1975. Turquoise mine and artifact correlation for Snaketown Site, Arizona. *Science*. *189*: 459–460.
Simpson, G. G. 1960. Notes on the measurement of faunal resemblance. *American Journal of Science, Bradley Volume*. *258–A*: 300–311.
Sinha, R. N. 1977. Uses of multivariate methods in the study of stored-grain ecosystems. *Environmental Entomology*. *6*: 185–192.
Sluss, T. P. Rockwood-Sluss, E. S., and Werner, F. G. 1977. Enzyme variation in semi-isolated populations of the mountain fly *Chamaemyia herbarum*. *Evolution*. *31*: 302–312.
Smith, D. W. 1969. A taximetric study of *Vaccinium* in northeastern Ontario. *Canadian Journal of Botany*. *47*: 1747–1759.
Sneath, P. H. A. 1964*a*. Computers in bacterial classification. *Advancement of Science*. *20*(88): 572–582.
Sneath, P. H. A. 1964*b*. New approaches to bacterial taxonomy: use of computers. In C. E. Clifton, S. Raffel, and M. P. Starr (Eds.), *Annual Review of Microbiology. Vol. 18*. Palo Alto, California: Annual Reviews, Inc.
Sneath, P. H. A. 1968. Goodness of intuitive arrangements into time trends based on complex pattern. *Systematic Zoology*. *17*: 256–260.
Sneath, P. H. A., and Sokal, R. R. 1973. *Numerical taxonomy*. San Francisco: W. H. Freeman.
Sokal, R. R. 1966. Numerical taxonomy. *Scientific American*. *215*(6): 106–116.
Sokal, R. R. 1974. Classification: purposes, principles, progress, prospects. *Science*. *185*: 1115–1123.
Sokal, R. R., and Michener, C. D. 1958. A statistical method for evaluating systematic relationships. *The University of Kansas Science Bulletin*. *38*: 1409–1438.
Sokal, R. R., and Sneath, P. H. A. 1963. *Principles of Numerical Taxonomy*. San Francisco: W. H. Freeman.
Solberg, H. E., Skrede, S., Elgjo, K., Blomhoff, J. P., and Gjone, E. 1976. Classification of liver diseases by clinical chemical laboratory results and cluster analysis. *The Scandinavian Journal of Clinical and Laboratory Investigation*. *36*: 81–85.
Späth, H. *Cluster Analysis Algorithms*. 1980. New York: Halsted Press.
SPSS X User's Guide. 1983. New York: McGraw-Hill Book Company.
Stevens, S. S. 1946. On the theory of scales of measurement. *Science*. *103*: 677–680.
Stocker, M., Gilbert, F. F., and Smith, D. W. 1977. Vegetation and deer habitat relations in southern Ontario: classification of habitat types. *Journal of Applied Ecology*. *14*: 419–432.
Stretton, R. J., Lee, J. D., and Dart, R. K. 1976. Relationships of *Aspergillus* species

based on their long chain fatty acids. *Microbios Letters. 2*: 89–93.

Swain, P. H. 1978. Fundamentals of pattern recognition in remote sensing. In P. H. Swain and S. M. Davis (Eds.), *Remote Sensing: The Quantitative Approach*. New York: McGraw-Hill Book Company.

Thilenius, J. F. 1972. *Classification of Deer Habitat in the Ponderosa Pine Forest of the Black Hills, South Dakota*. USDA Forest Service Research Paper RM-91. Fort Collins, Colorado: Rocky Mountain Forest and Range Experiment Station, May 1972.

Thompson, G. G. 1972. Palynologic correlation and environmental analysis within the marine Mancos Shale of southwestern Colorado. *Journal of Sedimentary Petrology. 42*: 287–300.

Tietjen, J. H. 1977. Population distribution and structure of the free-living nematodes of Long Island Sound. *Marine Biology. 43*: 123–136.

Tobler, W. R., Mielke, H. W., and Detwyler, T. R. 1970. Geobotanical distance between New Zealand and neighboring islands. *BioScience. 20*: 537–542.

Tsukamura, M., Mizuno, S., Tsukamura, S., and Tsukamura, J. 1979. Comprehensive numerical classification of 369 strains of *Mycobacterium*, *Rhodococcus*, and *Nocardia*. *International Journal of Systematic Bacteriology. 29*: 110–129.

Vander Kloet, S. P. 1978. The taxonomic status of *Vaccinium pallidum*, the hillside blueberries including *vaccinium vacillans*. *Canadian Journal of Botany. 56*: 1559–1574.

Vanneman, R. 1977. The occupational composition of American classes: results from cluster analysis. *American Journal of Sociology. 82*: 783–807.

Vasicek, Z., and Jicin, R. 1971. The problem of similarity of shape. *Systematic Zoology. 21*: 91–96.

Veldman, D. J. 1967. *Fortran Programming for the Behavioral Sciences*. New York: Holt, Rinehart and Winston.

Ward, R. H., Gershowitz, H., Layrisse, M., and Neel, J. V. 1975. The genetic structure of a tribal population, the Yanomama Indians. XI. Gene frequencies for 10 blood groups and the ABH-Le sector traits in the Yanomama and their neighbors; the uniqueness of the tribe. *American Journal of Human Genetics. 27*: 1–30.

Ward, R. H., and Neel, J. V. 1970. Gene frequencies and microdifferentiation among the Makiritare Indians. IV. A Comparison of a genetic network with ethnohistory and migration matrices; a new index of genetic isolation. *American Journal of Human Genetics. 22*: 538–561.

Ware, W. B., and Benson, J. 1975. Appropriate statistics and measurement. *Science Education. 59*: 575–582.

Webster, R. 1977. *Quantitative and Numerical Methods in Soil Classification and Survey*. Oxford: Clarendon Press.

West, N. E. 1966. Matrix cluster analysis of montane forest vegetation of the Oregon Cascades. *Ecology. 47*: 975–980.

Wharton, S. W., and Turner, B. J. 1981. ICAP: An interactive cluster analysis procedure for analyzing remotely sensed data. *Remote Sensing of Environment. 11*: 279–293.

Williams R. D., and Schreiber, M. M. 1976. Numerical and chemotaxonomy of the green foxtail complex. *Weed Science. 24*: 331–335.

Williams, W. T. 1967. Numbers, taxonomy, and judgment. *The Botanical Review. 33*: 379–386.

Williams, W. T., and Clifford, H. T. 1971. On the comparison of two classifications of the same set of elements. *Taxon. 20*, 519–522.

Williams, W. T., and Dale, M. B. 1965. Fundamental problems in numerical taxonomy. In R. D. Preston (Ed.), *Advances in Botanical Research. Vol. 2.* New York: Academic Press.

Wilson, E. O., and Bossert, W. H. 1971. *A Primer of Population Biology*. Stamford, Connecticut: Sinauer Associates, Inc., Publishers.

Wishart, D. 1981. *CLUSTAN User Manual*. Edinburgh, Scotland: Program Library Unit, Edinburgh University.

Woese, C. R. 1981. Archaebacteria. *Scientific American. 244*(6): 98–122.

Wright, R. L. 1977. On the application of numerical taxonomy in soil classification for land evaluation. *ITC Journal 1977-3. 3*: 482–510.

Younker, J. L., and Ehrlich, R. 1977. Fourier biometrics: harmonic amplitudes as multivariate shape descriptors. *Systematic Zoology. 26*: 336–342.

Zajicek, G., Maayan, C., and Rosenmann, E. 1977. The application of cluster analysis to glomerular histopathology. *Computers and Biomedical Research. 10*: 471–481.

Zelnio, R. N., and Simmons, S. A. 1981. Interpretation of research data: other multivariate methods. *American Journal of Hospital Pharmacy. 38*: 369–376.

Zonn, L. E. 1979. Housing and urban blacks: a social distance-residential distance view. *The Annals of Regional Science. 13*: 55–63.

Index

(continued on overleaf)

System Requirements. The Fortran IV source code for CLUSTAR/CLUSTID is available on 9-track (EBCDIC) magnetic tape. It is designed to compile and run on a mainframe (or 32-bit scientific desktop) computer that supports a Fortran IV or Fortran 77 compiler, with no changes, or at most minimal changes, to the source code. The programs, which are well documented, have specifically been tested on the following systems: VAX 11 series, IBM 4300 series, UNIVAC 1100 series.

Pricing and Documentation. The price of CLUSTAR/CLUSTID on tape is $175.00 and includes a User's Manual, documentation and listing of source code, and example test problems with output. The User's Manual is very detailed and contains specific instructions for implementation and use, including worked-out examples. Additional copies of the User's Manual can be purchased at $8.95 per copy.

Note: These programs contain all the methods described in *Cluster Analysis for Researchers*. The tapes, while complementary to the book, are not essential for learning and understanding the methods described in the book.

ORDER FORM

Please send me

_____ *CLUSTAR/CLUSTID: Computer Programs for Hierarchical Cluster Analysis* (0-534-03420-9). I enclose $175.00, which includes the price of the *User's Manual*, and cost of shipping/handling.

_____ additional copy/ies of *User's Manual for Clustar/Clustid Computer Programs* (0-534-3251-6) @ $8.95 each. Payment is enclosed (includes shipping/handling costs)

_____ copy/ies of *Cluster Analysis for Researchers* (0-534-03248-6) @ $36.00 each.

☐ Bill me. I'll return any books I don't want within the 15-day trial period without further obligation and the invoice will be cancelled. For books I keep, I'll pay the amount above plus my local sales tax and a small amount for postage and handling.

☐ Check enclosed. Publisher pays postage and handling. 15-day return privilege still applies.

☐ Charge my ☐ VISA ☐ MasterCard ☐ American Express

Card # ______________________ Exp. Date __________

Prices subject to change without notice. Offer valid in U.S. and Canada only. Residents of CA, KY, MA, MI, NC, NJ, NY, and WA, please add sales tax.

Please sign here: ______________________

We cannot process your order without your signature.

Please print to ensure proper delivery.

Name ______________________

Company ______________________

Address ______________________

City/State/Zip ______________________

Detach and mail to:

LIFETIME LEARNING PUBLICATIONS
10 DAVIS DRIVE
BELMONT, CA 94002